MATHEMATICAL MUSIC THEORY

Algebraic, Geometric, Combinatorial,
Topological and Applied Approaches to Understanding
Musical Phenomena

MATHEMATICAL MUSIC THEORY

Algebraic, Geometric, Combinatorial, Topological and Applied Approaches to Understanding Musical Phenomena

Editors

Mariana Montiel
Georgia State University, USA

Robert W Peck
Louisiana State University, USA

NEW JERSEY · LONDON · SINGAPORE · BEIJING · SHANGHAI · HONG KONG · TAIPEI · CHENNAI · TOKYO

Published by

World Scientific Publishing Co. Pte. Ltd.
5 Toh Tuck Link, Singapore 596224
USA office: 27 Warren Street, Suite 401-402, Hackensack, NJ 07601
UK office: 57 Shelton Street, Covent Garden, London WC2H 9HE

Library of Congress Cataloging-in-Publication Data
Names: Montiel, Mariana. | Peck, Robert William, 1963-
Title: Mathematical music theory : algebraic, geometric, combinatorial,
     topological and applied approaches to understanding musical phenomena /
     edited by: Mariana Montiel (Georgia State University, USA) and
     Robert W Peck (Louisiana State University, USA).
Description: New Jersey : World Scientific, 2018. | Includes bibliographical references.
Identifiers: LCCN 2018016235 | ISBN 9789813235304 (hc : alk. paper)
Subjects: LCSH: Music--Mathematics. | Music theory--Mathematics. | Musical analysis.
Classification: LCC ML3800 .M2456 2018 | DDC 781.01/51--dc23
LC record available at https://lccn.loc.gov/2018016235

British Library Cataloguing-in-Publication Data
A catalogue record for this book is available from the British Library.

Cover image: Dr. Maria Mannone

First published 2019 (Hardcover)
Reprinted 2020 (in paperback edition)
ISBN 978-981-122-138-5 (pbk)

Copyright © 2019 by World Scientific Publishing Co. Pte. Ltd.

For any available supplementary material, please visit
https://www.worldscientific.com/worldscibooks/10.1142/10858#t=suppl

Desk Editors: V. Vishnu Mohan/Tan Rok Ting

Typeset by Stallion Press
Email: enquiries@stallionpress.com

Introduction

Questions about variation, similarity, enumeration, and classification of musical structures have long intrigued both musicians and mathematicians. Mathematical models can be found for almost all levels of musical activities, from theoretical analysis to actual composition or sound production. Modern Music Theory has been incorporating more and more modern mathematical content during the last decades. One example is the application of methods from Algebraic Combinatorics, or Topology and Graph Theory, to the classification of different musical objects. However, these applications of mathematics in the context of the in-depth understanding of music also have led to interesting open problems in mathematics itself. The reach and depth of the contributions on mathematical music theory presented in this volume is significant. In general, the subjects fall in the mathematical classifications of: (i) Algebraic and Combinatorial Approaches (Douthett, Clampitt, and Carey; Douthett, Steinbach, and Hermann; Jedrzejewski; Kastine; Kochavi; Noll; Peck; and Plotkin); (ii) Geometric, Topological, and Graph-Theoretical Approaches (Hughes; Ivey; Mannone; Sivakumar and Tymoczko; Yust; and Milam); and (iii) Distance and Similarity Measures in Music (Gómez; Baker; and Shanahan). These three classifications form our sections and each contribution is a chapter of this volume.

In Chapter 1 of Section I, Douthett, Clampitt, and Carey first review some of the basic concepts introduced by Clough and

Myerson. Clampitt's extension of Clough and Myerson's work, particularly as it relates to cardinality equals variety for chords (CCV), is also discussed. In this work, the search for scales whose complements preserve properties such as CCV and well formedness is explored and identified. The generalization of similarities and differences between cardinality equals variety for lines (LCV) and CCV, as well as the connection between dual CCV and the twin primes conjecture, are also investigated.

In Chapter 2 of Section I, Douthett, Steinbach, and Hermann investigate m-cubes modulo n that are generated by musical dyads. In particular, they consider the subclass of toggling m-cubes, for which the pairwise intersections of the generating dyads are null, and the pcset used to generate the vertex set contains one and only one member from each generating dyad. They also explore a subclass that correspond to Cohn's maximally smooth cycles, leading to generalizations that relate certain Western and Hindustani musical systems. These cycles correspond to middle-layer graphs that are antipolar great cycles with strong diameters. The three-dimensional models that they study include Cohn's Cube and Hermann's Cubes. The higher-dimensional models include Douthett's Tesseract and Steinbach's Penteract.

In Chapter 3 of Section I, Jedrzejewski works with combinatorial designs which are pictured by graphs whose vertices represent the blocks. These designs can also be mapped to musical objects, where the substructure of each block highlights properties of uniqueness. Following the work of composer Tom Johnson, the author investigates the ways of representing block designs and presents special ones, called Catalan block designs, whose number of blocks is a Catalan number. They are depicted by a geometric construct called a *Stasheff polytope* or *associahedron*. As there are many families of combinatorial objects enumerated by the Catalan numbers the approach opens up many possibilities for applications to music by using the bijections between bracketings, Dyck paths, binary rooted trees, etc. Moreover, a generalized Catalan number is associated to each positive rational number, a process which leads to a new geometric structure called *rational associahedra*. Jedrzejewski states that these representations cover all combinatorial designs.

In Chapter 4 of Section I, Kastine seeks alternative methods to rhythmic tiling for composing monophonic canons which, he claims, will feature more interaction between the parts by allowing them to overlap at times. He approaches the goal of less monotony by allowing beats on which multiple parts overlap, beats of rest, unevenly distributed beats, a higher density of musical information in terms of note-to-length ratio (by definition, rhythmic tilings have a note-to-length ratio of exactly 1 note per beat) to arrive at more interesting (less regular) rhythms. The author creates what he calls *L-canons* and he takes them to the realm of melody as well. He proceeds from very strict mathematical definitions and, in the process, discovers a series of non-trivial problems to be solved. On the other hand, the author revisits the *S* and *G canons*, based on Sidon sets and Golomb rulers, which he had developed in previous work. In a similar way, as with the *L-canons*, a plethora of mathematical problems arise in the process of finding the ideal representatives of these constructions.

In Chapter 5 of Section I, Kochavi discusses the notion of rhythmic hierarchy in Luigi Nono's 1956 composition *Il canto sospeso*. Drawing on the notion of beat hierarchy, as typically manifest in the metrical structure of music with a time signature, Kochavi extends relevant concepts to this work by Nono, which was written without the conventional use of meter. Nevertheless, its durations are subject to compositional control that derives from the Fibonacci series. Specifically, Kochavi's analysis examines Nono's use of the Fibonacci series modulo n, including the Pisano period. Kochavi situates his analytical results in a discussion of the composition's extra-musical connotations as a post-war Italian resistance statement.

In Chapter 6 of Section I, Noll explains that Eric Regener's note interval system is the starting point for a number of refinements and extensions in connection with the study of modes. Two different kinds of transformations of the note space are presented: (1) signature morphisms and (2) Regener transformations. Tone systems such as the diatonic or the chromatic play a central role with respect to both types of transformations. They appear in the guise of signature morphisms: $\sigma : N \to D$ mediating between a note space N and a width/height-degree space D, both of which are isomorphic to Z_2. The image of the note space N within the degree space D is of index 7 in the diatonic case (or 12 in the chromatic, respectively). The carrier sets of modes in N are defined as preimages of a square-shaped principal diatonic domain in D. The diatonic Regener transformation turns the fifth/fourth coordinates of note intervals into major and minor step coordinates. These transformations have refinements as Special Sturmian morphisms, which serve to generate modal interval species. Modes are eventually described as contiguous paths of anchored intervals in the note space and are generated by lattice path transformations. The motto of the chapter — *One Note Samba* — points to the desire to control the variation of the meaning of a fixed note or note interval. Dual lattice path transformations are shown to be well-adapted for this purpose.

In Chapter 7 of Section I, Peck discusses the relationship between all-interval chords, as studied in musical pitch-class set theory, and the combinatorial theory of difference sets. This investigation rests on the notion of an interval as an element of a group (i.e., a generalized interval system), following the transformational theories of Lewin. Whereas an all-interval chord needs to possess at least one of every interval in an interval group by definition, some all-interval chords meet the stricter requirement of their possessing one and only one occurrence of each non-identity interval. These chords are isomorphic to planar difference sets. Such difference sets may be cyclic, (non-cyclic) abelian, or non-abelian, depending on the structure of the underlying group of intervals. Peck examines several examples of planar difference sets in musical works by Johann Sebastian Bach, Arnold Schoenberg, and Milton Babbitt.

In Chapter 8 of Section I, Plotkin looks for a way to formally define the opposite of a *parsimonious transformation*. Before he provides a definition, the author cites previous work related to the subject, including the contributions of Cohn, Childs, and Douthett and Steinbach who established rigorous criteria on the subject of parsimonious transformations. Plotkin explores a different approach this matter, in terms of scales instead of classes. At the same time, he argues for the augmented flexibility and reach of this approach, he also explains that this is the way his characterization of non-parsimonious transformations must be understood, through a technique he calls *inter-scale opposition.*

In Chapter 9 of Section II, Hughes states that one of the most intriguing and useful aspects of modeling chord spaces as orbifolds is that voice leadings take the form of paths in these spaces. He explores how viewing such paths as elements of the orbifold fundamental group or groupoid can help resolve voice leading ambiguities associated with pitch class doubling and voice crossings.

In Chapter 10 of Section II, Ivey takes as a starting point a sequence of publications by Tymoczko *et al.*, where they developed the notion of chord and voice-leading spaces constructed by applying musically relevant equivalence relations to cubic lattices representing n-dimensional pitch space. The quotient lattices are naturally embedded in topological spaces which are typically orbifolds. Ivey points out the difficulty of formulating a mathematically valid notion of geometry for these spaces that is also musically meaningful. Nonetheless, he continues, a mathematical examination of these constructions leads to several interesting avenues: the orbifold singularity points within these spaces often represent chords that maximize tonal ambiguity; many continuous chord spaces have non-trivial topology that has musical significance (e.g., homotopy generators in the orbifold lift to modulatory sequences of chords); and it may be possible to apply more sophisticated adjacency models to endow these spaces with a geometry that reflects the dominant features of certain stylistic corpora.

In Chapter 11 of Section II, Mannone states that musical performance starts from an indication of movement (a curve) hidden in the score, which then is transformed by the musician into a physical gesture (another curve), connecting the symbolic reality of the score to the physical reality of acoustics. Composition from improvisation follows the inverse path, from physical to symbolic. Symbolic gestures can be ideally transformed into physical ones via a connecting surface, known as a "world-sheet" in physics. This formalism can be applied to any musical instrument, including the voice. The relations between gestures on different musical instruments can be framed through category theory, allowing comparison within music itself, and between music and other fields. Experiments in which images and gestures in the visual arts have been transformed into music have their explanation in categorical terms, via gestural analogies and similarity. In fact, the mathematical definition of musical gestures, apart from explaining and modeling musical practice, may constitute a musical element itself in composition. Mannone's chapter also includes some examples of music from images, and an excerpt from an original orchestral piece, where gestural analogies and morphisms connect instruments and sounds.

In Chapter 12 of Section II, Sivakumar and Tymoczko state that voice leading is closely connected with homotopy, that is, the exploration of paths in higher-dimensional configuration spaces. Musicians explored these spaces centuries before mathematicians developed tools for describing them. In this chapter, they analyze the group structure of these contrapuntal paths, generalize traditional music-theoretical vocabulary for representing voice leading, and ask what proportion of the paths are realizable given our generalized vocabulary of musical possibilities.

In Chapter 13 of Section II, Yust takes a geometrical approach to generalized *Tonnetze*, drawing on previous work that includes Cohn's common-tone formulation and Tymoczko's voice-leading reformulation. Yust's geometric interpretation uses the common-tone approach to demonstrate that the toroidal spaces in which such *Tonnetze* are realized correspond to Fourier phase spaces. The result is that Yust is able to optimize a *Tonnetz* to a toroidal space (or vice versa) by using the discrete Fourier transform (DFT). Triangulating the 2-torus allows Yust to associate the resulting regions with members of a given trichordal set class. Accordingly, in extending this notion to the three-dimensional case, the *Tonnetz* becomes a network of tetrachordal set classes that share common trichordal subsets. Finally, Yust considers other *Tonnetz* types, particularly those with non-toroidal or bounded topologies.

In Chapter 14 of Section II, Milam presents a preliminary investigation of an original technique for generating musical material from mathematically deterministic games. Based on the premise that mathematical models underlie the logical and coherent relationships that organize all natural phenomena, Milam attempts to coordinate the outcome of purely mathematical processes with various elements of musical experience to produce deterministic musical compositions with inherent, but possibly unexpected, natural relationships.

In Chapter 15 of Section III, Gómez presents the efforts of COFLA (Computational Analysis of Flamenco Music), a group of researchers from several disciplines who study flamenco music. Their goal is to analyze flamenco music from different disciplines, incorporating music technology in that analysis. To accomplish this objective, COFLA is composed of an interdisciplinary team including exerts from areas such as Musicology, Ethnomusicology, History, Literature, Education, Sociology, but also Mathematics, Engineering, and Computer Science. The chapter includes a brief overview of the main musical features of flamenco music and three problems the COFLA group is currently working on. The problem of melodic similarity in flamenco a cappella songs is addressed from an interdisciplinary perspective.

In Chapter 16 of Section III, Baker designed an experiment with the goal of exploring how different computational and algorithmic measures can be used to mirror human behavior by employing a contextual paradigm. This goal lead to the use of a design that was not dependent on any sort of explicit understanding of Western musical notation. However, the similarity measures that were contrasted did depend on certain Charlie Parker jazz tunes that were selected.

In Chapter 17 of Section III, Shanahan proposes a corpus-study-based approach to studying prototypicality. He argues that, whereas corpus-based methods have mostly been used to analyze aspects of *usage*, they can also be applied to an examination of *typicality*. However, this approach involves more data, and presents certain problems with regard to the use of schema. The corpus that Shanahan investigates in this regard is the body of eighteenth-century

Italian melody from the RISM-World Dataset, which has previously been discussed in terms of schema and prototype, but not in a corpus-based analysis of those features. He divides his analysis into "high-level" and "low-level" features. He concludes that, in conducting such an analysis, many aspects of an entire composition should be encoded, including harmonic information. Using melodic information alone makes it difficult to examine tonal assimilation (and dissimilation) as "high-level" features. Ideally, such corpora will include more information, permitting the use of schema that are not used in traditional searching methods.

Mariana Montiel
Robert Peck

About the Editors

Mariana Montiel is Associate Professor at Georgia State University. Her research resolves around Mathematical Music Theory and frequently she carries out interdisciplinary work with music theorists and computer scientists. Simultaneously she cultivates an interest in pedagogical aspects and, in particular, in the use of abstract and symbolic representations, common to both disciplines, mathematics and music.

Robert Peck is Professor of Music Theory at Louisiana State University. An active researcher in the field of mathematical music theory, he is a founding co-Editor-in-Chief of the *Journal of Mathematics and Music,* and has co-organized seven special sessions on mathematics and music at meetings of the American Mathematical Society.

Contents

Section I

Chapter 1

From Musical Chords to Twin Primes

Jack Douthett[*], David Clampitt[†] and Norman Carey[‡]

University of New Mexico, Albuquerque, NM, USA
†The Ohio State University, Columbus, OH, USA
‡CUNY Graduate Center, New York, NY, USA

1.1. Introduction

A freshman music student is often taught by rote that each of the diatonic intervals (seconds through sevenths) come in two flavors: seconds and thirds in major and minor, fourths in perfect and augmented, fifths in diminished and perfect, and so on. The young musician learns to identify these intervals aurally and reproduce them in song. While this pairing of intervals for each scale degree might seem curious, within it lurks some non-trivial mathematics. Moreover, the musician might notice that within the diatonic scale there are three different triads (major, minor, and diminished) and every triad has three different notes, and that seventh chords come in four types and there are four notes in every seventh chord. But in general that is as far as it goes. And to try to attach meaning to the fact that the cardinalities of the diatonic scale (white keys) and its pentatonic complement (black keys) are twin primes would seem like weird numerology. But as it turns out, contemporary mathematical music theory draws these observations together. This chapter will trace these connections, collating some new results with some from the existing literature in music theory, mathematics, and physics.

1.2. Myhill's Property and Well-Formed Scales

Musical scales are defined to be finite sets of pitches ordered according to pitch height (increasing fundamental frequencies), and periodic at some interval (usually the octave, associated with the frequency ratio 2:1). We understand scales to be equivalent under musical transposition, i.e., translation of all its points by a constant. In music theory there are two ways of conceptualizing scales, which we may call *continuous background* and *discrete background*, or *continuous* and *discrete* for short (both are discrete sets). A continuous scale is an ordered set of real-valued points on a mod 1 circle. If we order these points from smallest to largest, then a scale is defined as this ordered set or a cyclic rotation thereof, to within transposition. In Fig. 1.1(a), the scale is $D = \{0, \log_2(9/8), \log_2(3/2)\}$, where the base 2 logarithms reflect the fact that there is a natural equivalence relation in music based on the octave. The discrete model is similar except that the circumference is divided into c integer parts labeled 0 through $c-1$ (equivalent to rational points $0, 1/c, \ldots, (c-1)/c$, but the clock face diagram is conventional in music theory). Reflecting their application in music theory, these numbers are called *pitch classes* (*pcs*). In Fig. 1.1(b), $c = 12$, and the scale is $D = \{0, 2, 7\}$ (according to music theory convention, the mod 12 pitch class 0 represents the note **C**, so D contains notes **C**, **D**, and **G**).

The scales in Fig. 1.1 have a property not generally found in an arbitrary scale. We call the intervals between adjacent points on the circle step intervals, and we measure generic intervals according to the number of step intervals they span. Thus step intervals have span 1, intervals that skip one intervening point have span 2, and so forth. Following Clough and Myerson [8, 9], and Clough and Douthett [7] we define the *spectrum* of a generic interval to be the set of specific interval sizes corresponding to the generic interval of a given span, and we designate this set by enclosing the span k in angle brackets. Note in Fig. 1.1(a) that the step intervals come in two sizes (the first spectrum is the doubleton $\langle 1 \rangle = \{\log_2(9/8), \log_2(4/3)\}$), as is the case for the spectrum of span 2 ($\langle 2 \rangle = \log_2(3/2), \log_2(16/9)\}$). This is also true of Fig. 1.1(b), where $\langle 1 \rangle = \{2, 5\}$ and $\langle 2 \rangle = \{7, 10\}$, the interval sizes understood to be modulo 12. If we let d be the cardinality of the

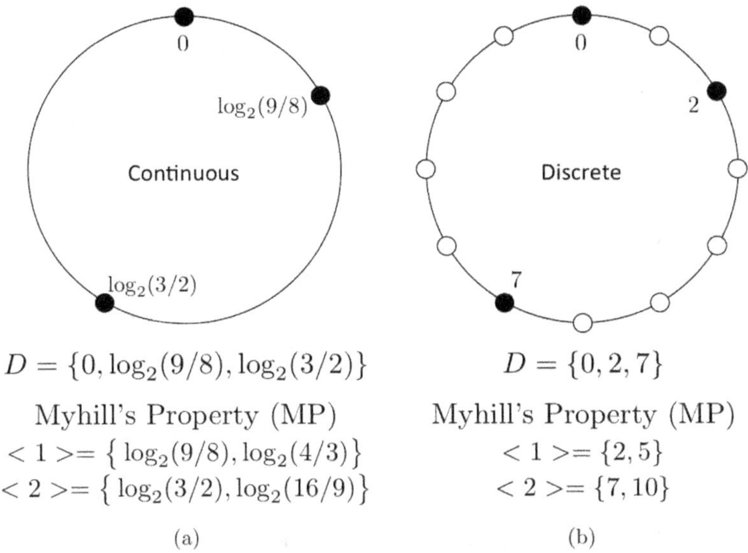

$$D = \{0, \log_2(9/8), \log_2(3/2)\} \qquad D = \{0, 2, 7\}$$

Myhill's Property (MP)   Myhill's Property (MP)

$$<1> = \{\log_2(9/8), \log_2(4/3)\} \qquad <1> = \{2, 5\}$$
$$<2> = \{\log_2(3/2), \log_2(16/9)\} \qquad <2> = \{7, 10\}$$

(a)   (b)

Fig. 1.1.   Well-formed scales.

discrete scale with background chromatic of cardinality c $(1 \leq d < c)$, then when each spectrum $\langle k \rangle, 1 \leq k \leq d - 1$, is a doubleton, we say the scale has *Myhill's Property (MP)*, as initially defined in [8, 9]. This is the property observed in the introduction that every non-zero diatonic interval comes in "two flavors"; it plays a key role in mathematical music theory. We adopt the terminology of Carey and Clampitt [2, 3], and call both discrete and continuous background scales with MP *well-formed (WF)* scales. Carey and Clampitt also admit *degenerate WF scales*, where the spectra are all singletons (i.e., the distinction between generic and specific interval collapses). We will assume in this chapter that all WF scales are non-degenerate.

A WF scale is a *generated scale*, that admits a *generating interval* or *generator* [2]. That is, with respect to the continuous background, there exists a real value $g > 0$, the generator, such that every element s of the scale is an integer multiple mod 1 of the generator: $s = ng \bmod 1$. If the generator is a rational value m/c, then with respect to the continuous background the elements of the generated scale of cardinality d are $\{0, (m/c) \bmod 1, 2(m/c) \bmod 1, \ldots, (d-1)(m/c) \bmod 1\}$, or, equivalently, with respect to the discrete background, mod c residues $\{0, m \bmod c, 2m \bmod c, \ldots,$ $(d-1)m \bmod c\}$. The property that defines a WF scale is that the span of its generating interval must be a constant: there must be a generic interval whose spectrum contains two intervals, one of multiplicity $d-1$ and one of multiplicity 1.

There is one last observation to make in Fig. 1.1. In the discrete case, we may reasonably define the complementary scale to be the scale formed by the pcs not included in D. There is no natural way to do that in the continuous case. In what follows, complementary scales are going to play a key role in our observations, so from this point on we will assume all scales are discrete.

We begin with a simple question. Are complements of WF scales WF? Figure 1.2(a) reproduces Fig. 1.1(b) (the WF scale $D = \{0, 2, 7\}$), and Fig. 1.2(b) represents its complement ($D' = \{1, 3, 4, 6, 8, 9, 10, 11\}$). Since several of the spectra in Fig. 1.2(b) are triples, it is clear that the general answer to our question above is "no". The next obvious question is: is the complement of a WF scale ever WF?

1.3.   Complementary WF Scales

As seen in Fig. 1.2, the complement of a WF scale need not be WF. Now we go on the hunt for complementary scales that are both WF.

In 1991, inspired by the work of Clough and Myerson, Clough and Douthett wrote their paper on *maximally even (ME)* scales [7]. In an ME scale, every spectrum consists of one or two consecutive integers, and several important chords and scales have this property: augmented triad, fully diminished seventh chord, black key pentatonic scale, whole-tone scale, diatonic scale and the half-step/whole-step octatonic scale. In the mid-90s it was discovered that the formalism of ME scales also applied to the Ising model: ME configurations minimize average configurational energy [11]. This approach was extended to determine the phase transitions and fractal dimensions of the Ising model [14]. This was followed by two papers that gave a more mathematically suitable definition for ME sets (scales), which were shown to be equivalent to the definition adopted by Clough and Douthett [10, 12, 13]. Other treatments of ME sets include [1, 15].

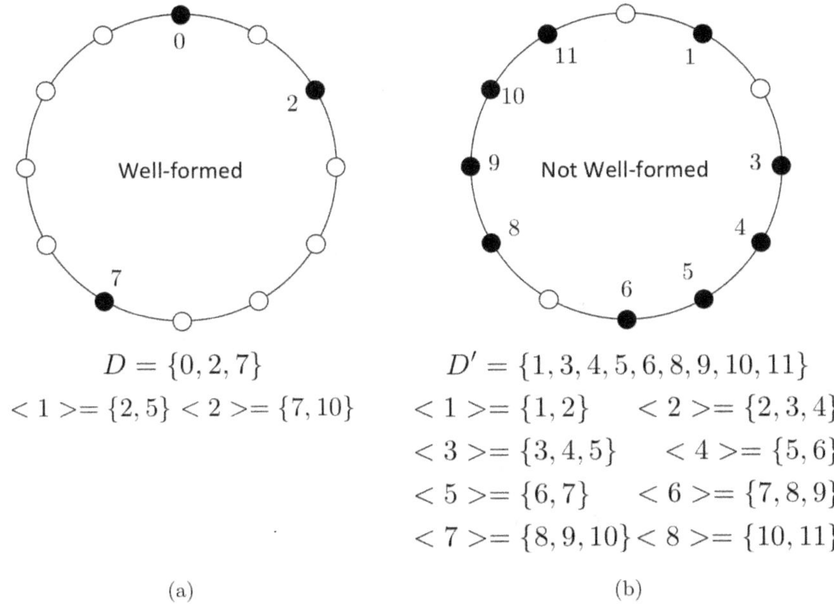

$$D = \{0, 2, 7\}$$
$$< 1 >= \{2, 5\} \quad < 2 >= \{7, 10\}$$

$$D' = \{1, 3, 4, 5, 6, 8, 9, 10, 11\}$$
$$< 1 >= \{1, 2\} \qquad < 2 >= \{2, 3, 4\}$$
$$< 3 >= \{3, 4, 5\} \qquad < 4 >= \{5, 6\}$$
$$< 5 >= \{6, 7\} \qquad < 6 >= \{7, 8, 9\}$$
$$< 7 >= \{8, 9, 10\} \quad < 8 >= \{10, 11\}$$

(a)   (b)

Fig. 1.2. Complementary scales (a) 3-in-12, (b) 9-in-12.

Clough and Douthett divided ME scales into three classes: those whose spectra are singletons (e.g., whole-tone scales), those whose spectra contain both singletons and doubletons (e.g., octatonic scales), and those whose spectra are all doubletons (e.g., pentatonic and diatonic scales). Since scales in the last of these classes are WF, we refer to these scales as *WFME* scales. Clough and Douthett [7] showed that the complement of a WFME scale is also a WFME scale. So, WFME scales are the first example of complementary WF scales. Figure 1.3 shows the complementary black-key pentatonic and white-key diatonic scales with their doubleton spectra listed below.

Next we explore the antithesis of ME scales, *minimally even* (*me*) scales. The term *me* first appears in the physics literature [11]. As mentioned above, ME configurations minimize average configurational energy [10–12]. It is also shown in [10, 11] that me configurations maximize average configurational energy. In musical terms, a me scale is simply a chord cluster, and it is not hard to see that all me scales are WF, as illustrated in Figs. 1.4(a) and 1.4(b).

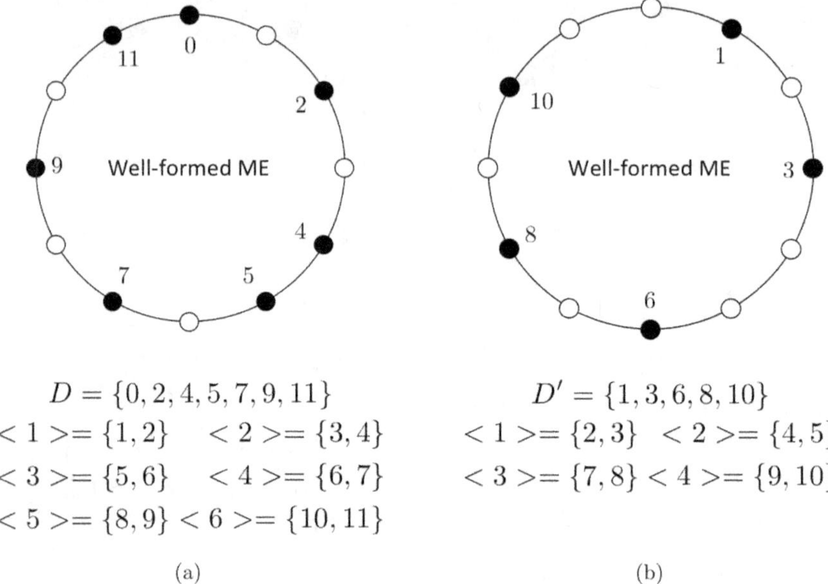

$$D = \{0, 2, 4, 5, 7, 9, 11\}$$
$$<1>= \{1,2\} \quad <2>= \{3,4\}$$
$$<3>= \{5,6\} \quad <4>= \{6,7\}$$
$$<5>= \{8,9\} <6>= \{10,11\}$$

(a)

$$D' = \{1, 3, 6, 8, 10\}$$
$$<1>= \{2,3\} \quad <2>= \{4,5\}$$
$$<3>= \{7,8\} <4>= \{9,10\}$$

(b)

Fig. 1.3. Complementary WFME scales (a) 7-in-12, (b) 5-in-12.

For consistency in our use of the term WF, we will call these scales WFme scales (despite the redundancy). So, we now have two classes of WF scales whose complementary scales are also WF. Are there others?

To search for other complementary WF scales we make the following definitions. We consider a scale D of cardinality $d > c/2$, where the chromatic background set is of cardinality c. (In the case of scales D where c is even and $d = c/2$, it is clear that the cluster scale of cardinality d and its complement are both essentially the same WFme scale. We will see that this is the only pair of WF complements for this cardinality.) We do not assume that D is WF, but only that its step spectrum has two elements. All WF scales D are therefore included in these definitions:

$$\langle 1 \rangle = \{x, y\}, x < y,$$
$$d' = c - d,$$
$$a = \text{the number of step intervals of size } x,$$
$$b = \text{the number of step intervals of size } y.$$

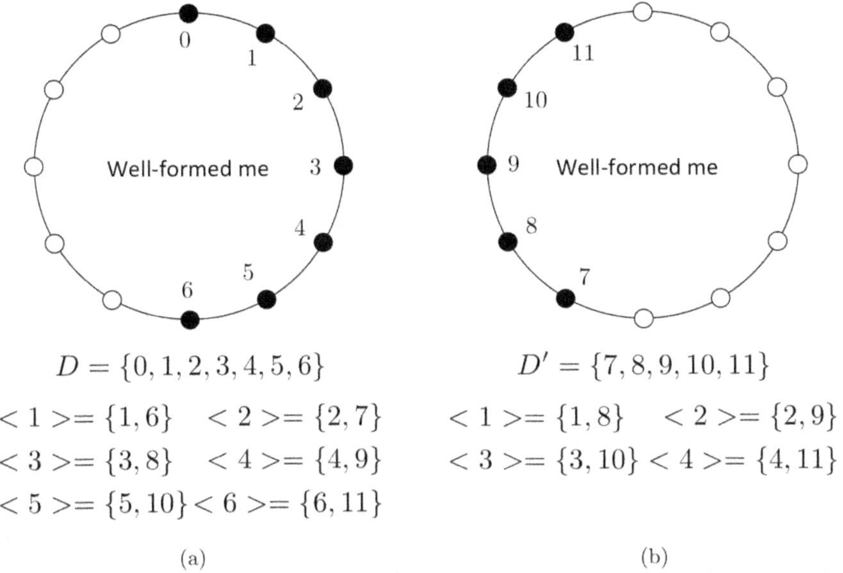

$$D = \{0, 1, 2, 3, 4, 5, 6\}$$

$$< 1 >= \{1, 6\} \quad < 2 >= \{2, 7\}$$
$$< 3 >= \{3, 8\} \quad < 4 >= \{4, 9\}$$
$$< 5 >= \{5, 10\} < 6 >= \{6, 11\}$$

(a)

$$D' = \{7, 8, 9, 10, 11\}$$

$$< 1 >= \{1, 8\} \quad < 2 >= \{2, 9\}$$
$$< 3 >= \{3, 10\} < 4 >= \{4, 11\}$$

(b)

Fig. 1.4.   Complementary WFme scales (a) 7-in-12, (b) 5-in-12.

This leads us to two linear Diophantine equations, $a, b \neq 0$:

$$a + b = d,$$

$$ax + by = c.$$

To employ the parameters in these equations to discuss the WF scales of cardinality d and the existence or non-existence of complementary WF scales of cardinality d', we draw on a lemma from well-formed scale theory [3–5].

Lemma 1. *A generated scale D of cardinality d with generator real number θ, $0 < \theta < 1$, is WF if and only if d is the denominator of a convergent or semi-convergent in the continued fraction representation of θ. Furthermore, if D is WF, the span of the generator is the multiplicative inverse $\mathrm{mod}\ d$ of one of the step interval multiplicities, without loss of generality $b^{-1} \bmod d = \mathbf{b}$, taking the least positive residue $\mathrm{mod}\ d$. Then $\mathbf{b}/d$ is the required (semi-) convergent to θ.*

Theorem 1. *Let D be a scale of cardinality d, with parameters defined as above. Then $b|d'$.*

Proof. Because $d > c/2$ and $y > 1$, we have $x = 1$. This is clear from the pigeonhole principle, but also falls out of the following argument: If both integers $x, y > 1$, $c = ax + by \geq 2(a + b) = 2d > c$, contradiction. Thus we have $x = 1$. Then subtracting $a + b = d$ from $a + by = c$, we have $b(y - 1) = c - d = d'$. It follows that $b|d'$.   □

Theorem 2. *The scale D is WF if and only if $b|d'$ and $(b, d) = 1$.*

Proof. We begin with the converse. If there exists a WF scale D of cardinality d, it follows from Theorem 1 that $b|d'$. That $(b, d) = 1$ follows from the lemma.

Now we assume $b|d'$ and $(b, d) = 1$, and demonstrate the existence of WF scale D. Then $b^{-1} \bmod d$ exists, and we again let $\boldsymbol{b}$ be the least positive residue $b^{-1} \bmod d$. We have $(b, c) = 1$: $b|(c - d)$, so $c \equiv d \bmod b$, and d is coprime with b.

Consider first two special cases, $b = 1$, and $b = d'$. If $b = 1$, by the lemma, the span of the generating interval is 1, and since $d > c/2$, the generator is 1 (or $1/c$ with respect to the continuous background). Then the generated scale D is WFme, and so is its complement.

If $b = d'$, it follows that $(c, d) = 1$: Here, $b = c - d$, or $d = c - b$, and so $c \equiv b \bmod d$. From $(b, d) = 1$, we have $(c, d) = 1$. Let n be the positive integer such that $|\boldsymbol{b}d' - dn| = 1$ (from the definition of $\boldsymbol{b}$). Because $(c, d) = 1$ we can also find the unique positive integer m that satisfies $|c\boldsymbol{b} - md| = 1$. Then by continued fraction theory, $\boldsymbol{b}/d$ and n/d' are full convergents to m/c, and therefore the complementary scales of cardinalities d and d' are WF, by the lemma.

Let us now stipulate that we do not have the trivial case $b = d' = 1$, which was covered above. Now we can say that we have constructed the unique pair of WF complements with cardinalities d and d': given $b = d'$, we have generator m/c, and n/d' and $\boldsymbol{b}/d$ are the unique pair of fractions that have m/c as their mediant: $n/d' \oplus \boldsymbol{b}/d = (n + \boldsymbol{b})/(d' + d) = m/c$, subject to the equations above and also to $|nc - md'| = 1$: $|nc - md'| = |nc - (n + \boldsymbol{b})b| = |n(c - b) - b\boldsymbol{b}| = |nd - b\boldsymbol{b}| = 1$. We therefore have the unique divisor b and unique generator mod c, such that both d and d' are denominators of convergents to the generator, and are thereby WF.

Moreover, these complementary scales are both WFME. Since $b = d', a + d' = d, a + d'y = c$, and subtracting the former equation from the latter, $c - d = d' = d'(y - 1)$, which implies $y = 2$, and D is WFME. Then the complement is also WFME ([7]: the complement of an ME set is ME).

Now consider the case where b is a proper divisor of d': $bw = d'$, with $1 < b, w < d'$. Then solve for positive n such that $|bb - nd| = 1$. It follows that n/b is a convergent to b/d, which in turn is a semi-convergent to $(b + wn)/(d + wb) = (b + wn)/c$. Note that

$$b(b + wn) - n(d + wb) = (bb + bwn) - (nd + nwb)$$
$$= bb - nd$$
$$= \pm 1.$$

Let $b + wn = g$. By the lemma, the discrete set

$$D = \{0, g \bmod c, 2g \bmod c, \ldots, (d - 1)g \bmod c\},$$

is well-formed, by construction.

Although D is WF, it is not ME: by applying Theorem 1, we know that the size of the larger step interval $y = w + 1 > 2$, so the sizes of the step intervals are not consecutive integers. Moreover, the complement D' is not well-formed. As we showed above, there is a unique pair of WF complements with cardinalities $d = c - b$ and $d' = b$. $\qquad\square$

Corollary 1. *From the foregoing, there is a one-to-one correspondence between divisors of d' that are coprime with d and distinct WF scales of cardinality d.*

Corollary 2. *If $(d, d') = 1$, the only complementary WF scales of cardinalities d and d' are either both WFme or both WFME. If $(d, d') \neq 1$, the only complementary WF scales are WFme.*

Proof. There exist complementary WFme scales for all complementary cardinalities d and d', including the case $d = d' = c/2$, excluded from the above theorems. From Theorem 2, the unique case of complementary WF scales is when $b = d'$, $(d, d') = 1$. In this case, the complementary WF scales are WFME. $\qquad\square$

Example. Consider $c = d + d' = 11 + 9$. Since $d' = 9$ has three divisors (1, 3, and 9), all coprime with d (since $(d, d') = 1$), there are three WF scales of cardinality $d = 11 \bmod 20$.

For $b = 1$ and $y = 10$, $D_1 = \{0, 1, 2, 3, 4, 5, 6, 7, 8, 9, 10\} \bmod 20$ (WFme), which is a WFme scale since there is only one step interval greater than 1 (chromatic cluster). Thus its complement is WFme.

For $b = 9$ and $y = 2$, $D_2 = \{0, 1, 3, 5, 7, 9, 10, 12, 14, 16, 18\} \bmod 20$ (WFME). $9^{-1} \bmod 11 = 5$, and we find a convergent to 5/11 with denominator 9 by solving for $5 \cdot 9 - n \cdot 11 = 1$. Then $n = 4$, and 4/9 and 5/11 are by construction convergents to generator 9/20, the mediant of 4/9 and 5/11. We note that 9/20 generates complementary WF scales: taking 11 consecutive multiples of 9 mod 20, we have in generation order $\{0, 9, 18, 7, 16, 5, 14, 3, 12, 1, 10\}$, which is D_2. The literal complement is $D_2' = \{2, 4, 6, 8, 11, 13, 15, 17, 19\}$. D_2' transposed up by 1 chromatic step (add 1 mod 20 to each value) yields the first 9 consecutive multiples of 9 mod 20 in generation order, the WF scale of cardinality $d' = 9$: $\{0, 3, 5, 7, 9, 12, 14, 16, 18\}$.

For $b = 3$, with $y = 4$, $D_3 = \{0, 1, 2, 3, 7, 8, 9, 10, 14, 15, 16\} \bmod 20$ (WF, not ME nor me). To construct this scale, note that $3^{-1} \bmod 11 = 4$, we find a convergent to 4/11 with denominator 3 by solving $4 \cdot 3 - n \cdot 11 = 1$. Then $n = 1$, and 1/3 is a convergent to 4/11, which in turn is a semi-convergent to $(4 + 3 \cdot 1)/(11 + 3 \cdot 3) = 7/20$. Taking 11 consecutive multiples of 7 mod 20, we have in generation order $\{0, 7, 14, 1, 8, 15, 2, 9, 16, 3, 10\}$, which is D_3. The complementary set has a step spectrum which is a triple, so it is not WF.

The reader is invited to test the case where $c = 25$, $d = 15$, where among the divisors of $d' = 10$ we have 2 which is coprime with 15. Applying the theorem, one should arrive at a well-formed scale $\{0, 1, 2, 3, 4, 5, 6, 7, 13, 14, 15, 16, 17, 18, 19\}$ with generator 13/25. On the other hand, with the divisor 5, not coprime with 15, there is a scale that is ME but not WF: $\{0, 1, 2, 5, 6, 7, 10, 11, 12, 15, 16, 17, 20, 21, 22\}$.

1.4. Cardinality Equals Variety

There are two kinds of cardinality equals variety (CV): *line cardinality equals variety (LCV)* and *chord cardinality equals variety (CCV)*.

A *line* is defined to be an ordered set of pitch classes. Consider a scale D of cardinality $d, 1 \leq d < c$, in a chromatic universe U_c of cardinality c, $D = \{d_0, d_1, \ldots, d_{d-1} | d_i \in U_c, d_i < d_{i+1}, 0 \leq i < d-1\}$. Let a line L of length k be $L = \langle d(m_1), d(m_2), \ldots, d(m_k) \rangle$, $d(m_i) \in D$, with $0 \leq m_i < d$, $1 \leq k \leq d$.

The line genus, or generic description of L, is $gL = \langle m_2 - m_1, m_3 - m_2, \ldots, m_k - m_{k-1}, m_1 - m_k \rangle$, all differences modulo d. The line species, or specific description of line L, is $sL = \langle d(m_2) - d(m_1), d(m_3) - d(m_2), \ldots, d(m_k) - d(m_{k-1}), d(m_1) - d(m_k) \rangle$, all differences modulo c. The angle brackets mean these are strict sequences: rotated sets are considered inequivalent. We say that D has LCV if for all lines of length k with generic description gL there are k distinct specific descriptions sL, for $1 \leq k \leq d$.

A *chord* is defined to be an unordered set of pitch classes. D and U_c are defined as above. Then let $\chi \subset D$ be a chord of cardinality k, such that $\chi = \{d(m_1), d(m_2), \ldots, d(m_k)\}$, but in order to compare descriptions, we choose to order the set such that $m_i < m_{i+1}$, for $1 \leq i < k$. The generic description of the chord is the ordered set $g\chi = (m_2 - m_1, m_3 - m_2, \ldots, m_k - m_{k-1}, m_1 - m_k)$, all differences modulo d, and the specific description is the ordered set $s\chi = (d(m_2) - d(m_1), \ldots, d(m_k) - d(m_{k-1}), d(m_1) - d(m_k))$, all differences modulo c, but now the round parentheses mean that all cyclic rotations of these ordered sets are considered to be equivalent. We say that D has CCV if for all chords of cardinality k with generic description $g\chi$ there are k distinct, inequivalent specific descriptions $s\chi$, for $1 \leq k < d$. Note that the case $k = d$ must now be excluded: this is generic description $(1, 1, \ldots, 1)$, and there is exactly 1 specific description up to rotation.

Clough and Myerson [8] showed that the requirement for LCV is very simple.

Theorem 3 (Clough and Myerson). *A scale has LCV if and only if the scale is WF.*

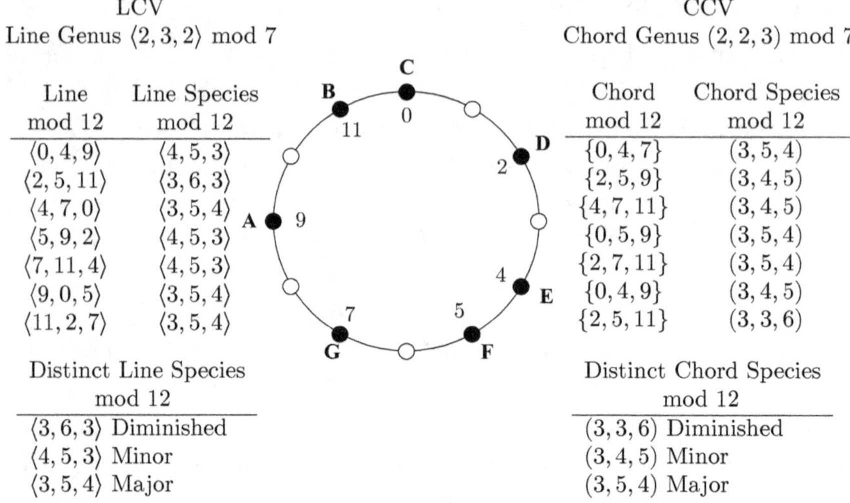

Line mod 12	Line Species mod 12
$\langle 0,4,9 \rangle$	$\langle 4,5,3 \rangle$
$\langle 2,5,11 \rangle$	$\langle 3,6,3 \rangle$
$\langle 4,7,0 \rangle$	$\langle 3,5,4 \rangle$
$\langle 5,9,2 \rangle$	$\langle 4,5,3 \rangle$
$\langle 7,11,4 \rangle$	$\langle 4,5,3 \rangle$
$\langle 9,0,5 \rangle$	$\langle 3,5,4 \rangle$
$\langle 11,2,7 \rangle$	$\langle 3,5,4 \rangle$

Chord mod 12	Chord Species mod 12
$\{0,4,7\}$	$(3,5,4)$
$\{2,5,9\}$	$(3,4,5)$
$\{4,7,11\}$	$(3,4,5)$
$\{0,5,9\}$	$(3,5,4)$
$\{2,7,11\}$	$(3,5,4)$
$\{0,4,9\}$	$(3,4,5)$
$\{2,5,11\}$	$(3,3,6)$

Distinct Line Species mod 12	
$\langle 3,6,3 \rangle$	Diminished
$\langle 4,5,3 \rangle$	Minor
$\langle 3,5,4 \rangle$	Major

Distinct Chord Species mod 12	
$(3,3,6)$	Diminished
$(3,4,5)$	Minor
$(3,5,4)$	Major

Fig. 1.5. CV for Lines and CV for Chords in ME 7-in-12.

Figure 1.5 illustrates the difference between LCV and CCV. In the usual diatonic, we consider the line genus of an arpeggiated first inversion triad, $\langle 2,3,2 \rangle$. Since this is a sequence, the rotation cannot be changed. On the other hand, as a chord (set), genus $(2,3,2)$ is the same as $(3,2,2)$ is the same as $(2,2,3)$. In Column 1 under LCV we list the literal triadic lines beginning from A minor in first inversion, in mod 12 pcs, $\langle 0,4,9 \rangle$, and then the other members of the line class following scale order. Column 2 under LCV indicates the associated specific line descriptions (species). We see that there are 3 specific descriptions: $\langle 4,5,3 \rangle$, $\langle 3,5,4 \rangle$, and $\langle 3,6,3 \rangle$. In Column 1 under CCV we list the triads in scale order according to their roots, C major, d minor, etc., and since they are considered to be unordered sets, in Column 2 under CCV we may choose to list pcs of the chord (set) in ascending order as natural numbers. When specific descriptions are taken, we choose the rotation that begins with the common smallest value, 3. Again we see that there are 3 specific descriptions: $(3,4,5)$, $(3,5,4)$, and $(3,3,6)$.

Since in this case both generic lines and chords come in three species, one might think LCV and CCV are beholden to the same

LCV
Line Genus $\langle 2, 2, 2, 2 \rangle$ mod 8

Not CCV
Chord Genus $(2, 2, 2, 2)$ mod 8

Line mod 13	Line Species mod 13	Chord mod 13	Chord Species mod 13
$\langle 0, 3, 6, 9 \rangle$	$\langle 3, 3, 3, 4 \rangle$	$\{0, 3, 6, 9\}$	$(3, 3, 3, 4)$
$\langle 1, 4, 8, 11 \rangle$	$\langle 3, 4, 3, 3 \rangle$	$\{1, 4, 8, 11\}$	$(3, 3, 3, 4)$
$\langle 3, 6, 9, 0 \rangle$	$\langle 3, 3, 4, 3 \rangle$	$\{0, 3, 6, 9\}$	$(3, 3, 3, 4)$
$\langle 4, 8, 11, 1 \rangle$	$\langle 4, 3, 3, 3 \rangle$	$\{1, 4, 8, 11\}$	$(3, 3, 3, 4)$
$\langle 6, 9, 0, 3 \rangle$	$\langle 3, 4, 3, 3 \rangle$	$\{0, 3, 6, 9\}$	$(3, 3, 3, 4)$
$\langle 8, 11, 1, 4 \rangle$	$\langle 3, 3, 3, 4 \rangle$	$\{1, 4, 8, 11\}$	$(3, 3, 3, 4)$
$\langle 9, 0, 3, 6 \rangle$	$\langle 4, 3, 3, 3 \rangle$	$\{0, 3, 6, 0\}$	$(3, 3, 3, 4)$
$\langle 11, 1, 4, 8 \rangle$	$\langle 3, 3, 4, 3 \rangle$	$\{1, 4, 8, 11\}$	$(3, 3, 3, 4)$

Distinct Line Species mod 13	Distinct Chord Species mod 13
$\langle 3, 3, 3, 4 \rangle$	$(3, 3, 3, 4)$
$\langle 3, 3, 4, 3 \rangle$	
$\langle 3, 4, 3, 3 \rangle$	
$\langle 4, 3, 3, 4 \rangle$	

Fig. 1.6. No CV for Chords in ME 8-in-13.

rules. But consider the line $\langle \mathbf{C}, \mathbf{D}, \mathbf{E}, \mathbf{F}, \mathbf{G}, \mathbf{A}, \mathbf{B} \rangle$. This produces the seven diatonic modes as lines (LCV holds) but only one chord (this is why CCV must exclude the complete scale D, as noted above). While Clough and Myerson's focus was LCV, they did comment on CCV: "The situation for chords is a bit more complicated. In the usual diatonic scale, CV holds except for $k = 7$ — there is only one seven-note chord" [8]. So one might think that by requiring the chord be a proper subset of the scale, the problem would be solved. This actually works for the diatonic scale, but in general it does not.

Consider the scale $D = \{0, 1, 3, 4, 6, 8, 9, 11\}$ mod 13 and the line and chord genus $\langle 2, 2, 2, 2 \rangle$, as shown in Fig. 1.6. Although we have a proper subset, there are four line species but only one chord species. This is because the generic description is an equal division of the cardinality of the scale, and the line species are equivalent under rotation.

LCV
Line Genus $\langle 1,2,2\rangle$ mod 5

Line mod 7	Line Species mod 7
$\langle 0,1,4\rangle$	$\langle 1,3,3\rangle$
$\langle 1,2,5\rangle$	$\langle 1,3,3\rangle$
$\langle 2,4,0\rangle$	$\langle 2,3,2\rangle$
$\langle 4,5,1\rangle$	$\langle 1,3,3\rangle$
$\langle 5,0,2\rangle$	$\langle 2,2,3\rangle$

Distinct Line Species mod 7
$\langle 1,3,3\rangle$
$\langle 2,2,3\rangle$
$\langle 2,3,2\rangle$

Not CCV
Chord Genus $(1,2,2)$ mod 5

Chord mod 7	Chord Species mod 7
$\{0,1,4\}$	$(1,3,3)$
$\{1,2,5\}$	$(1,3,3)$
$\{0,2,4\}$	$(2,2,3)$
$\{1,4,5\}$	$(1,3,3)$
$\{0,2,5\}$	$(2,2,3)$

Distinct Chord Species mod 7
$(1,3,3)$
$(2,2,3)$

Fig. 1.7. No CV for Chords in ME 5-in-7.

As it turns out not even this is restrictive enough. Consider the scale $D = \{0,1,2,4,5\}$ mod 7 and the genus $\langle 1,2,2\rangle$ in Fig. 1.7. Once again LCV holds, but CCV does not. So what are the conditions for CCV? The answer comes in [6].

Theorem 4 (Clampitt). *A scale of cardinality d in a chromatic universe of cardinality c has CCV if and only if the following conditions hold:*

(1) *the scale is WF,*
(2) *d is prime,*
(3) *$d \le \lfloor c/2 \rfloor + 1$.*

The reference to the proper subset in Clampitt's definition takes care of Clough and Myerson's first observation on CCV; part (2) of his theorem takes care of their second concern, and part (3) takes care of the problem not observed by Clough and Myerson (Fig. 1.7). That is, the maximally even 5-in-7 scale shown in Fig. 1.7 satisfies the first two conditions, but not the third: it is WF and its cardinality is prime, but it does not exhibit CCV because $5 > \lfloor 7/2 \rfloor + 1$.

1.5. Dual CCV

As mentioned above, the diatonic scale has CCV, as does its pentatonic complement. This can be checked with Clampitt's theorem, or the scales are small enough that they could be checked manually. This observation takes us to our next definition.

Definition (Clampitt–Douthett). A chromatic universe with cardinality c is said to support dual CCV if there exist complementary scales that both have CCV.

It is this definition that illustrates the unusual connections between musical chords and primes. If a scale D of cardinality d and its complement of cardinality d' have CCV, then by part (3) of Clampitt's theorem, for $c > 5$

$$\frac{c}{2} - 1 \le d', d \le \frac{c}{2} + 1,$$

and d, d' must both be prime by part (2). (If $c > 5$ were odd, then the complementary cardinalities would be $\lfloor c/2 \rfloor$ and $\lfloor c/2 \rfloor + 1$, at least one of which would be non-prime, so to satisfy CCV, c must be even.) Moreover, the cardinalities must both be $c/2$ or one is $\frac{c}{2} - 1$ and its complement is $\frac{c}{2} + 1$.

If the complementary scales have cardinality $c/2$, they are WFme, by the remark preceding Theorem 1 that dispatched the $c/2$ case.

Otherwise they can be either WFme or WFME. If the complementary WF scales that both have CCV are WFme, then we say the universe U_c supports dual me CCV, and if the complementary WF scales that both have CCV are ME, we say the universe U_c supports dual ME CCV. In either case, we say U_c supports dual CCV.

Theorem 5. *If U_c with $c = d + d'$ is a universe that supports dual CCV where d and d' are the cardinalities of the complementary pairs of scales with CCV, then d and d' are primes that differ at most by 2. Moreover,*

(1) *if $d = d'$, then the complementary scales are WFme;*
(2) *if $d \ne d'$, then d and d' are 3 and 2, respectively, or twin primes and there exist complementary WFME and WFme scale pairs.*

This leads to our last theorem and a conjecture.

Theorem 6. *There are infinitely many universes U_c that support dual me CCV.*

Proof. There are infinitely many primes, and every chromatic universe U_{2p}, p a prime, supports dual me CCV. $\square$

Conjecture 1. *There are infinitely many universes U_c that support dual ME CCV (equivalent to the twin primes conjecture).*

Bibliography

[1] E. Amiot, *Music Through Fourier Space: Discrete Fourier Transform in Music Theory* (Springer, 2016).

[2] N. Carey and D. Clampitt, Aspects of well-formed scales, *Music Theory Spectrum* **11** (1989) 187–208.

[3] N. Carey and D. Clampitt, Self-similar pitch structures, their duals, and rhythmic analogues, *Perspect. New Music* **34**(2) (1996) 62–87.

[4] N. Carey and D. Clampitt, Two theorems concerning rational approximations, *J. Math. Music* **6**(1) (2012) 61–66.

[5] N. Carey and D. Clampitt, Addendum to 'Two theorems concerning rational approximations', *J. Math. Music* **11**(1) (2017) 61–63.

[6] D. Clampitt, Cardinality equals variety for chords in well-formed scales, with a note on the twin primes conjecture, in *Music Theory and Mathematics: Chords, Collections, and Transformations* (eds.) J. Douthett, M. Hyde and C. Smith (University of Rochester Press, 2008), pp. 9–22.

[7] J. Clough and J. Douthett, Maximally even sets, *J. Music Theory* **35** (1991) 93–173.

[8] J. Clough and G. Myerson, Variety and multiplicity in diatonic systems, *J. Music Theory* **29**(2) (1985) 249–270.

[9] J. Clough and G. Myerson, Musical scales and the circle of fifths, *Amer. Math. Monthly* **93** (1986) 695–701.

[10] J. Douthett, The theory of maximally and minimally even sets, the one-dimensional antiferromagnetic Ising model, and the continued fraction compromise of musical scales, Ph.D. dissertation, University of New Mexico (1999).

[11] J. Douthett and R. Krantz, Energy extremes and spin configurations for the one-dimensional antiferromagnetic Ising model with arbitrary-range interaction, *J. Math. Phys.* **37** (1996) 3334–3353.

[12] J. Douthett and R. Krantz, Maximally even sets and configurations: Common threads in mathematics, physics, and music. *J. Comb. Optim.* **14** (2007) 47–70.

[13] J. Douthett and R. J. Krantz, Dinner tables and concentric circles: a harmony of mathematics, music, and physics, *College Math. J.* **39**(3) (2008) 203–211.

[14] R. J. Krantz, J. Douthett and S. D. Doty, Maximally even sets and the devil's staircase phase diagram for the one-dimensional Ising antiferromagnet with arbitrary-range interaction, *J. Math. Phys.* **39** (1998) 4675–4682.

[15] G. T. Toussaint, The Euclidean algorithm generates traditional musical rhythms, in *Proc. BRIDGES: Mathematical Connections in Art, Music and Science*, Banff, Alberta, Canada (2005), pp. 47–56.

Chapter 2

Hypercubes and the Generalized Cohn Cycle

Jack Douthett[*,§], Peter Steinbach[†,¶] and Richard Hermann[‡,‖]

*University of New Mexico,
Albuquerque, NM, USA

†School of Math, Science, and Engineering,
Central New Mexico Community College,
Albuquerque, NM, USA

‡Department of Music,
University of New Mexico,
Albuquerque, NM, USA

§douthett@comcast.net
¶psteinbach@cnm.edu
‖harhar@unm.edu

2.1. Introduction

This is an exploratory chapter based on previous presentations by Jack Douthett, Richard Hermann, and Robert Peck on musical connections to hypercubes. Hermann and Douthett explored musical connections to the 3-cube symmetry group at the 2003 *Society for Music Theory Meeting* in Madison, WI and at the 2005 *AMS Central Sectional Meeting* in Evanston, IL. Peck and Douthett presented similar work that included the 4-cube at the 2011 *Joint Mathematics Meeting* in New Orleans. Finally in 2016 Douthett focused on the musical connections to *dyad generated hypercubes* at the 2016 *AMS Southeastern Sectional Meeting*, in Athens, GA.

In this chapter we explore *dyad generated m-cubes mod n* and focus on a subclass of this collection we call *toggling m-cubes*. This subclass has many musical implications that we briefly discuss. We also explore a subclass of Richard Cohn's *maximally smooth cycles* that we will call *Cohn's cycles* for reference. We discover that Cohn's cycles as subgraphs of their corresponding 3-cubes are *middle layer graphs*, have *strong diameters*, and are *antipolar great cycles*. These observations lead us to three generalizations of Cohn's cycles that relate to both Western and Hindustani musical systems. For referential ease we will name several hypercubes we explore after the persons who first constructed them or who inspired their constructions.

2.2.   Dyad Generated m-Cubes Mod n

We start by defining a mod n universe $U_n = \{0, 1, 2, \ldots, n - 1\}$ and its power set $\mathcal{V}_{U_n} = \mathcal{P}(U_n)$ where its members, called *pitch-class sets* (*pcsets*), are objects of a permutation group. The group of *transformations* acting on the pcsets is $(\mathcal{I}_{U_n}, \oplus)$ where $\mathcal{I}_{U_n} = \mathcal{P}(U_n)$ and $\oplus$ is the symmetric difference. The action $i \in (\mathcal{I}_{U_n}, \oplus)$ of on $v \in \mathcal{V}_{U_n}$ is defined as follows:

$$i(v) = i \oplus v \in \mathcal{V}_{U_n}.$$

To build an m-cube, we choose a pcset $v \in \mathcal{V}_{U_n}$ and chose m *generating dyads* (doubletons) from $(\mathcal{I}_{U_n}, \oplus)$. For this chapter we will assume the members of each generating dyad are chromatic (consecutive integers mod n). We define $(\mathcal{I}_{m\,(\mathrm{mod}\,n)}, \oplus)$ as the subgroup of $(\mathcal{I}_{U_n}, \oplus)$ generated by the generating dyads and let $\mathcal{V}_{m\,(\mathrm{mod}n)}$ be the orbit of v induced by the action of $(\mathcal{I}_{m\,(\mathrm{mod}\,n)}, \oplus)$ on v. Two pcsets $v_1, v_2 \in \mathcal{V}_{m\,(\mathrm{mod}n)}$ are adjacent if v_1 and v_2 are exchanged by one of the generating dyads. The m-cube induced by this action is called a *dyad generated m-cube mod n* and denoted $Q_{m\,(\mathrm{mod}n)}$. For example, for the pcset $\{1, 3\} \in \mathcal{V}_{U_{12}}$ and generating dyads $\{0, 1\}, \{1, 2\} \in (\mathcal{I}_{U_{12}}, \oplus)$, the subgroup generated by $\{0, 1\}$ and

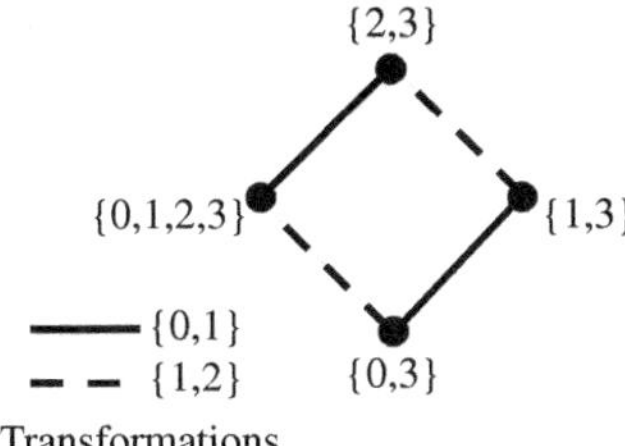

Fig. 2.1. Dyad generated 2-cube mod 12.

$\{1, 2\}$ is

$$(\mathcal{I}_{2\,(\text{mod }12)}, \oplus) = \{\{\}, \{0, 1\}, \{1, 2\}, \{0, 2\}\},$$

and the orbit of $\{1,3\}$ when acted on by the members of $(\mathcal{I}_{2\,(\text{mod }12)}, \oplus)$ is

$$\mathcal{V}_{2\,(\text{mod }12)} = \{\{0, 3\}, \{1, 3\}, \{0, 1, 2, 3\}, \{2, 3\}\}.$$

This leads to the dyad generated 2-cube mod 12 $(Q_{2\,(\text{mod }12)})$ in Fig. 2.1.

From this point on we will assume $n = 12$ and leave off the mod n. In addition when we refer to hypercube structures in terms of their usual binary codes, the value of n is irrelevant.

Although not called as such, Douthett introduced dyad generated m-cubes at the First Clough Symposium (then called the *SUNY Buffalo Working Group*) in 1993 at Buffalo, NY. The idea was inspired by Richard Cohn's [2] *maximally smooth cycles.*[a] One such cycle appears in Brahms' Concerto for Violin, Cello, and Orchestra in A Minor, Op. 102 ("Double Concerto"), mm. 270–278.

---

[a]Cohn shared his observations and more with John Clough and Jack Douthett at the 1992 *Society of Music Theory Meeting* in Kansas City. This led to a flurry of letter exchanges discussing Cohn's work that, in turn, led to the First Clough Symposium in 1993. For those interested in learning more about Cohn's maximally smooth cycles and other Neo-Riemannian topics, one can start with articles by Cohn [2–6] and Lewin [14]. The Riemann referred to in *Neo-Riemannian* is the late nineteenth–early twentieth century music theorist Hugo Riemann, not the mathematician Bernhard Riemann.

The cycle in Brahms' Double Concerto consists of three voices that *toggle* by half-steps through six harmonic triads as illustrated below:

Top Voice :	0(C)	$\to$	11(B)	$\to$	11(B)	$\to$	11(B)	$\to$	0(C)	$\to$	0(C)	$\to$	0(C)
Middle Voice :	8(A$\flat$)	$\to$	8(G$\sharp$)	$\to$	8(G$\sharp$)	$\to$	7(G)	$\to$	8(G)	$\to$	8(G)	$\to$	8(A$\flat$)
Lower Voice :	3(E$\flat$)	$\to$	3(D$\sharp$)	$\to$	4(E)	$\to$	4(E)	$\to$	4(E)	$\to$	3(E$\flat$)	$\to$	3(E$\flat$)
Triad :	A$^\flat$	$\to$	g$^\sharp$	$\to$	E	$\to$	e	$\to$	C	$\to$	c	$\to$	A$^\flat$

This illustration comes from Cohn's [2, p. 15] score reduction of Brahms' Double Concerto, mm. 270–278. The triads are shown in the bottom row. The top voice toggles between pcs 11 and 0 (notes B and C); the middle voice toggles between 7 and 8 (G and G$\sharp$ = A$\flat$), and the lower voice toggles between 3 and 4 (E$\flat$ = D$\sharp$ and E). We will call this cycle and its transpositions (rotations mod 12) *Cohn's cycles*. Cohn calls the collection of these six triads along with certain group operators not related to this paper a *hexatonic system*, not because there are six triads but because the union of the triads is the hexatonic set

$$U_{\text{Hex}} = \{0, 3, 4, 7, 8, 11\}.$$

Since there are 4 Cohn cycles, there are 4 hexatonic systems.

Douthett observed that Cohn's cycle is a subgraph of the 3-cube in Fig. 2.2(a) where Cohn's cycle is the black edged subgraph. For three voices that are either high in the toggle or a half-step lower, there are 3C0 (3 choose 0) = 1 way of choosing zero high voices (bottom vertex or zeroth level in Fig. 2.2(a), 3C1 = 3 ways of choosing one high voice (First level), 3C2 = 3 ways of choosing two high voices (second level), and 3C3 = 1 way of choosing three high voices (third level or top vertex). Voice leading in a Cohn cycle takes care of the two middle levels. If we include the other two levels with the same rules for adjacency (triads are adjacent if they are exchanged by precisely one toggling pair), we get the 3-cube in Fig. 2.2(a) that includes two augmented triads (the augmented triads are also subsets of U_{Hex}). We call this cube and its three transpositions *Cohn's cubes*. The binary vertex codes in Fig. 2.2 will be discussed later.

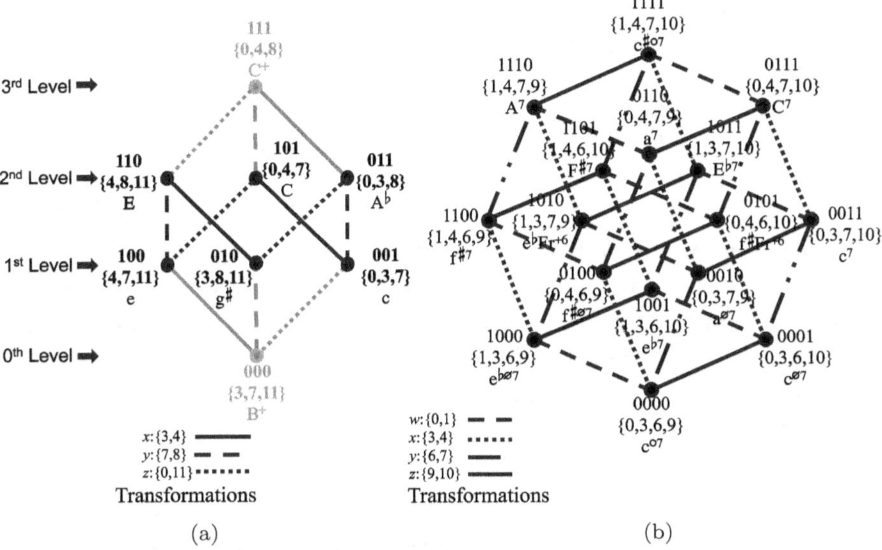

Fig. 2.2. Cohn's cube and Douthett's tesseract.

While this was the thought process Douthett went through at
the time, one can also arrive at Cohn's cube by mimicking the
procedure for Fig. 2.1. Pick any triad in Cohn's cycle, say $\{0,4,7\}$.
There are three generating dyads (sets that toggle the pcs) that define
the voice leading in Cohn's cycle. They are $\{0,11\}$, $\{3,4\}$, and $\{7,8\}$.
The subgroup $(\mathcal{I}_{U_{12}}, \oplus)$ these generate is

$$(\mathcal{I}_{\mathrm{Cohn}}, \oplus) = \{\{\}, \{0,3,4,11\}, \{0,3,4,7,8,11\}, \{0,7,8,11\}, \{0,11\},$$
$$\{3,4\}, \{3,4,7,8\}, \{7,8\}\}$$

and the orbit of $\{0,4,7\}$ is

$$\mathcal{V}_{\mathrm{Cohn}} = \left\{ \{3, \underset{001}{\overset{c}{7}}, 0\}, \{3, \underset{011}{\overset{A^\flat}{8}}, 0\}, \{4, \underset{101}{\overset{C}{7}}, 0\}, \{4, \underset{111}{\overset{C^+}{8}}, 0\}, \{3, \underset{000}{\overset{B^+}{7}}, 11\}, \right.$$
$$\left. \{3, \underset{010}{\overset{g^\sharp}{8}}, 11\}, \{4, \underset{100}{\overset{e}{7}}, 11\}, \{4, \underset{110}{\overset{E}{8}}, 11\} \right\}.$$

While pcsets need not be ordered, for clarity we have ordered
them in $\mathcal{V}_{\mathrm{Cohn}}$ to reflect their binary codes below them. If the pc is

low in the toggle, the corresponding binary coordinate (bit) is 0; if high, the bit is 1. The map that takes the binary code $b_1 b_2 b_3$ to the corresponding pcset is

$$b_1 b_2 b_3 \rightarrow \{b_1 + 3, \ b_2 + 7, \ b_3 + 11\}.$$

Using the symmetric difference adjacency rule given for dyad generated m-cubes again leads to Cohn's cube in Fig. 2.2(a). There are four such Cohn cubes in the mod 12 system, and their union, hinging at their common augmented triads, is called *Cube Dance*. These graphs and others can be found in [9].

Cohn [2] called pairs of antipodal triads in Cohn's cycle *hexatonic poles*. However for toggling m-cubes it will be convenient to simply call Cohn's poles, pairs of *antipodal triads* and reserve the term *pole* for the two vertices in Cohn's cube that are not in his cycle. The pcset that includes the low chromatic members from the generating dyads (corresponding to 000) will be called the *LO pole* and the pcset that includes all high members (111), the *HI pole*. Care must be taken here, since lower does not necessarily imply the smaller number. For example, in the chromatic sense the note **B** is lower than **C**, which implies 11 is lower than 0. The LO pole in Cohn's cube is B⁺ ($\{3, 7, 11\}$), and the HI pole is C⁺ ($\{0, 4, 8\}$).

Douthett also introduced the 4-cube in Fig. 2.2(b) based on the fully diminished seventh chord $\{0, 3, 6, 9\}$ at this symposium. Again Douthett's thought process was the combinatorial approach described above for Cohn's cube, but we can also get this 4-cube by mimicking the approach in Fig. 2.1 with the generating dyads $\{0, 1\}$, $\{3, 4\}$, $\{6, 7\}$, and $\{9, 10\}$. The union of the pcsets in this 4-cube is the octatonic set

$$U_{\text{Oct}} = \{0, 1, 3, 4, 6, 7, 9, 10\}.$$

We call this 4-cube *Douthett's tesseract*, and its poles are c°⁷ (0000) and c♯°⁷ (1111). The map that takes the binary codes $b_1 b_2 b_3 b_4$ to the corresponding pcsets is

$$b_1 b_2 b_3 b_4 \rightarrow \{b_1 + 0, \ b_2 + 3, \ b_3 + 6, \ b_4 + 9\}.$$

There are three Douthett tesseracts in the mod 12 system, and their union, hinged at their common fully diminished seventh chords, is called the 4-*Cube Trio*.[b,c]

Close scrutiny of Figs. 2.1 and 2.2 exposes some curious differences in their structures. In Fig. 2.2 the members of the generating dyads determine which pcs toggle. For example, the transformation $\{0, 7, 8, 11\} \in (\mathcal{I}_{\text{Cohn}}, \oplus)$ exchanges the triads $\text{C}(\{0, 4, 7\})$ and $\text{E}(\{4, 8, 11\})$. But when we rewrite this transformation in terms of the composition of generating dyads we get $\{0, 7, 8, 11\} = \{0, 11\} \oplus \{7, 8\}$. These generating dyads tell us that the triads C and E are exchanged by toggling the notes $\text{B}(11)$ and $\text{C}(0)$ and toggling the notes $\text{G}(7)$ and $\text{G}^{\sharp}(8)$. The note $\text{E}(4)$ remains fixed. Note also that pcsets need not be ordered; it is the generating dyads that determine the voice leading. This is not the case in Fig. 2.1 where the generating dyad $\{1, 2\}$ exchanges $\{0, 3\}$ and $\{0, 1, 2, 3\}$. No such toggling exists. The toggling in Fig. 2.2 occurs for two reasons:

1. The pairwise intersections of the generating dyads are null.
2. The pcset used to generate the vertex set contains one and only one member from each generating dyad.

When these two conditions are met, we call the m-cube a *toggling* m-*cube*. Thus Fig. 2.1 is a dyad generated 2-cube that does not toggle

---

[b]At the First Clough Symposium in 1993 Douthett named each 4-cube an *OctaTower* and called their union *Power Towers*. However in the Neo-Riemannian edition of the *Journal of Music Theory*, Douthett and Steinbach [9] left out the French sixth chords in the towers so these graphs would only included seventh chords. In that article, End Note 12 discusses this change and renames what was called *Power Towers* at the Clough symposium the 4-*Cube Trio*. Several years later while developing his theory of voice leading and geometry, Dmitri Tymoczko independently rediscovered this graph [17, p. 106].

[c]While we focus on the construction of m-cubes and their subgraphs that have musical implications, we do not explore musical applications in depth. For those interested in exploring the music analytic aspects of our graphs, we recommend a couple of important twenty first century musical treatises: Dmitri Tymoczko [17] and Richard Cohn [7]. These authors employ Cohn's cycles, m-cubes, cube dance, the 4-cube trio, and other constructs in their analyses of the works of Brahms, Chopin, Schubert, Shostakovich, Wagner, and many others.

while Cohn's cube and Douthett's tesseract in Fig. 2.2 are toggling m-cubes.

Toggling m-cubes have several properties that are not generally shared by those in the larger class of dyad generated m-cubes. For a toggling m-cube all the pcsets in the vertex set have the same cardinalities, and each pcset contains precisely one member from each of the generating dyads. If immutable tones (tones that do not toggle) are in one pcset vertex they must be in all of them. This will become clear at the end of Sec 2.4. In addition the subsets of mutable tones in the polar pcsets are anhemitonic (have no semitones); else at least two of the generating dyads would overlap.

The distance between two pcsets in a toggling m-cube is half the cardinality of the transformation that exchanges them. Equivalently, this distance is half the cardinality of the symmetric difference of the pcsets. That is if v_1 and v_2 are pcsets in a toggling m-cube and i is the transformation that exchanges them, then the distance between v_1 and v_2 is

$$d(v_1, v_2) = \frac{1}{2}|v_1 \oplus v_2| = \frac{1}{2}|i|.$$

Thus, the edge set of a toggling m-cube is

$$\varepsilon_m = \{v_1 v_2 : \frac{1}{2}|v_1 \oplus v_2| = 1\}.$$

It follows that the triads **C** and **e** are adjacent since

$$d(\mathrm{C}, \mathrm{e}) = d(\{0, 4, 7\}, \{4, 7, 11\}) = \frac{1}{2}|\{0, 4, 7\} \oplus \{4, 7, 11\}|$$

$$= \frac{1}{2}|\{0, 11\}| = 1.$$

For chords (or scales) this means that the fewer notes they have in common, the further apart they are in the m-cube.

We say two pcsets are *dyad-coupled* if one pcset contains one member from each generating dyad and the other pcset contains the other dyad member. It follows that two pcsets are dyad-coupled in a toggling m-cube if and only if they are a distance m apart (i.e., they are a pair of antipodal pcsets). We note that in m-cubes two pcsets are dyad-coupled if and only if their corresponding binary codes are complements.

There is one last observation about Fig. 2.2. Note that the pcsets used to generate the vertex sets include no pcs outside those in the union of their generating dyads (U_{Hex} and U_{Oct}, respectively). When this is the case, pairs of antipodal pcsets are complements of the union of the generating dyads.

2.3. The Cohn and Hermann Cubes

A *toggling k-subcube* of an arbitrary dyad generated m-cube Q_m is a toggling k-cube $Q_k, k \leq m$, whose k dyad generators are members of the set of the m-cube dyad generators. A *toggling subgraph of Q_m* is a subgraph of a toggling k-subcube. All subgraphs in a toggling m-cube are clearly toggling. But as we shall see, toggling subgraphs can also live in dyad generated m-cubes that do not toggle.

Figure 2.3 shows two 3-cubes, Cohn's cube and *Hermann's cube*. We have already discussed Cohn's cube, but there is one small difference between this Cohn's cube and the one in Fig. 2.2(a). This

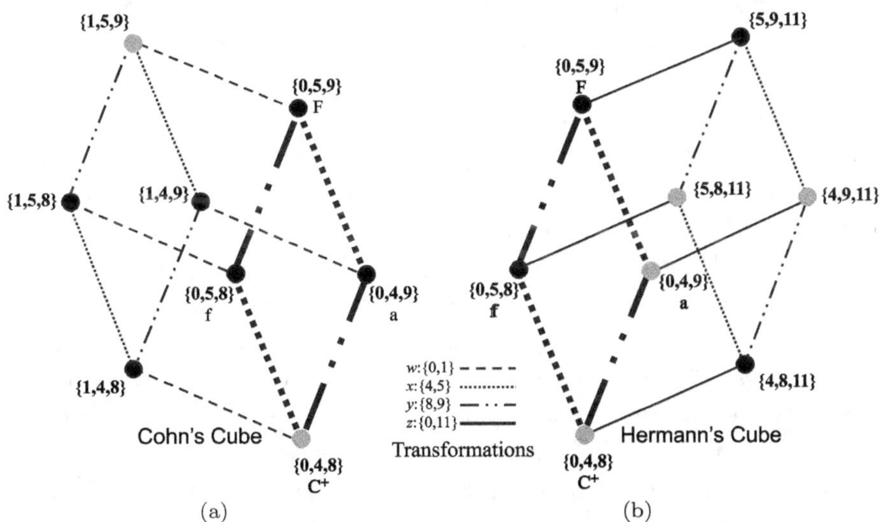

Fig. 2.3. Cohn's cube and Hermann's cube.

one is transposed up by a half-step. It is in the wxy-space, and its generating dyads are $\{0,1\}$, $\{4,5\}$, and $\{8,9\}$. We do this so Cohn's and Hermann's cubes have a common face, which gives us a means for comparison.

Hermann's cube is in the xyz-space and evolved from Richard Hermann's analysis of the chromatic voice leading between trichords found in Berlioz's "Villanelle" from *Nuit d'été* for Mezzo-Soprano or Tenor and Piano Op. 7, no. 1 H81A, mm. 28–31. In this case the generators are $\{0,11\}$, $\{4,5\}$, and $\{8,9\}$. Thus

$$(\mathcal{I}_{\text{Hermann}}, \oplus) = \{\{\}, \{0,4,5,11\}, \{0,4,5,8,9,11\}, \{0,8,9,11\}, \{0,11\},$$
$$\{4,5\}, \{4,5,8,9\}, \{8,9\}\},$$

The orbit of any one of the pcsets in Hermann's cube in Fig. 2.3(b) is

$$\mathcal{V}_{\text{Hermann}} = \{\{0,4,8\}, \{0,4,9\}, \{0,5,8\}, \{0,5,9\}, \{4,8,11\}, \{4,9,11\},$$
$$\{5,8,11\}, \{5,9,11\}\}.$$

The black vertices in Fig. 2.3(a) indicate the triads transposed up a half-step employed by Brahms in his Double Concerto. In Fig. 2.3(b) the black vertices indicate the trichords employed by Berlioz in his Villanelle.

Like Cohn's cube, the pairs of antipodal pcsets of Hermann's cube are dyad-coupled. Both cubes are toggling 3-cubes. In addition the dyads $\{4,5\}$ and $\{8,9\}$ are common to both cubes and define a toggling 2-cube in the xy-space.

Since in both cubes there is a total of four dyad generators, we can embed these cubes in the dyad generated 4-cube in Fig. 2.4. The union of Cohn's and Hermann's cubes is the black edged subgraph, and the rest of the 4-cube is in gray. This 4-cube does not toggle, since the intersection of the generating dyad $\{0,1\}$ in Cohn's cube and the dyad $\{0,11\}$ in Hermann's cube overlap. It follows that non-toggling dyad generated m-cubes can include toggling k-subcubes. Any subgraph of either Cohn's cube or Hermann's cube is a toggling subgraph of the 4-cube. But the union of these cubes *is not* a toggling

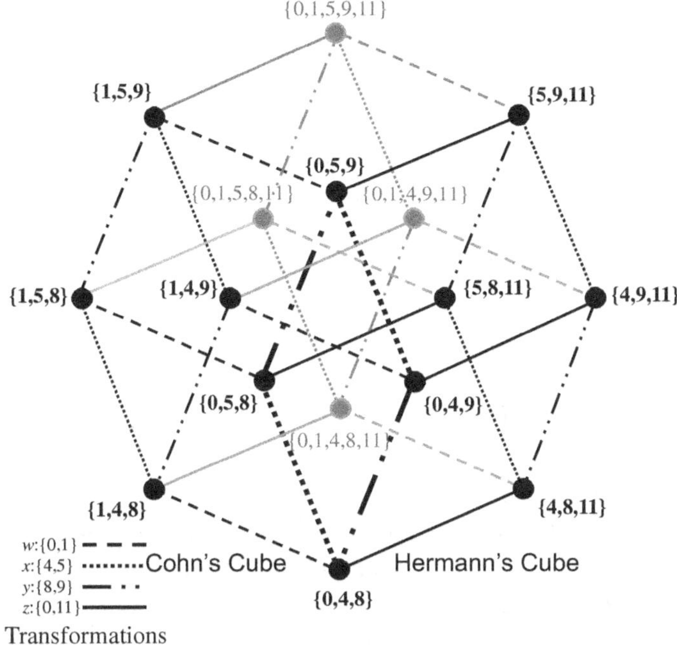

Fig. 2.4. Cohn's Cube and Hermann's cube embedded in a 4-cube.

subgraph since two of the generating dyads defining the edges of this subgraph overlap (i.e., $\{0, 11\} \cap \{0, 1\} = \{0\}$).

2.4. Revisiting the Cohn Cycle

Given our previous discussion on Cohn's cycle and Brahms' Double Concerto, one might reasonably wonder if Brahms employed some kind of mathematics to construct his cycle of triads. This is highly unlikely. More likely is that Brahms experimented with harmonic triads until he found a way of moving from one to the next by moving a single tone by a half-step. After the sixth move he arrived back at the first triad. While there is no evidence to support this speculation, given the emphasis on chromatic voice leading in the late nineteenth century, it is not an unreasonable speculation. At any rate, because of the tight structure of this cycle, it inherits many interesting mathematical properties, and we explore those properties in this section. In addition, we will offer three ways to generalize Cohn's cycle. But it is first necessary to establish some new terminology.

Consider an m-cube Q_m with the usual binary coded vertices, and let the ith level of Q_m be the *antichain* of mCi vertices whose binary codes have i 1s.[d] The partitioning of the vertices into these $m + 1$ levels will be called the $mC0 - mC1 - mC2 - \cdots - mCm$ partition. Thus the level partition of Cohn's cube in Fig. 2.2(a) is the 1-3-3-1 partition.[e] We borrow Gregor and Škrekovski's [11] notation, and let G_m^k be the subgraph of Q_m induced by the vertices in the levels between k and $m - k$ for a fixed k, $0 \le k < m/2$, and odd $m \ge \max\{3, 2k + 1\}$. Thus $G_m^0 = Q_m$, and G_3^1 is Cohn's cycle.

The first half of Cohn's cycle is A♭-g♯-E-e, which is a *chain* of maximal length 3 (to go a step further would give a triad at a distance 2 from A♭).[d] To complete a 6-cycle, it is necessary to find another chain of maximal length 3 from e to A♭ that has no triads in common with the first half of Cohn's cycle. Actually there are many. Two such chains are e-B$^+$c-A♭ and the second half of the Cohn cycle e-C-c-A♭. These two return chains give rise to the two 6-cycles below:

<pre>
        g♯   E   g♯   E
     {3,8,11} - {4,8,11}   {3,8,11} - {4,8,11}
        010   —   110   010   —   110
        /   \   /   \
A♭ {3,8,0} 011   100 {4,7,11} e   A♭ {3,8,0} 011   100 {4,7,11} e
        \   /   \   /
        001   —   000   001   —   101
     {3,7,0} - {3,7,11}   {3,7,0} - {4,7,0}
        c   B⁺   c   C
</pre>

Both cycles have diameter 3, but not all diameters are the same. Note that the antipodal pairs in Cohn's *cycle* on the right are also

---

[d]A *chain* is a totally ordered subset in a partially ordered set. For our purposes a *chain* is a path in Q_m of strictly increasing distance. An *antichain* is a set of vertices that have no edges in common (all chains have length 0). Thus each level in Q_m is an antichain.

[e]Since our rendering of Douthett's tesseract emphasizes the w-, x-, y-, and z-axes, it may be difficult to identify the different levels. For a rendering of a 4-cube whose levels are clear, see Cohn's 4-cube trio [7, p. 158].

antipodal pairs in Cohn's *cube*. However in the cycle on the left 010 ($g^\sharp$) and 000 (B^+) are *antipodes* in the *cycle* but are *adjacent* in Cohn's *cube*. If all antipodal pairs in a cycle in Q_m are also antipodal pairs in Q_m, we say the diameter of the cycle is *strong*. So both cycles above have a diameter of 3, but only the diameter of Cohn's cycle is strong. If the diameter of a cycle in a toggling m-cube is strong and its length is equal to the length of the diameter of the m-cube, we call the cycle a *great cycle*. Thus great cycles have length $2m$-cycle. For Cohn's cube, Cohn's cycle is one of four great cycles, and all four are listed below:

$$
\begin{array}{rccccccc}
\text{Antipolar Great Cycle}: & A^\flat & g^\sharp & E & e & C & c & A^\flat \\
\text{Meridian}: & A^\flat & C^+ & E & e & B^+ & c & A^\flat \\
\text{Meridian}: & A^\flat & g^\sharp & B^+ & e & C & C^+ & A^\flat \\
\text{Meridian}: & a^\flat & E & C^+ & C & c & B^+ & a^\flat
\end{array}
$$

Great cycles that go through the poles are called *meridians*; otherwise they are called *antipolar great cycles*. So Cohn's cube has 3 meridians and 1 antipolar great cycle, and the antipolar great cycle is Cohn's cycle. Great cycle's antipodal binary (pcsets) pairs are always complements (dyad-coupled).

It is not difficult to construct great cycles in all m-cubes, $m \geq 2$. Figure 2.5(a) shows two great cycles in Douthett's tesseract. One is an antipolar great cycle (the *star* cycle inside), and the other is a meridian (the *circle* cycle outside). These two great 8-cycles include all sixteen chords in Douthett's tesseract, and they have no chords in common.

While all great cycles have strong diameters, not all cycles with strong diameters are great cycles. Figure 2.5(b) shows a 16-cycle with a strong diameter in Douthett's tesseract. Neither of the graphs in Fig. 2.5 is unique. There are many pairs of great cycles in the 4-cube that have no vertices in common, and there are many Hamiltonian cycles that have strong diameters but are not great cycles.

Now observe the 1-3-3-1 partition of Cohn's cube in Fig. 2.2(a). The graph induced by the two middle levels (the 3-3 part of the partition) is called a *middle layer graph*. For Cohn's cube, the middle

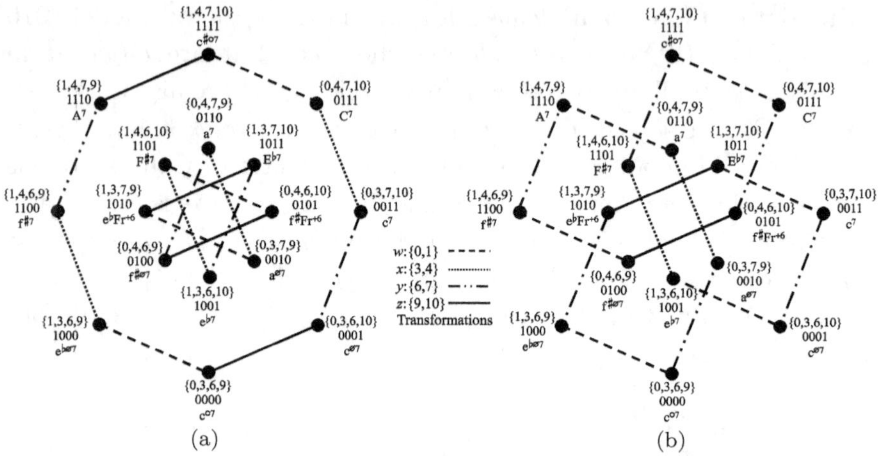

Fig. 2.5.   Strong diameter cycles in Douthett's tesseract.

layer graph is the Cohn cycle. Phrased this way, Cohn's cycle is an example of a problem in mathematics, known as the *middle levels conjecture* [12] that remained unsolved until 2014.

Middle Levels Conjecture: Let Q_m be an m-cube where $m = 2k + 1$ and $k \geq 1$. The middle layer graph (the graph induced by middle levels k and $k + 1$) is Hamiltonian (has a Hamiltonian cycle).

Torsten Mütze's [16] proved this conjecture and took it a step further by establishing a lower bound for the number of Hamiltonian cycles in a middle layer graph.

Theorem (Mütze). Let Q_m be an m-cube for odd $m \geq 3$, and let $\alpha = 2^{[(m+1)/4]}$. The middle layer graph has at least $2^{\alpha-2}$ Hamiltonian cycles.

While the problem is simple for the 3-cube, it is far from simple for larger odd values of m.

There is also a *generalized middle levels conjecture* [8, 13].

Generalized Middle Levels Conjecture. The graph G_m^k is Hamiltonian for each k, $0 \leq k < m/2$, and odd $m \geq \max\{3, 2k + 1\}$.

It is well known that $G_m^0 = Q_m$ is Hamiltonian for all $m \geq 2$, even or odd. The graph G_m^1 is Hamiltonian for odd $m \geq 3$ [10, 15]: $G_m^2 \ldots$ is

Hamiltonian for odd $m > 5$ [11], and the middle layer graph $G_m^{(m-1)/2}$ is Hamiltonian for $m \geq 3$ [16].

We now offer three ways to generalize Cohn's cycle:

G1. The cycle is an antipolar great cycle.
G2. The cycle has a strong diameter.
G3. The cycle is a G_m^k Hamiltonian cycle for fixed k, $0 \leq k < m/2$, and odd $m^3 \geq \max\{3, 2k + 1\}$.

Generalization G1 includes antipolar great cycles as in the star in Fig. 2.5(a) but excludes meridians. Generalization G2 includes meridians as in the circle in Fig. 2.5(a) and also cycles with strong diameters that are not great cycles as in Fig. 2.5(b). Generalization G3 is the only one that does not require a strong diameter, and we will look closer at cycles that satisfy only G3 in the next section.

To find a cycle in a 3-cube other than the Cohn cube that satisfies all 3 generalizations we need go no further than Hermann's cube. In this case the polar pcsets are $\{4, 8, 11\}$ and $\{0, 5, 9\}$. Its antipolar great cycle is

$$\{5, 8, 11\} - \{5, 9, 11\} - \{4, 9, 11\} - \{0, 4, 9\} - \{0, 4, 8\}$$
$$- \{0, 5, 8\} - \{5, 8, 11\}.$$

This cycle is also a middle layer (G_3^1) Hamiltonian cycle with a strong diameter, satisfying all three generalizations. However for musical reasons, perhaps a more interesting great cycle in Hermann's cube is the meridian

$$\{\mathbf{0}, \mathbf{5}, \mathbf{8}\} - \{\mathbf{0}, \mathbf{5}, \mathbf{9}\} - \{\mathbf{5}, \mathbf{9}, \mathbf{11}\} - \{\mathbf{4}, \mathbf{9}, \mathbf{11}\} - \{\mathbf{4}, \mathbf{8}, \mathbf{11}\}$$
$$- \{0, 4, 8\} - \{\mathbf{0}, \mathbf{5}, \mathbf{8}\}.$$

This meridian includes all the trichords employed by Berlioz in his Villanelle (bold).

In both Cohn's and Hermann's cubes, the polar pcsets include only mutable tones. As mentioned in Sec. 2.3, immutable tones can be included in the pcsets in the vertex set. But if pcs are included in one pcset, they must be included in all of them. If the pc 3 is included in the pcsets in Hermann's cube, the meridian above becomes

$$\{0,3,5,8\} - \{0,3,5,9\} - \{3,5,9,11\} - \{3,4,9,11\} - \{3,4,8,11\}$$
$$- \{0,3,4,8\} - \{0,3,5,8\}.$$

The polar pcsets are $\{3,4,8,11\}$ and $\{0,3,5,9\}$. However, the subsets of LO and HI *mutable* tones are as before, $\{4,8,11\}$ and $\{0,5,9\}$. When the polar pcsets include immutable pcs, the intersections of pairs of antipodal pcsets will yield the set of fixed pcs. For example, the pcsets $\{0,3,5,8\}$ and $\{3,4,9,11\}$ are antipodal pairs in the above cycle, and

$$\{0,3,5,8\} \cap \{3,4,9,11\} = \{3\}.^{\text{f}}$$

Since the last two cycles include the polar pcsets and are not Hamiltonian cycles in G_3^k for any k, they satisfy only G2.

2.5. The 32 Scales and Steinbach's Penteract

Yet another good example of the toggling m-cube is inspired by the Ragamala (*garland of melodies*) form of Hindustani and Carnatic music theory. We can relate and compare musical scales from around the world if, first, we postulate 12 tones to the octave. Whether we use the Western equal temperament, some just intonation, or the exquisite tunings of Hindustani or Balinese music, we may assume that a heptatonic scale is chosen from these loosely defined 12 tones. Choose them so that the first and fifth degrees — tonic and dominant — are fixed. The mutable scale degrees 2, 3, 4, 6, and 7 occur in one of two states: *low and high*. Within this construction

---

[f]Including immutable tones in this way can be seen in Tymoczko's scale analysis of Shostakovich's F♯ *Minor Prelude and Fugue*, Op. 87 [17, pp. 329–332]. The unison, third, and sixth degrees of the F♯ natural minor scale are fixed, and the other four scale degrees are allowed to toggle a half-step lower. This leads to a toggling 4-cube structure of heptatonic scales. A similar toggling 5-cube structure of heptatonic scales will be explored in the next section.

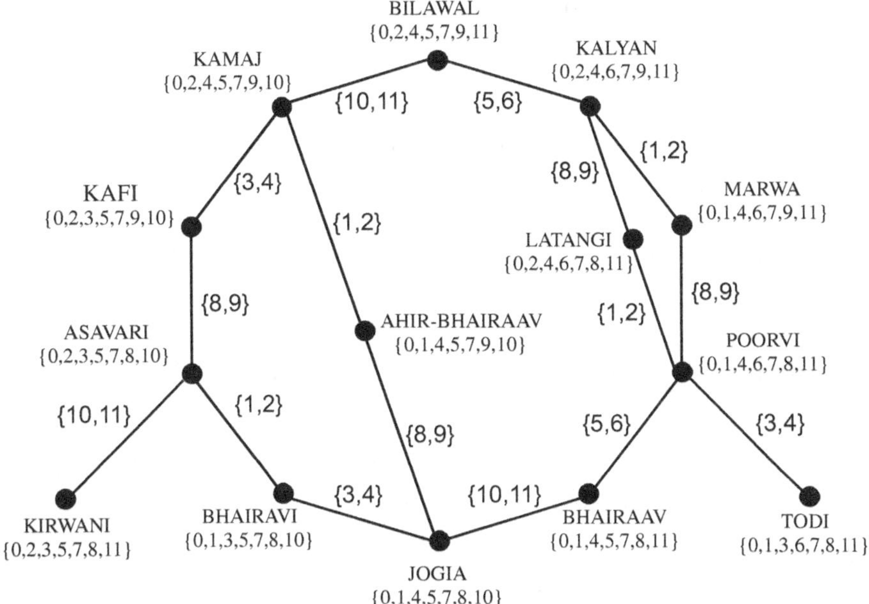

Fig. 2.6. The 14 septak scales.

and ignoring slight differences in tuning, we can think of "low and high" as toggling notes determined by generating dyads.

Since the first and fifth degrees (pcs 0 and 7) are fixed, the generating dyads are $\{1,2\}$, $\{3,4\}$, $\{5,6\}$, $\{8,9\}$, and $\{10,11\}$. With 5 generating dyads there are $2^5 = 32$ possible scales. While Hindustani music theory recognizes 32 heptatonic scales (thaat) *in theory*, practice ignores most of them. A century ago, Bhatkhande [1] identified ten of the 32 scales as the Hindustani Canon, but today musicians use as many as 14 of the scales, which are shown in Fig. 2.6. The *septak* (heptatonic) scales represented by the vertices shift one to another by half-step note changes. We take some liberty with the term *pc* in that the number 0 does not necessarily represent the note C, but rather indicates the first note of the scale. In that sense we think of these scales as *ordered* pcsets.

The 32 theoretical scales are shown in the *Table of 32 Possible Scales* below in binary code and as pcsets, and we have given some of them commonly used names. The map that takes the binary codes to their corresponding pcsets is

$$b_1 b_2 b_3 b_4 b_5 \rightarrow \{0,\; b_1 + 1,\; b_2 + 3,\; b_3 + 5,\; 7,\; b_4 + 8,\; b_5 + 10\}.$$

Since the first and fifth degrees are fixed, 0 and 7 are fixed in the map and appear in every scale.

Septak scales with mutable 2nd, 3rd, 4th, 6th, and 7th degrees

---

$00000 \rightarrow \{0,1,3,5,7,8,10\}$
Bhairavi (Phrygian)

$00001 \rightarrow \{0,1,3,5,7,8,11\}$
(Neapolitan Minor)

$00010 \rightarrow \{0,1,3,5,7,9,10\}$

$00011 \rightarrow \{0,1,3,5,7,9,11\}$
(Neapolitan)

$00100 \rightarrow \{0,1,3,6,7,8,10\}$

$00101 \rightarrow \{0,1,3,6,7,8,11\}$
Todi

$00110 \rightarrow \{0,1,3,6,7,9,10\}$

$00111 \rightarrow \{0,1,3,6,7,9,11\}$

$01000 \rightarrow \{0,1,4,5,7,8,10\}$
Jogia (Spanish Phrygian)

$01001 \rightarrow \{0,1,4,5,7,8,11\}$
Bhairaav

$01010 \rightarrow \{0,1,4,5,7,9,10\}$
Ahir-Bhairaav

$01011 \rightarrow \{0,1,4,5,7,9,11\}$

$10000 \rightarrow \{0,2,3,5,7,8,10\}$
Asavari (Natural Minor)

$10001 \rightarrow \{0,2,3,5,7,8,11\}$
Kirwani (Harmonic Minor)

$10010 \rightarrow \{0,2,3,5,7,9,10\}$
Kafi (Dorian)

$10011 \rightarrow \{0,2,3,5,7,9,11\}$
(Melodic Minor)

$10100 \rightarrow \{0,2,3,6,7,8,10\}$
(Gypsy Minor 7)

$10101 \rightarrow \{0,2,3,6,7,8,11\}$
(Hungarian Minor)

$10110 \rightarrow \{0,2,3,6,7,9,10\}$
(Ukrainian)

$10111 \rightarrow \{0,2,3,6,7,9,11\}$

$11000 \rightarrow \{0,1,4,5,7,8,10\}$
(Romanian)

$11001 \rightarrow \{0,2,4,5,7,8,11\}$
(Harmonic Major)

$11010 \rightarrow \{0,2,4,5,7,9,10\}$
Kamaj (Mixolydian)

$11011 \rightarrow \{0,2,4,5,7,9,11\}$
Bilawal (Ionian-Major)

(Continued)

(*Continued*)

$01100 \to \{0,1,4,6,7,8,10\}$	$11100 \to \{0,2,4,6,7,8,10\}$ (Whole-Tone Plus 1)
$01101 \to \{0,1,4,6,7,8,11\}$ Poorvi	$11101 \to \{0,2,4,6,7,8,11\}$ Latangi
$01110 \to \{0,1,4,6,7,9,10\}$	$11110 \to \{0,2,4,6,7,9,10\}$ (Acoustic)
$01111 \to \{0,1,4,6,7,9,11\}$ Marwa	$11111 \to \{0,2,4,6,7,9,11\}$ Kalyan (Lydian)

Table of 32 Possible Scales

Paths of neighboring scales in Fig. 2.6 make it possible to modulate from one scale to the next via a single degree change (single generating dyad). The poles are $\{0,1,3,5,7,8,10\}$ (Phrygian) and $\{0,2,4,6,7,9,11\}$ (Lydian), and the Hindustani names for these scales are Bhairavi and Kalyan, respectively. They differ at all 5 mutable scale degrees, implying they are a distance of 5 apart. This is consistent with our distance formula in Sec. 2.2:

$$d(\text{Bhairavi}, \text{Kalyan})$$
$$= \frac{1}{2}|\{0,1,3,5,7,8,10\} \oplus \{0,2,4,6,7,9,11\}| = 5.$$

The graph that connects single-change neighbors among the 32 normal scales is a toggling 5-cube, which can be conveniently embedded on the surface of the sphere in Fig. 2.7. The pcsets have been left out to avoid cluttering the spherical representation. The Phrygian (Bhairavi) 00000 ($\{0,1,3,5,7,8,10\}$) and Lydian (Kalyan) 11111 ($\{0,2,4,6,7,9,11\}$) are at the poles, a distance of 5 from each other. Recall that to say the binary codes are complements is to say their corresponding scales are dyad-coupled. We call this 5-cube *Steinbach's Penteract*. The 32 scales in Steinbach's penteract have a 1-5-10-10-5-1 partition, corresponding to the fifth row of Pascal's triangle. The LO pole (00000) is at the zeroth level. There are

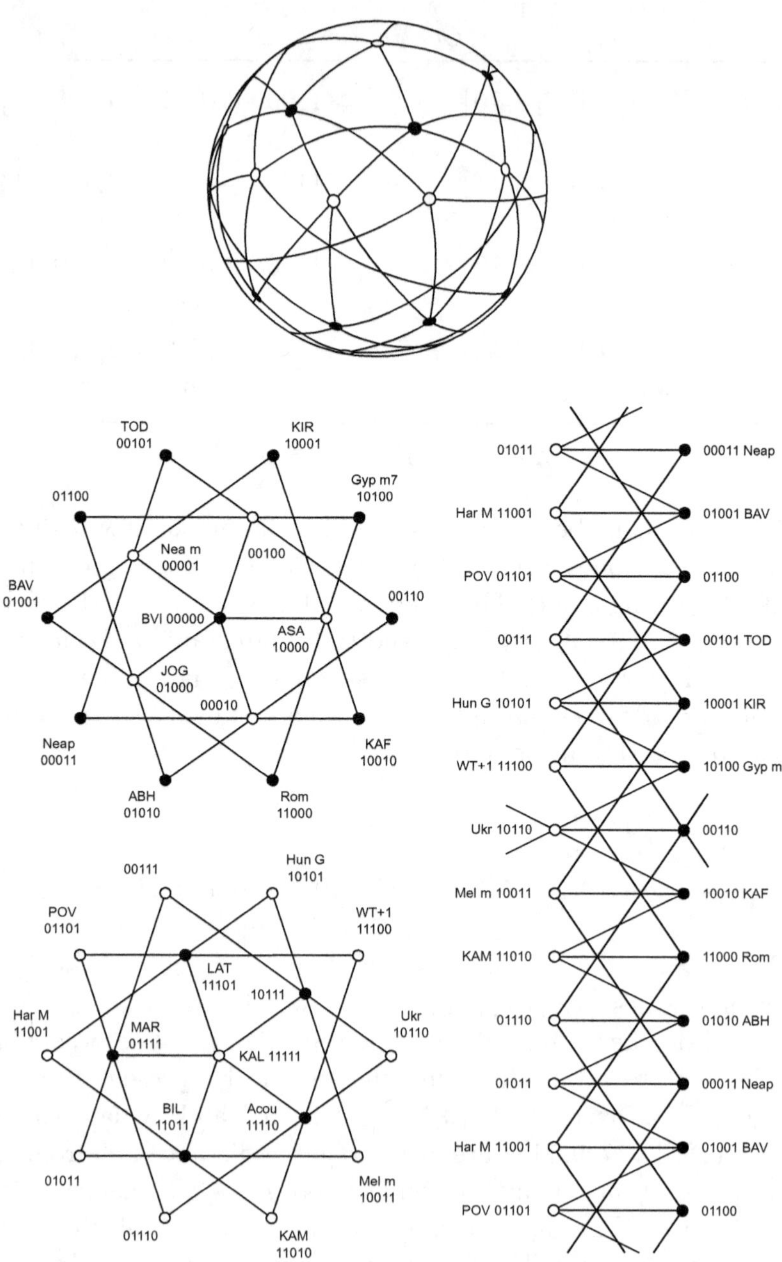

Fig. 2.7. Steinbach's penteract.

5 vertices at the first level at a distance of 1 from the LO pole,
10 vertices at distance 2 at level 2, etc.

Next consider the chain of maximal length 5 from Kamaj to its
dyad-coupled antipodal scale Todi. The septak scales are listed below
their binary representations:

$$11010 \quad — \quad 01010 \quad — \quad 01000$$
$$\{0,2,4,5,7,9,10\} - \{0,1,4,5,7,9,10\} - \{0,1,4,5,7,8,10\}$$

$$— \quad 01001 \quad — \quad 00001 \quad — \quad 00101$$
$$-\{0,1,4,5,7,8,11\} - \{0,1,3,5,7,8,11\} - \{0,1,3,6,7,8,11\}$$

We return to Kamaj via the complementary binary (dyad-coupled
pcset) chain below:

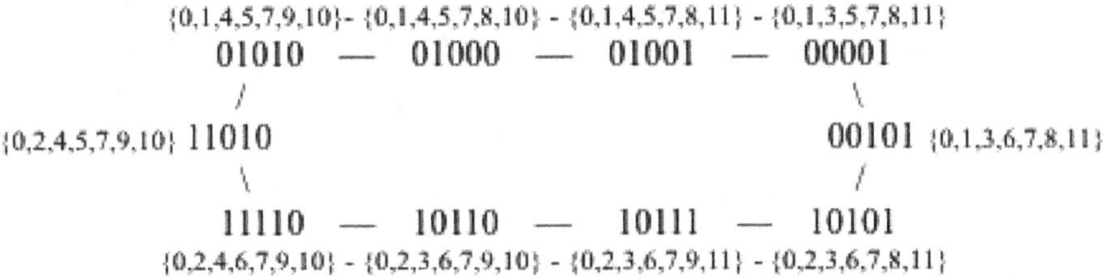

```
        {0,1,4,5,7,9,10}- {0,1,4,5,7,8,10} - {0,1,4,5,7,8,11} - {0,1,3,5,7,8,11}
           01010  —  01000  —  01001  —  00001
             /                               \
{0,2,4,5,7,9,10} 11010                           00101 {0,1,3,6,7,8,11}
             \                               /
           11110  —  10110  —  10111  —  10101
        {0,2,4,6,7,9,10} - {0,2,3,6,7,9,10} - {0,2,3,6,7,9,11} - {0,2,3,6,7,8,11}
```

Since all the antipodal pairs of scales in the cycle are also
antipodal pairs in Steinbach's penteract, the cycle has a strong
diameter of 5. Moreover the poles are not included. It follows that
the cycle satisfies G1 in Sec. 2.4, implying the cycle is an antipolar
great cycle. Unlike Cohn's cycle, which is unique in the 3-cube, there
are 132 antipolar great cycles in Steinbach's penteract.

It is easy to modify the above great cycle so that it passes through
the poles, making it a meridian, satisfying G2 but not G1:

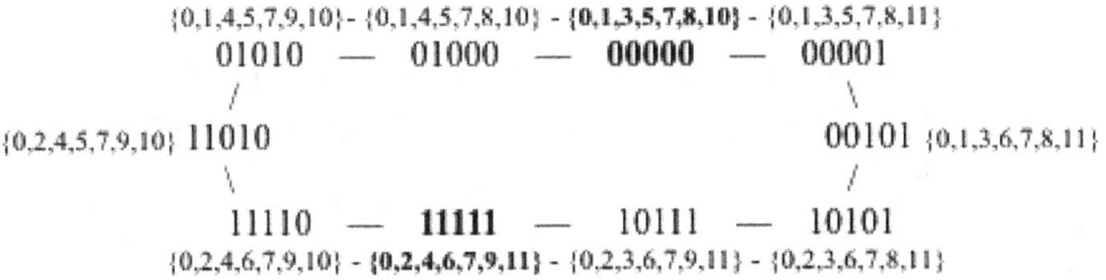

```
        {0,1,4,5,7,9,10}- {0,1,4,5,7,8,10} - {0,1,3,5,7,8,10} - {0,1,3,5,7,8,11}
           01010  —  01000  —  00000  —  00001
             /                               \
{0,2,4,5,7,9,10} 11010                           00101 {0,1,3,6,7,8,11}
             \                               /
           11110  —  11111  —  10111  —  10101
        {0,2,4,6,7,9,10} - {0,2,4,6,7,9,11} - {0,2,3,6,7,9,11} - {0,2,3,6,7,8,11}
```

Steinbach's penteract contains 192 great cycles in total, and 60 of them are meridians (actually, given any fixed pair of antipodes, there are $5!/2 = 60$ great cycles passing through them). Four-cubes have 24 great cycles including 12 meridians; 3-cubes have only 4 great cycles that includes 3 meridians, and the 2-cube is a meridian and hence has no antipolar great cycles. We offer the conjecture here that an m-cube has $2^{m-2}(m-1)!$ great cycles.

A special compositional structure known as Ragamala is a way for a musician to move from raag to raag, often changing scales as well. (The reader is encouraged to listen to *Ragamala in Jogia* or *Rag Mishra Bhairavi*, both by Ram Narayan.) This was an inspiration to try to map all possible scale changes with the penteract. We do not suggest that Fig. 2.6 has been used in Ragamala composition, but we point out that it is nearly a meridian on the sphere. There are several ways the reader can find to make it a meridian. One way is to note that starting on $\{0, 1, 3, 5, 7, 8, 10\}$ (Bhairavi) and proceeding clockwise to $\{0, 2, 4, 6, 7, 9, 11\}$ (Kalyan), the generating dyads defining the edges are in order $\{1, 2\}$, $\{8, 9\}$, $\{3, 4\}$, $\{10, 11\}$, and $\{5, 6\}$. This path is a maximal chain (length 5). To get a great cycle we repeat this dyad pattern until we return to Bhairavi. Two of these dyads, $\{1, 2\}$ and $\{8, 9\}$, are repeated in Fig. 2.6, but then the pattern changes. If $\{0, 1, 4, 5, 7, 8, 11\}$ (Bhairaav) is replaced by $\{0, 1, 3, 6, 7, 8, 11\}$ (not named in the Table of 32 Possible Scales) and $\{0, 1, 4, 5, 7, 8, 11\}$ (Jogia) is replaced by $\{0, 1, 3, 6, 7, 8, 10\}$ (also not named), the 10-cycle becomes a meridian with poles Bhairavi and Kalyan.

Next we examine the G3 Cohn's cycle generalizations in Steinbach's penteract. We start with G_5^1. We know G_5^1 Hamiltonian cycles exist [10, 15], but do any of them have a strong diameter? The answer is "Yes", and Fig. 2.8 shows such a cycle read from top to bottom and from left to right.

Thus this cycle satisfies both G2 and G3. The pairs of complementary diameter endpoints (cycle antipodes) can be seen in every third column.

00001	10100	10011	11110	01011	01100
10001	00100	10010	01110	11011	01101
10101	00110	10000	01010	11001	01111
11101	10110	11000	00010	01001	00111
11100	10111	11010	00011	01000	00101

Fig. 2.8. Strong diameter G_5^1 Hamiltonian cycle.

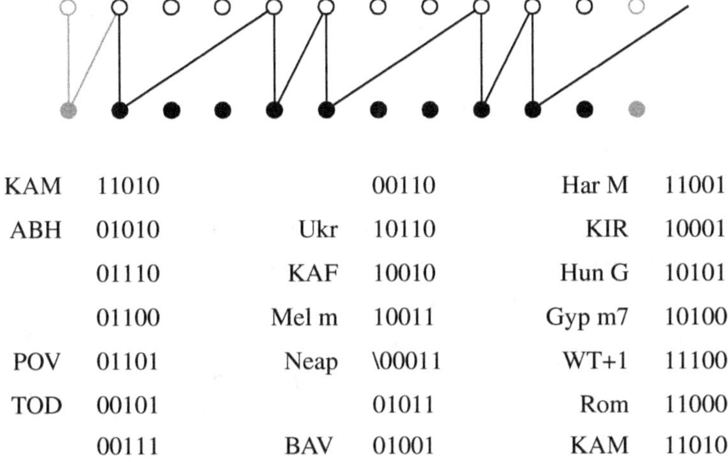

KAM	11010		00110	Har M	11001
ABH	01010	Ukr	10110	KIR	10001
	01110	KAF	10010	Hun G	10101
	01100	Mel m	10011	Gyp m7	10100
POV	01101	Neap	\00011	WT+1	11100
TOD	00101		01011	Rom	11000
	00111	BAV	01001	KAM	11010

Fig. 2.9. G_5^2 (middle layer) Hamiltonian cycle.

Figure 2.9 is a Hamiltonian cycle of the G_5^2 (middle layer graph) shown in the lower right of Fig. 2.7. In both figures the black vertices are in level 2 and the white in level 3. However, this cycle does not have a strong diameter, and in fact no strong diameter G_5^2 Hamiltonian cycle exists. This can be shown with a simple *parity argument*. Observing the 1-5-10-10-5-1 partition of Steinbach's penteract, the length of a middle layer Hamiltonian cycle is $10 + 10 = 20$. It follows that its diameter has length 10. Moreover a vertex is in level 2 if and only if its binary complement is in level 3. Since each level is an antichain, a path between two vertices is even if and only if both vertices are in the same level. Since the diameter

00000	01100	11000	10100
00001	01101	11001	10101
00011	01111	11011	10111
00010	01110	11010	10110
00110	01010	11110	10010
00111	01011	11111	10011
00101	01001	11101	10001
00100	01000	11100	10000

Fig. 2.10. $G_5^0 = Q_5$ Hamiltonian cycle.

of a Hamiltonian cycle is 10 (even), the endpoints of a diameter are in the same level, and cannot be complements (or dyad coupled). It follows that the cycle does not have a strong diameter.

Figure 2.10 shows a $G_5^0 = Q_5$ Hamiltonian cycle. As is the case for G_5^2 (Fig. 2.9), its diameter is not strong, and no strong diameter G_5^2 Hamiltonian cycle exists. An argument similar to the parity argument above can illustrate this. A path between the LO vertex (00000) and another vertex in Q_5 is even if and only if the other vertex is in an even level (levels 0, 2, or 4). Since the complement of the LO vertex is the HI vertex (11111) in level 5 and the diameter of a Q_5 Hamiltonian cycle is $\frac{1}{2} \cdot 2^5 = 16$ (even), there cannot be a strong diameter Q_5 Hamiltonian cycle. Hence G_5^2 and Q_5 Hamiltonian cycles can only satisfy G3. This parity argument can be used in general to show that no strong diameter G_m^k Hamiltonian cycle exists if the cycle diameter is even.

We have a few last comments. What can be said about the higher dimensional toggling hypercubes in the mod 12 system? First there is only one hypercube of higher dimension — the 6-cube. More than six generating dyads would force dyad overlapping. In addition the poles of a toggling 6-cube must be whole-tone scales; else generating dyads would overlap. The 6-cube has 64 vertices and 192 edges, which makes it difficult to get a coherent rendering in which all the vertices are labeled. We leave it to the reader to make this construction and label its vertices. There are also microtonal systems that have many more than 12 divisions. For a mod n universe (n divisions), the highest dimensional toggling m-cube would have dimension $m = \lfloor n/2 \rfloor$.

In addition there are many pcsets in the mod 12 system that are anhemitonic, and any one of them can serve as a mutable subset of a polar pcset of an m-cube. Might any of these m-cubes based on other anhemitonic pcsets have musical relevance?

Furthermore we have restricted the generating dyads to *chromatic* dyads. What might happen if we loosen this restriction? Finally what applications might non-toggling dyad generated m-cubes have?

Acknowledgments

Richard Hermann wishes to acknowledge the assistance provided by MISHA-Labex GREAM, Groupe de Rescherches Expérimentales sur l'Acte Musical based at the University of Strasbourg, Prof. Alessandro Arbo, Director.

Bibliography

[1] V. N. Bhatkhande, *Hindustani Sangeet Paddhati* (Sangeet Karyalaya, 1990).

[2] R. Cohn, Maximally smooth cycles, hexatonic systems, and the analysis of late romantic triadic progressions, *Music Anal.* **15**(1) (1996) 9–40.

[3] R. Cohn, Neo-riemannian operations, parsimonious trichords, and their 'Tonnetz' Representations, *J. Music Theory* **41**(1) (1997) 1–66.

[4] R. Cohn, Introduction to Neo-Riemannian theory: A survey and historical prospective, *J. Music Theory* **42**(2) (1998) 167–180.

[5] R. Cohn, Square dances with cubes, *J. Music Theory* **42**(2) (1998) 283–296.

[6] R. Cohn, Weitzmann's regions, my cycles, and Douthett's dancing cubes, *Music Theory Spectrum* **22**(1) (2000) 89–103.

[7] R. Cohn, *Audacious Euphony: Chromatic Harmony and the Triad's Second Nature* (Oxford University Press, 2012).

[8] I. J. Dejter, W. Cedeno and V. Jauregui, A note on Frucht diagrams, boolean graphs, and Hamilton cycles, *Discrete Math.* **114** (1993) 131–315.

[9] J. Douthett and P. Steinbach, Parsimonious graphs: A study in parsimony, contextual transformations, and modes of limited transposition, *J. Music Theory* **42**(2) (1998) 241–263.

[10] M. El-Hashash and A. Hassan, On the Hamiltonicity of two subgraphs of the hypercube, *Congr. Numer.* **148** (2001) 7–32.

[11] P. Gregor and R. Škrekovski, On generalized middle-level problem, *Information Sci.* **180**(12) (2010) 2448–2457.

[12] I. Havel, Semipaths in directed cubes, in *Graphs and Other Combinatorial Topics*, ed. M. Fiedler (Teubner, Leipzig, 1983), pp. 101–108.

[13] G. Hurlbert, The antipodal layer problem, *Discrete Math.* **128** (1994) 237–245.

[14] D. Lewin, Cohn functions, *J. Music Theory* **40**(2) (1996) 181–216.

[15] S. C. Locke and R. Strong, Spanning cycles in hypercubes: 10892, *Amer. Math. Monthly* **110** (2003) 440–441.

[16] T. Mütze, Proof of middle level conjecture, preprint (2014), http:arxiv.org/pdf/1404.4442.pdf.

[17] D. Tymoczko, *A Geometry of Music: Harmony and Counterpoint in the Extended Common Practice* (Oxford University Press, 2011).

Chapter 3

Associahedra, Combinatorial Block Designs and Related Structures

Franck Jedrzejewski

University of Paris-Saclay,
CEA-INSTN, 91191-Gif-sur-Yvette, France
franckjed@gmail.com

3.1. Combinatorial Designs

A *combinatorial design* is a (k, b)-matrix of positive integers such that each column is called a block, and satisfying nice substructure properties. Tom Johnson has used combinatorial designs in music since 2003 [5]. Formally, a *t-design* $t\text{-}(v, k, \lambda)$ is a pair $D = (X, \mathcal{B})$ where X is a v-set $(X = \mathbb{Z}_v)$ and $\mathcal{B}$ a collection of k-subsets of X called blocks such that every t-subset of X is contained in exactly λ blocks. D is simple if it has no repeated block. The simple example is the Fano plane or design $(7, 3, 1)$. The design includes 7 blocks of 3 digits, taken through the set of the first numbers $X = \{0, 1, \ldots, 6\}$. Each pair of numbers appears in only one block: $\{0, 1\}$ appears in the first block, $\{0, 2\}$ appears in the second block, $\{0, 3\}$ appears in the first block, and so on, until $\{5, 6\}$ which appears in the fifth block. Each element of X appears three times in three different blocks. These properties guarantee the variety and the interest that one can have by transposing these numbers to musical objects.

$$
\begin{array}{ccccccc}
0 & 0 & 0 & 1 & 1 & 2 & 3 \\
1 & 2 & 4 & 2 & 5 & 3 & 4 \\
3 & 6 & 5 & 4 & 6 & 5 & 6
\end{array}
$$

Block designs have different names depending upon their parameter values: $2\text{-}(v, k, \lambda)$ are called *Balanced Incomplete Block Design* (BIBD) and simply denoted by (v, k, λ), for $\lambda = 1$, $t\text{-}(v, k, 1)$ are *Steiner Systems*, $t\text{-}(v, 3, 1)$ are *Triple Systems* (TS), $2\text{-}(v, 3, 1)$ are *Steiner Triple Systems* (STS) and $2\text{-}(v, 4, 1)$ are called *Steiner Quadruple System* (SQS) [2]. There are no known examples of non-trivial t-designs with $t \geq 6$. For $t = 5$, the design $5\text{-}(24, 8, 1)$ is a well-known Steiner system. A *symmetric design* is a BIBD (v, k, λ) with $b = v$. Of course, block designs are considered up to isomorphism. Two t-designs $(X_1, \mathcal{B}_1)$ and $(X_2, \mathcal{B}_2)$ are *isomorphic* if there is a bijection $\varphi : X_1 \to X_2$ such that $\varphi(\mathcal{B}_1) = \mathcal{B}_2$. The number of blocks of a $t\text{-}(v, k, \lambda)$ design is given by the formula

$$
b = \lambda \frac{v!}{(v - t)!} \frac{(k - t)!}{k!}.
$$

Setting

$$
r = \lambda \frac{(v - 1)!}{(v - t)!} \frac{(k - t)!}{(k - 1)!}
$$

leads to the famous simple relation

$$
bk = vr.
$$

The complement of $t - (v, k, \lambda)$ design $D = (X, \mathcal{B})$ is the design $D^c = (X, X \backslash \mathcal{B})$ of parameters $t - (v, v - k, \mu)$ with

$$
\mu = \lambda \binom{v - t}{k} \bigg/ \binom{v - t}{k - t} = \lambda \frac{(v - k)!}{(v - t - k)!} \frac{(k - t)!}{k!}
$$

D and D^c have the same number of blocks. For $t = 2$, the block design D with b blocks has a complement D^c with b blocks and parameters $(v, v - k, b - 2r + \lambda)$. For instance, the complement of $(7, 3, 1)$ is $(7, 4, 2)$ with blocks $\{0, 1, 2\}^c = \{3, 4, 5, 6\}$, etc.

Another important class of designs are *resolvable designs*. A *parallel class* in a design is a set of blocks that partition the point set. A design (v, k, λ) is *resolvable* if its blocks can be partitioned into parallel classes. For instance, $(9, 3, 1)$ is resolvable as shown in the following table. The 12 blocks of this design can be arranged by groups of three blocks such that the union of these blocks (a column in the following table) is the point set.

$(0,1,2)$	$(0,3,6)$	$(0,4,8)$	$(0,5,7)$
$(3,4,5)$	$(1,4,7)$	$(1,5,6)$	$(1,3,8)$
$(6,7,8)$	$(2,5,8)$	$(2,3,7)$	$(2,4,6)$

In 1850, Reverend Thomas Penyngton Kirkman (1806–1895) posed the so-called schoolgirls problem "*Fifteen young ladies in a school walk out abreast for seven days in succession: it is required to arrange them daily, so that no two walk twice abreast*". Kirkman's ladies can be considered the beginning of block designs distributed into 35 blocks of three elements, with each pairs coming together exactly once. It has been shown that this problem leads to a resolvable Steiner Triple system 2-$(v, 4, 1)$ if and only if $v \equiv 3$ (mod 6). The Kirkman problem is solved for $v = 15$. There are 7 non-isomorphic solutions. One solution is given in the following table.

Monday	$(0,1,2)$	$(3,9,11)$	$(4,7,13)$	$(5,8,14)$	$(6,10,12)$
Tuesday	$(0,3,4)$	$(1,8,10)$	$(2,10,14)$	$(5,7,11)$	$(6,9,13)$
Wednesday	$(0,5,6)$	$(1,7,9)$	$(2,11,13)$	$(3,12,14)$	$(4,8,10)$
Thursday	$(1,3,5)$	$(0,10,13)$	$(2,7,12)$	$(4,9,14)$	$(6,8,11)$
Friday	$(1,4,6)$	$(0,11,14)$	$(2,8,9)$	$(3,7,10)$	$(5,12,13)$
Saturday	$(2,3,6)$	$(0,7,8)$	$(1,13,14)$	$(4,11,12)$	$(5,9,10)$
Sunday	$(2,4,5)$	$(0,9,12)$	$(1,10,11)$	$(3,8,13)$	$(6,7,14)$

Composer Tom Johnson has used this design in a piece called *Kirkman's ladies*. If one wonders if it is possible for the ladies to continue their walks for a complete semester of 13 weeks, so as to include all $\binom{15}{3} = 455$ possible blocks leads to a huge structure called a $(15, 3, 1)$ large designs. This is this structure Tom Johnson used in his work. He also composed a *Septet II* for 2 flutes, oboe, clarinet, 2 violins and viola with the Fano plane $(7, 3, 1)$; *Block Design for piano* based on 4-$(12, 6, 10)$ built on 30 base blocks and the automorphism that adds 1 to each point except 11 which remains the same; *Vermont Rhythms* a 42×11 rhythms based on design $(11, 6, 3)$; *5 Chords* (2009) for organ, 23 minutes of organ music all derived from a $(11, 4, 6)$ block design. This last piece raises the question of the graphical representation of a block design. Johnson gives several graphs of the same design: *Cosmological view* in which every single chord has no notes in common with exactly four chords; *Pentagonal view* where each chord has one pair of notes in common with one chord, the other pair in common with one other chord, and no notes in common with the adjacent chords; *Spider web view* linking chords with 3 notes in common and *Starfish view* where three pairs of notes combine to form 3 chords (see [6]). This leads us to the question of how to draw a block design.

The easy case is when blocks are constructed from generators $\mathcal{B} = \langle B \mid T_1^v(B) \equiv 1 \rangle$ with action of the cyclic group. These designs are represented by cyclic representations. A special case are the designs

$$2 - \left(\frac{p^m - 1}{p - 1}, \frac{p^{m-1} - 1}{p - 1}, \frac{p^{m-1} - 1}{p - 1} \right)$$

of the projective geometry $\mathrm{PG}(m - 1, p)$ depicted in the following table. The last column indicates the generator block of the design.

$(7, 3, 1)$	$\mathrm{PG}(2, 2)$	$(0, 1, 3)$
$(13, 4, 1)$	$\mathrm{PG}(2, 3)$	$(0, 1, 3, 9)$
$(21, 5, 1)$	$\mathrm{PG}(2, 4)$	$(0, 1, 4, 14, 16)$
$(31, 6, 1)$	$\mathrm{PG}(2, 5)$	$(0, 1, 3, 8, 12, 18)$
$(57, 8, 1)$	$\mathrm{PG}(2, 7)$	$(0, 1, 3, 13, 32, 36, 43, 52)$
$(73, 9, 1)$	$\mathrm{PG}(2, 8)$	$(0, 1, 3, 7, 15, 31, 36, 54, 63)$
$(91, 10, 1)$	$\mathrm{PG}(2, 9)$	$(0, 1, 3, 9, 27, 49, 56, 61, 77, 81)$

Another case is given by *Netto's theorem* (1893): Let p prime, $n \geq 1$, $p^n \equiv 1 \pmod 6$. Let $\mathbb{F}_{p^n}$ be a finite field on X of size $p^n = 6t + 1$ with 0 as its zero element and α a primitive root of unity. The sets $B_i = \{\alpha^i, \alpha^{i+2t}, \alpha^{i+4t}\} \bmod p^n$ for $i = 1, 2, \ldots, t - 1$ are generators $(T_j(B) = j + B \bmod p^n)$ of the set blocks of an $\mathrm{STS}(p^n)$ on X.

But in the general case, there is no canonical way to draw a t-design.

3.2. Catalan Block Designs

It is well known that they are many families of combinatorial objects enumerated by the Catalan numbers

$$\mathrm{Cat}(n) = \frac{(2n)!}{(n+1)!n!}.$$

These families include non-crossing partitions, binary rooted trees, triangulations of a convex $(n + 2)$-gon by diagonals, bracketing of a non-associative products of $(n + 1)$ factors, Dyck paths and much more. The bijections between these structures are fully described in some lectures or textbooks (see for example [3, 11]). Since binary rooted trees and Dyck paths play an increasing role in music theory, we will be interested in Catalan block designs, that is block designs t-(v, k, λ) whose number of block b is a Catalan number for some n:

$$b = \lambda \frac{v!}{(v-t)!} \frac{(k-t)!}{k!} = \frac{(2n)!}{(n+1)!n!}.$$

For the first values of $n \geq 4$, Catalan block designs exist. For $n = 4$, designs with 14 blocks are for instance $(7, 3, 2)$, $(8, 4, 3)$ and the Steiner system 3-$(8, 4, 1)$. For $n = 5$, designs $(7, 3, 6)$, $(8, 4, 9)$, $(15, 5, 4)$, $(21, 5, 2)$, $(21, 6, 3)$, and 3-$(8, 4, 3)$, etc. have 42 blocks. As the number of blocks b and the number of symbols v increase very quickly, large block designs interest musicians who work with computers. For instance, for $n = 6$, block designs have 132 blocks: $(33, 8, 7)$, $(33, 9, 9)$, $(121, 11, 1)$, 4-$(11, 5, 2)$, 4-$(12, 6, 4)$, 5-$(12, 6, 1)$, and for $n = 7$, designs $(66, 6, 3)$ and $(286, 20, 2)$ have 429 blocks.

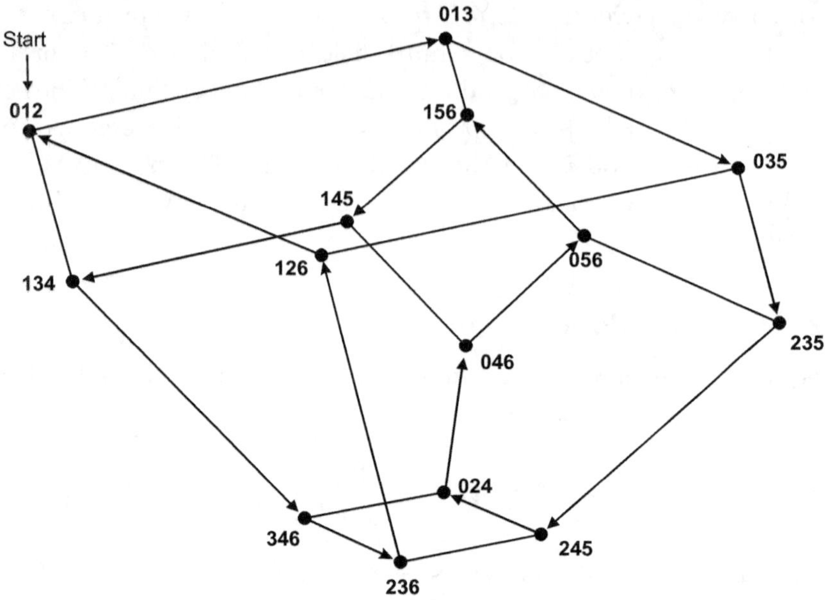

Fig. 3.1. Associahedron of the design $(7, 3, 2)$.

Catalan structures can be represented by some associahedra. Associahedron appears in Dov Tamari's 1951 thesis [12] (without the name) as the realization of a lattice of bracketings of a word with n factors (see [4, 10]). An associahedron is an $(n-1)$-dimensional convex polytope in which each vertex is a non-crossing partition, or a Dyck path, or one of the structures seen above [13]. For our purpose, vertices will be the blocks of a Catalan block design. For instance, the block design $(7, 3, 2)$ of 14 blocks can be displayed on the three-dimensional associahedron (see Fig. 3.1). We can even find a Hamiltonian cycle which passes through all vertices (following the arrows). Two vertices linked by an arrow have two points in common. From now on, blocks are written in their compact forms: 012 means the set $\{0, 1, 2\}$.

3.3. Design 3-(8,4,1)

Let us continue with another example: the Steiner system $3\text{-}(8, 4, 1)$. In a first step, we explain how to build it from the design $(7, 3, 1)$.

Add the component 7 to each block of $(7, 3, 1)$:

$$
\begin{array}{ccccccc}
0 & 1 & 2 & 3 & 0 & 1 & 0 \\
1 & 2 & 3 & 4 & 4 & 5 & 2 \\
3 & 4 & 5 & 6 & 5 & 6 & 6 \\
7 & 7 & 7 & 7 & 7 & 7 & 7
\end{array}
$$

Next, for each block, consider the supplementary block and add it to the structure. For example, the block 0137 has the complement 2456. This leads to the 3-$(8, 4, 1)$ design of 14 blocks. In this design, each pair of notes appears exactly three times.

$$
\begin{array}{cccccccccccccc}
0 & 1 & 2 & 3 & 0 & 1 & 0 & 2 & 0 & 0 & 0 & 1 & 0 & 1 \\
1 & 2 & 3 & 4 & 4 & 5 & 2 & 4 & 3 & 1 & 1 & 2 & 2 & 3 \\
3 & 4 & 5 & 6 & 5 & 6 & 6 & 5 & 5 & 4 & 2 & 3 & 3 & 4 \\
7 & 7 & 7 & 7 & 7 & 7 & 7 & 6 & 6 & 6 & 5 & 6 & 4 & 5
\end{array}
$$

By construction, each sub-block of 3 digits is unique. One way to represent it is to plot three concentric heptagons, whose vertices are numbered from 0 to 6 (in dotted lines on Fig. 3.2). On this figure, the two yellow (gray) blocks (0317 and 2465) are supplementary blocks. They have no point in common. The same thing is repeated by a rotation of $2\pi/7$: blocks 1427 and 5036 have no point in common, and so on.

This representation seems to be better than the representation using the associahedra. The blocks show their structure more easily.

3.4. Design $(21, 6, 3)$

Consider now the design $(21, 6, 3)$. It has two generators

$$u = (0, 1, 3, 11, 16, 20), \qquad v = (0, 1, 7, 12, 15, 19)$$

and can be constructed by adding 1 mod 21 to each generator. The design $(21, 6, 3)$ has 42 blocks. It is shown in Fig. 3.3. Each block has 6 elements taken in an alphabet of 21 symbols. The outer circle starts with the block u at the north pole. By adding 1 mod 21 and turning clockwise, its translates form 21 blocks. The inner circle starts with the block v and its translates are also arranged clockwise. Each pair

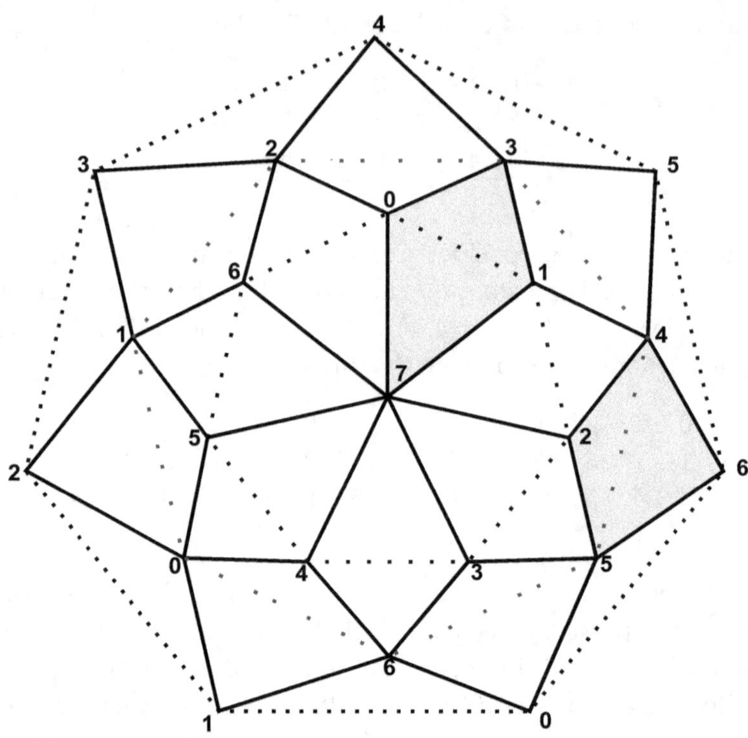

Fig. 3.2. Representation of the design 3-$(8, 4, 1)$.

$(n, n + 1)$ appears in exactly 3 blocks. The vertices of the triangles between the two circles with two points in common are denoted $(n, n+1)$ for $n = 0, \ldots, 20$. These are the blocks $(u+n, u+n+1, v+n)$. Unmarked triangles are blocks $(u + n, v - 1 + n, v + n)$.

On the inner circle, if we look at blocks having 3 points in common, and draw the graph, we see that there are three orbits each forming a complete graph K_7 (see Fig. 3.4). Each orbit is generated from an initial block: $(0, 3, 7, 9, 10, 16)$, $(2, 5, 9, 11, 12, 18)$ and $(1, 4, 8, 10, 11, 17)$. The vertices of the complete graph K_7 are denoted by the initial block and translated by 3 (mod 21). The complete graph K_7 is surrounded by blocks taken from the outer circle of the $(21, 6, 3)$ design with three points in common. The block $(0, 3, 6, 9, 10, 16)$ is linked to the block $(0, 1, 3, 11, 16, 20)$ by the points $(0, 3, 16)$. On the outer heptagon, by turning clockwise, one passes

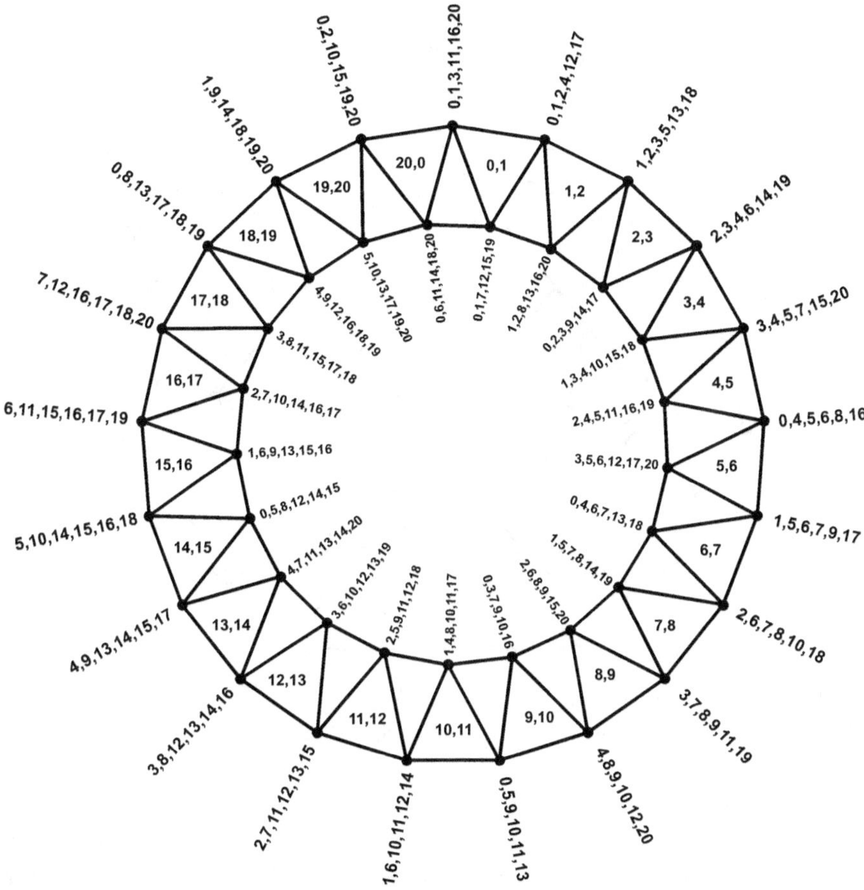

Fig. 3.3. Representation of design $(21, 6, 3)$.

from one block to the next, by adding 3 to each element as for the inner heptagon. Adjacent blocks of the outer heptagon have 2 points in common.

Let us now try to draw all the blocks having three points in common around the inner heptagon. We have seven structures having pairwise three points in common (see Fig. 3.5). Each structure is composed of 6 blocks. Three of these blocks, belonging to the inner circle of Fig. 3.3, are connected by a triangle. The three other blocks belong to the outer circle.

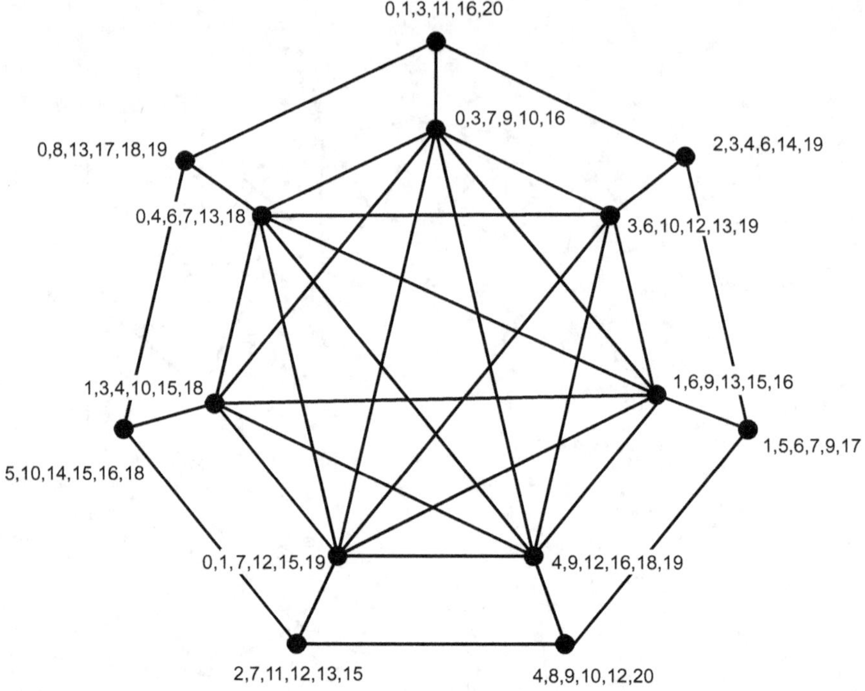

Fig. 3.4. Orbit of design $(21, 6, 3)$.

These 7 structures having three points in common can be placed at the top of a tiling of hexagons (see Fig. 3.8). In the tiling, three vertices of each hexagon have one of these structures. The other three vertices are free. The edges of each hexagon are labeled by blocks of the outer circle of Fig. 3.3. The three blocks that meet in a free vertex have no point in common. In this tiling, we have either three points in common, or no point in common.

3.5. Rational Associahedra

Rational associahedra have been introduced by Drew Armstrong [1]. A rational associahedron is a simplicial complex based on a generalization of Catalan numbers called *rational Catalan numbers*, and defined for all rational number $x \in \mathbb{Q}\backslash[-1, 0]$ written uniquely as

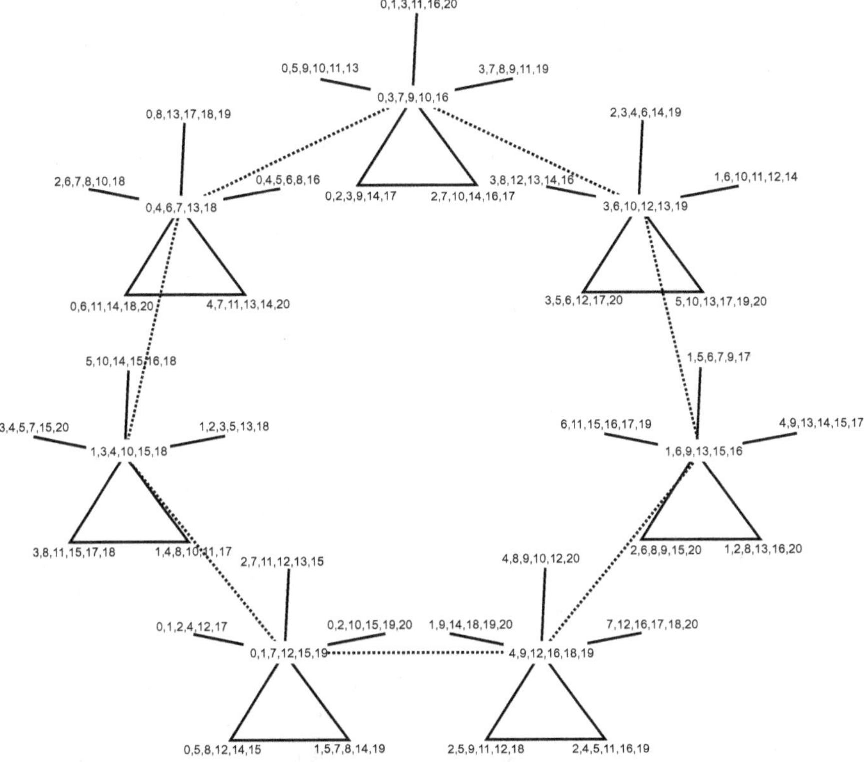

Fig. 3.5. Design $(21, 6, 3)$. Inner heptagon with 3 points in common.

$x = \frac{r}{s-r}$ where r and s are coprime positive integers $r \neq s$.

$$\mathrm{Cat}(x) = \mathrm{Cat}(r, s) = \frac{(r + s - 1)!}{r! s!}.$$

For a positive integer $x = n$, we recover the classical Catalan numbers.

$$\mathrm{Cat}(n) = \mathrm{Cat}(n, n + 1).$$

If we define a *rational Catalan design* as a combinatorial design whose number of blocks b is a *rational Catalan number* $\mathrm{Cat}(x)$ for some x positive rational number, we see that all combinatorial designs can be considered as a rational Catalan design. The reason is we can always find x such that $b = \mathrm{Cat}(x)$. It is enough to take $x = 2/(2b - 1)$,

$(r = 2, s = 2b - 1)$. That is to say, with this definition, we can always reach a combinatorial design of $b = (x+1)/2x$ blocks. The difference between the definition of a rational design with a Catalan number and this new definition with a rational Catalan number is that with the latter we can have several values of x for a given b. For instance, the rational Catalan number of 14 blocks is obtained for the integer $x = 4$, with $r = 4$, $s = 5$ (the classical case), and also for $x = 2/25$, with $r = 2$ and $s = 27$. Note that for a given b, one value of x is not necessarily an integer, for instance: $\mathrm{Cat}(x) = 12$, x is $x = 2/21$ and $x = 3/4$. In some cases, we get only one solution: $\mathrm{Cat}(x) = 16$, $x = 2/29$.

Furthermore, *derived Catalan Numbers* are defined by:

$$\mathrm{Cat}'(x) = \mathrm{Cat}\left(\frac{1}{x-1}\right) = \mathrm{Cat}\left(\frac{x}{1-x}\right).$$

An important property derives from that $\frac{1}{1/x-1} = \frac{x}{1-x}$ and hence

$$\mathrm{Cat}'(x) = \mathrm{Cat}'(1/x).$$

Another important property is that rational Catalan numbers are in relation with Dyck paths. An (r, s)-*Dyck path* is a staircase walk on the integer lattice $\mathbb{Z}^2$ from $(0, 0)$ to (r, s) that lies strictly above (but may touch) the diagonal $y = rx/s$. A Dyck path can be identified with its *Dyck word* being a two letter sequence over an alphabet $\{x, y\}$ with x's representing up steps and y's representing right steps. Originally Dyck words were defined as "well-parenthesized words" over the alphabet $\Sigma = \{(,)\}$, that is pairs of parentheses which are correctly matched. If for a word $\omega \in \Sigma^*$, we define $\mathrm{imb}(\omega) = |\omega|_(- |\omega|_)$, the number of occurrences of the left parenthesis minus the number of occurrences of the right parenthesis. ω is a Dyck word if and only if $\mathrm{imb}(\omega) = 0$ and $\mathrm{imb}(u) \geq 0$ for all prefix u of ω. It has been shown that the number of (r, s)-Dyck paths is $\mathrm{Cat}(r, s)$. We recover a classical result that the number of (classical) Dyck paths (staircase from $(0, 0)$ to (n, n)) is the Catalan number $\mathrm{Cat}(n, n + 1) = \mathrm{Cat}(n)$.

Associated with these rational Catalan numbers exists a geometric structure that D. Armstrong called *rational associahedron*

$\mathrm{Ass}(r,s)$ which has properties similar to the classical case of associahedron. In particular, there exist bijections between the set of homogeneous (r,s)-non-crossing partitions and the set of (r,s)-Dyck paths. A *rational associahedron* $\mathrm{Ass}(r,s)$ is the simplicial complex consisting of all non-crossing dissections of the regular s-gone $\mathbb{P}_{s+1}$, $s \geq 2$. The facets are collections $F(D)$ of admissible diagonals of $\mathbb{P}_{s+1}$, where D is a (r,s)-Dyck path (see [1] for details). The number of facets in $\mathrm{Ass}(r,s)$ is $\mathrm{Cat}(r,s)$. A diagonal of $\mathbb{P}_{s+1}$ which separates i vertices from $s - i - 1$ vertices appears as a vertex of $\mathrm{Ass}(r,s)$ if and only if $i \in S(r,s)$

$$S(r,s) = \left\{ \left\lfloor \frac{is}{r} \right\rfloor, 1 \leq i < r \right\},$$

where $\lfloor \xi \rfloor$ is floor(ξ), the greatest integer $\leq \xi$. For instance, if $x = 3/2$, $r = 3$ and $s = 5$, the rational associahedron $\mathrm{Ass}(3,5)$ has 6 vertices, since $S(3,5) = \{1,3\}$. $\mathrm{Ass}(3,5)$ has 7 facets as there are $\mathrm{Cat}(3,5) = 7$ Dyck paths. The associahedron is drawn on Fig. 3.6. Each vertex represents an admissible dissection of $\mathbb{P}_6$. On this graph, a facet links two vertices. Each facet corresponds to a $(3,5)$-Dyck path, a staircase from $(0,0)$ to $(3,5)$. Symbol x represents up step and y represents right step. The simplicial complex $\mathrm{Ass}(r,s)$ has dimension $r - 2$. The h-vector is the integer sequence $(h_{-1}, h_0, \ldots, h_{r-2})$ where h_{i-2} is the number of (r,s)-Dyck paths with i non-trivial vertices runs. A *vertical run* in a Dyck path is the maximal sequence of $(0,1)$ steps. h_{i-2} is also the number of Dyck paths with $(i-1)$ "valleys", according to the shape of Dyck path. It is also the number of monomial y^m in the Dyck word minus 1. It is the Narayana number given by

$$h_{i-2} = \mathrm{Nar}(r,s,i) = \frac{1}{r} \binom{r}{i} \binom{s-1}{i-1}.$$

For the non-crossing partitions point of view, the number of (r,s)-Dyck paths with α_j vertices runs of length j is the Kreweras number:

$$\mathrm{Krew}(r,s,\alpha) = \frac{(s-1)!}{\alpha_0! \alpha_1! \ldots \alpha_r!}.$$

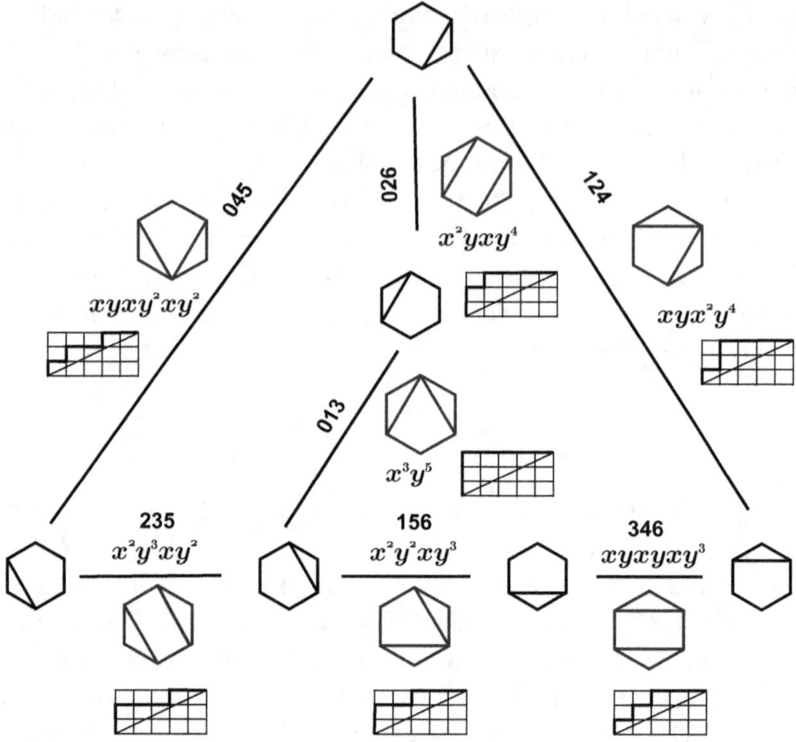

Fig. 3.6. Rational associahedron Ass$(3, 5)$.

The f-vector is the integer sequence $(f_{-1}, f_0, \ldots, f_{r-2})$ of Ass(r, s) with $f_{-1} = 1$, and f_i is the number of i-dimensional faces $0 \leq i \leq r - 2$, also given by the Kirkman numbers:

$$f_{i-2} = \mathrm{Kir}(r, s, i) = \frac{1}{r} \binom{r}{i} \binom{s+i-1}{i-1}.$$

These numbers verify the relation

$$\sum_{i=-1}^{r-2} f_i(t-1)^{r-2-i} = \sum_{i=-1}^{r-2} h_i t^{r-2-i}$$

and the *Reduced Euler Characteristic* defined from the f-vector by the alternate sum of its components is equal to the derived Catalan number:

$$\chi = \sum_{i=-1}^{r-2} (-1)^i f_i = (-1)^r \mathrm{Cat}'(r,s).$$

For instance, for the Ass$(3,5)$ of Fig. 3.6, h-vector $= (1,4,2)$, f-vector $= (1,6,7)$ and the reduced Euler characteristic $\chi = -2$ which is equal to $(-1)^3 \mathrm{Cat}'(3/2) = -\mathrm{Cat}(2) = -2$.

Dyck paths are related by plactic relations. Plactic relations were introduced by Knuth [7], and later studied by Lascoux and Schützenberger [8, 9]. On an alphabet with two letters $A = \{x, y\}$, with $x < y$, plactic relations are

$$xyx \equiv xxy, \quad yxy \equiv xyy.$$

The plactic monoid is the quotient of the free monoid by the Knuth congruences $A^*/\equiv$. Dyck paths of Ass$(3,5)$ is a class of this monoid with 3 letters x and 5 letters y, namely

$$x^3 y^5 \equiv x^2 yxy^4 \equiv xyx^2 y^4 \equiv xyxyxy^3 \equiv x^2 y^2 xy^3$$
$$\equiv x^2 y^3 xy^2 \equiv xyxy^2 xy^2.$$

The blocks of the $(7,3,1)$ design are represented on Fig. 3.6. One passes from one block to another by adding 2 to each element of the block.

Let us take another example. The design $(9,3,1)$ has 12 blocks and 4 parallel classes:

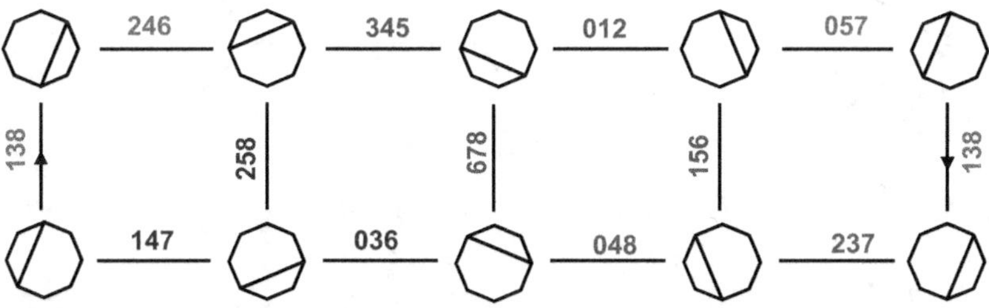

Fig. 3.7. Möbius strip of design $(9,3,1)$.

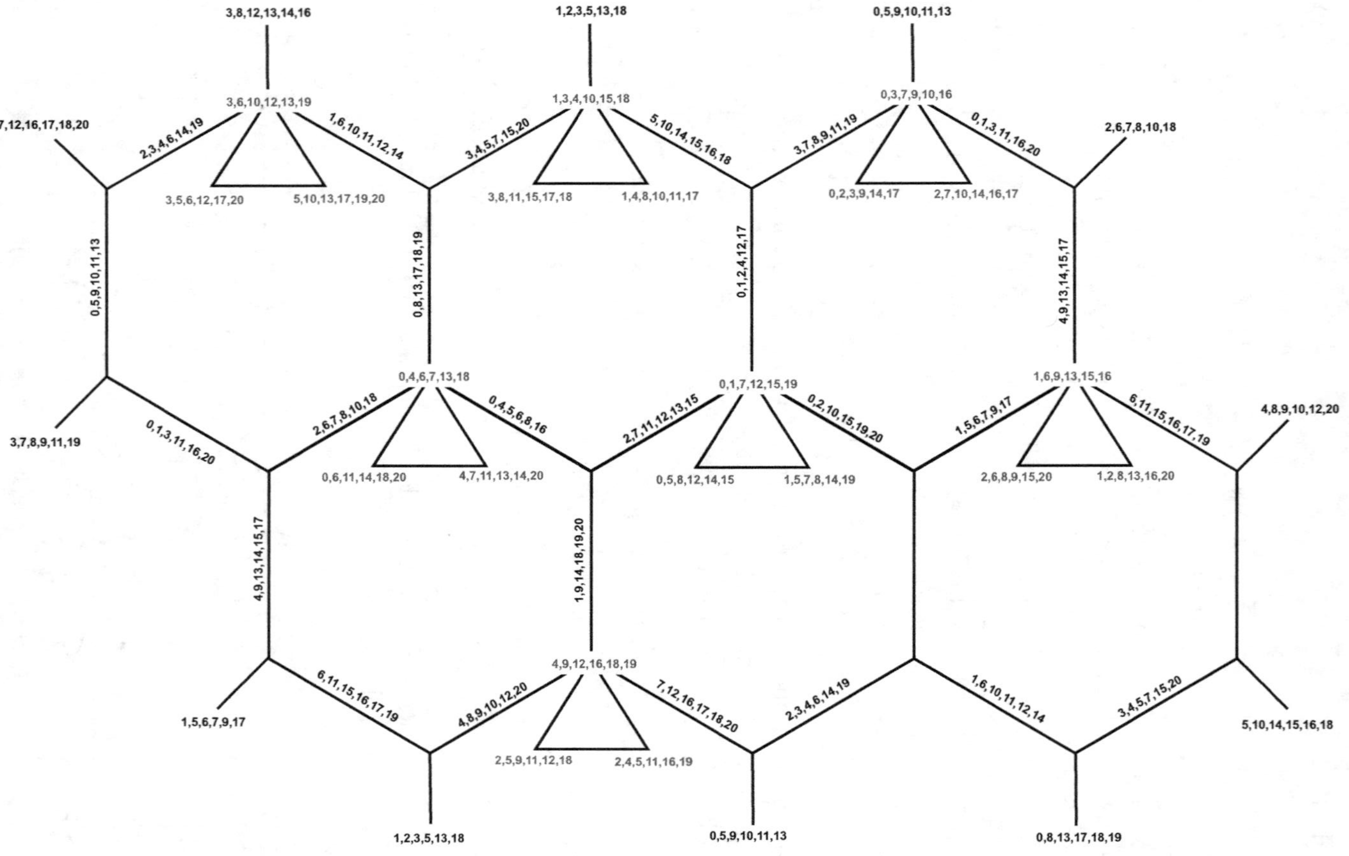

Fig. 3.8. Periodic tiling of design $(21, 6, 3)$.

000011122236
134534534547
268787676858

Since $12 = \text{Cat}(3,7)$, the rational associahedron $\text{Ass}(3,7)$ with 8 vertices and 12 facets could be used to represent the design $(9,3,1)$. On the dissection plane $\mathbb{P}_8$, each vertex separates 2 vertices from 4 vertices. Dyck paths lead to 12 facets. Each facet is associated with a block of the design. These blocks are depicted on Fig. 3.7. The figure is a Möbius strip (glue the ribbon with respect to the arrows). At each vertex there are three blocks which form a parallel class (a partition of $\mathbb{Z}_9$), except when the ribbon wraps.

Bibliography

[1] D. Armstrong, B. Rhoades and N. Williams, Rational associahedra and noncrossing partitions, *Electron. J. Combin.* **20**(3) (2013) 1–27.

[2] C. J. Colburn and J. H. Dinitz, *Handbook of Combinatorial Designs*, 2nd edn. (Chapman & Hall, 2007).

[3] S. Fomin and N. Reading, Root systems and generalized associahedra, in *Geometric Combinatorics*, eds. E. Miller, V. Reiner and B. Sturmfelds, IAS/Park City, Mathematis Series, Vol. 13, (American Mathematical Society, 2007).

[4] C. Hohlweg, Permutahedra and associahedra, generalized associahedra from the geometry of finite reflection groups, in *Associahedra, Tamari Lattices and Related Structures*, eds. F. Müller-Hoissen *et al.* (Birkhäuser, 2012), pp. 129–159.

[5] T. Johnson, *Combinatorial Designs in My Music* (Editions 75, 2009).

[6] T. Johnson and F. Jedrzejewski, *Looking at Numbers* (Birkhäuser, 2014).

[7] D. E. Knuth, Permutations, matrices, and generalized Young tableaux, *Pacific J. Math.* **34**(1970) 709–727.

[8] A. Lascoux and M. P. Schützenberger, Le monoïde plaxique, in *Non-commutative structures in Algebra and Combinatorics* (Quaderni della Ricerca Scientifica del CNR, Roma, 1981).

[9] A. Lascoux, B. Leclerc and J. Y. Thibon, The plactic monoid, in *Algebraic Combinatorics on Words*, ed. M. Lothaire (Cambridge University Press, 2002), p. 164–196.

[10] F. Müller-Hoissen, J. M. Pallo and J. Stasheff (eds.), *Associahedra, Tamari Lattices and Related Structures*, Tamari Memorial Festschrift (Birkhäuser, 2012).

[11] R. P. Stanley, *Enumerative Combinatorics*, Vol. 2 (Cambridge University Press, 1999).

[12] D. Tamari, Monoïdes préordonnés et chaînes de Malcev, Thèse de Mathématique, Paris (1951).

[13] J. Yust, The geometry of melodic, harmonic, and metrical hierarchy, in *Mathematics and Computation in Music, Second Int. Conf. MCM 2009*, eds. E. Chew, A. Childs and C.-H. Chuan, Springer Communications in Computer and Information Science, Vol. 38 (Springer, 2009), pp. 180–192.

Chapter 4

Rhythmic and Melodic L-canons

Jeremy Kastine

*Georgia Highlands College,
Rome, GA, USA*

4.1. Introduction

In this chapter we will explore techniques for composing canons with a composite monophonic texture (one pitch sounding at a time). One technique for composing monophonic canons is based on rhythmic tiling. Figure 4.1 gives an example of a rhythmic tiling.

Regardless of what melody and rule for melodic imitation that we choose to apply to a rhythmic tiling, the resulting canon will be monophonic due to the disjointness of the parts. Figure 4.2 gives an example of one possible melodization of the rhythmic tiling in Fig. 4.1.

The canon in Fig. 4.2 can be written in a more compact tabular form as follows.

0	1	2	3	4	5	6	7	8	9	10	11	
0	.	7	.	.	4	.	5	.	.	.	.	:||
.	.	.	.	4	.	11	.	.	8	.	9	:||
.	11	.	12	.	.	.	.	7	.	14	.	:||

Rhythmic tiling has been studied in some detail and has been applied to musical composition in various ways [1, 2]. Using rhythmic

Fig. 4.1. This is an example of a rhythmic tiling since each beat is occupied by exactly one note. The second and third parts imitate the first, starting on the fifth and ninth beats, respectively.

Fig. 4.2. This is one possible melodization of the rhythmic tiling in Fig. 4.1. The second and third parts are transposed a minor third higher and a perfect fifth higher, respectively, with respect to the first part.

tiling to compose monophonic canons has some drawbacks though. Most rhythmic tilings exhibit a high degree of regularity (predictability) in the rhythmic theme and/or in the set of translations of that rhythm. For instance, in the example above, the entries of the voices are equally spaced. Vuza canons, rhythmic tilings in which neither the rhythmic theme nor the set of translations are periodic, provide an exception to this rule, but they quite rare. In fact, the smallest Vuza canon is 72 beats long [4].

Another objection that we might have to using rhythmic tiling to compose canons is that there is often no sense of interaction among the parts, since the parts are disjoint. Also, the fixed density of musical information (exactly one note per beat, where the beats are evenly spaced) can become rather monotonous.

In this chapter we will seek alternative methods for composing monophonic canons which will often feature

- more interaction between the parts, by allowing them to overlap at times,
- less monotony, by allowing beats on which multiple parts overlap, beats of rest, and unevenly distributed beats,
- higher density of musical information, in terms of their note-to-length ratio (by definition, rhythmic tiling have a note-to-length of exactly 1 note per beat), and
- more interesting (less regular) rhythms.

Many open problems regarding the construction of these canons will be given.

4.2. Rhythmic L-canons

Let us first give a formal definition of rhythmic tiling.

Definition 1. Let X and A be subsets of $\mathbb{Z}_n$. Then (X, A) is a rhythmic tiling provided that

- $(X + c_1) \cap (X + c_2) = \emptyset$ for distinct $c_1, c_2 \in A$, and
- $\bigcup_{c \in A}(X + c) = \mathbb{Z}_n$.

In the example from the introduction, $n = 12$, $X = \{0, 2, 5, 7\}$, and $A = \{0, 4, 8\}$. Now let us introduce the following generalization.

Definition 2. Let X be a finite subset of $\mathbb{R}$, and let A be a subset of $\mathbb{R}^2$. We will say that (X, A) is a rhythmic L-canon provided that

- $|(b_1 X + c_1) \cap (b_2 X + c_2)| \leq 1$ for distinct $(b_1, c_1), (b_2, c_2) \in A$, and
- $|x_1 - x_2| \geq 1$ for distinct $x_1, x_2 \in \bigcup_{(b,c) \in A}(bX + c)$.

In some cases, we will insist that all b_i be equal, in which case we will have a rhythmic L-canon without augmentation.

This definition generalizes rhythmic tiling in the following ways.

- The parts (the sets of the form $bX + c$) are allowed to overlap once.
- Rhythmic augmentation is allowed.

- X and A are real-valued.
- There is no "beat-filling" requirement.
- If A is finite, then the resulting canon will be finite, as opposed to the endlessly repeating nature of rhythmic tiling.

If we let $X = \{0, 4, 7, 8\}$ and $A = \{(2, 0), (1, 1), (1.75, 9)\}$, then we have the following.

$$2X + 0 = \{0, 8, 14, 16\},$$
$$1X + 1 = \{1, 5, 8, 9\},$$
$$1.75X + 9 = \{9, 16, 21.25, 23\}.$$

Note that the intersection of any two of these sets is a singleton and that the distance between any two elements of their union, $\{0, 1, 5, 8, 9, 14, 16, 21.25, 23\}$, is at least 1. So (X, A) is a rhythmic L-canon. The following is its score.

0	1	5	8	9	14	16	21.25	23
x	.	.	x	.	x	x	.	.
.	x	x	x	x	.	.	.	.
.	.	.	.	x	.	x	x	x

$$(4.1)$$

As stated in the introduction, one of our goals is to compose canons with a relatively high note-to-length ratio. This canon, with a note-to-length ratio of $\frac{12}{24} = 0.5$, performs rather poorly in this respect. The following example, with $X = \{0, 8, 13, 16\}$ and $A = \{(0.72, 0), (0.36, 1.08), (0.585, 6.84)\}$, is little bit better, with a note-to-length ratio of $\frac{12}{17.2} \approx 0.7$.

0	1.08	3.96	5.76	6.84	9.36	11.52	14.445	16.2
x	.	.	x	.	x	x	.	.
.	x	x	x	x	.	.	.	.
.	.	.	.	x	.	x	x	x

$$(4.2)$$

Notice that (4.1) and (4.2) are similar, in the sense that they both begin with (part 1, note 1), followed by (part 2, note 1) and (part 2, note 2), and then (part 1, note 2) and (part 2, note 3) at the same time, and so on. In other words, they are both of the following form.

t_0	t_1	t_2	t_3	t_4	t_5	t_6	t_7	t_8
×	·	·	×	·	×	×	·	·
·	×	×	×	×	·	·	·	·
·	·	·	·	×	·	×	×	×

$$(4.3)$$

If we would like to find the shortest possible rhythmic L-canon of this form, then we must solve the following optimization problem.

$$
\begin{aligned}
\text{minimize} \quad & t_8 \\
\text{subject to} \quad & t_0 = 0 \\
& t_{i+1} - t_i \geq 1 \\
& \operatorname{rank}
\begin{bmatrix}
t_3 - t_0 & t_5 - t_3 & t_6 - t_5 \\
t_2 - t_1 & t_3 - t_2 & t_4 - t_3 \\
t_6 - t_4 & t_7 - t_6 & t_8 - t_7
\end{bmatrix} = 1
\end{aligned}
$$

$$(4.4)$$

The constraint that the matrix have rank 1 forces the rows of the matrix to be proportional, so that the rhythms are proportional. This is a non-convex problem, which makes it somewhat difficult to solve. Numerical methods produce the following solution.

0	1	3.61	4.72	5.72	6.72	8.53	9.72	10.79
×	·	·	×	·	×	×	·	·
·	×	×	×	×	·	·	·	·
·	·	·	·	×	·	×	×	×

$$(4.5)$$

This canon has a note-to-length ratio of $\frac{12}{11.79} \approx 1.02$, which is much better than those of the two preceding examples and even slightly better than that of tiling canons (which have a note-to-length ratio of exactly 1).

Problem 1. What is the most efficient numerical algorithm for solving problems like (4.4)? Do they always have a unique solution? Do local minima occur? Is it possible to solve them analytically?

The following definition formalizes and generalizes the type of similarity that we noted between the preceding examples of rhythmic L-canons.

Definition 3. We will say that two L-canons (X, A) and (X', A') are similar provided that there exist one-to-one correspondences $f : X \to X'$ and $g : A \to A'$ such that for all $x_1, x_2 \in X$ and $(b_1, c_1), (b_2, c_2) \in A$,

$$b_1 x_1 + c_1 \leq b_2 x_2 + c_2 \quad \text{if and only if} \quad b_1' x_1' + c_1' \leq b_2' x_2' + c_2',$$

where $x_1' = f(x_1)$, $x_2' = f(x_2)$, $(b_1', c_1') = g(b_1, c_1)$, and $(b_2', c_2') = g(b_1, c_1)$.

Consider the set of all rhythmic L-canons (X, A) with fixed $|X|$, $|A| \in \mathbb{N}$. While there are uncountably many such rhythmic L-canons, each similarity class corresponds to a certain ordering of its notes that is unique (up to permutation of the notes and parts). For example, (4.3) gives the note order corresponding to all of the preceding rhythmic L-canons. The following is another example of two similar rhythmic L-canons.

0	2	4	6	8	10	11	13	15
x	x	·	x	·	·	x	·	·
·	·	x	x	·	x	·	·	x
·	x	x	·	x	·	·	x	·

0	1	2	3	4	5	6	7	8
x	x	·	x	·	·	x	·	·
·	·	x	x	·	x	·	·	x
·	x	x	·	x	·	·	x	·

Both of these rhythmic L-canons, as well any other rhythmic L-canon in the same similarity class, have the following note order.

t_0	t_1	t_2	t_3	t_4	t_5	t_6	t_7	t_8
x	x	·	x	·	·	x	·	·
·	·	x	x	·	x	·	·	x
·	x	x	·	x	·	·	x	·

Given a possible note order, there may or may not exist a corresponding rhythmic L-canon. Consider the following note order.

t_0	t_1	t_2	t_3	t_4
x	·	x	·	x
·	x	·	x	x

In this case, there exists a corresponding rhythmic L-canon.

0	2	4	5	8
x	·	x	·	x
·	x	·	x	x

Now consider the following note order.

t_0	t_1	t_2	t_3	t_4	t_5
x	·	x	·	x	
·	x	x	·	·	x

In this case, no corresponding rhythmic L-canon exists. Indeed, comparing the first two notes of each part, we would conclude that part 1 possesses a greater factor of augmentation. Comparing the last two notes of each part, we would conclude that part 2 has a greater factor of augmentation. This contradiction implies that no solution exists.

Problem 2. Given a possible note order, consider the problem of determining whether a corresponding rhythmic L-canon exists, and if so, finding one. What is the most efficient numerical algorithm for solving this non-convex feasibility problem? Is it possible to solve this problem analytically? Are there any useful necessary or sufficient conditions for the existence of a solution?

Problem 3. For fixed $|X|, |A| \in \mathbb{N}$, the number of similarity classes is less than or equal to the number of possible note orders, which is finite. For relatively small $|X|, |A|$, we would like to enumerate all similarity classes. What is the most efficient algorithm for doing so? For larger $|X|, |A|$, when this problem becomes intractable, what bounds can we give on the number of similarity classes?

Problem 4. For each fixed $|X|, |A| \in \mathbb{N}$, what is the shortest rhythmic L-canon? If we can solve the previous problem, then we can find an optimal solution for each similarity class, and then pick the shortest. However, this method will be quite slow, and will become intractable for even moderate values of $|X|$ and $|A|$. How can we search for the solution to this problem more efficiently?

4.3. Melodic L-canons

Suppose that we want to melodize (4.5). Let the sequence $\{y_1, y_2, y_3, y_4\}$ represent the pitch values of the melody, and let $\{z_1, z_2, z_3\}$ represent the degree to which each part is translated up or down. In other words, the score will be of the following form.

0	1	3.61	4.72	5.72	6.72	8.53	9.72	10.79
(y_1+z_1)	$\cdot$	$\cdot$	(y_2+z_1)	$\cdot$	(y_3+z_1)	(y_4+z_1)	$\cdot$	$\cdot$
$\cdot$	(y_1+z_2)	(y_2+z_2)	(y_3+z_2)	(y_4+z_2)	$\cdot$	$\cdot$	$\cdot$	$\cdot$
$\cdot$	$\cdot$	$\cdot$	$\cdot$	(y_1+z_3)	$\cdot$	(y_2+z_3)	(y_3+z_3)	(y_4+z_3)

The following equations must hold in order to maintain monophony.

$$y_2 + z_1 = y_3 + z_2,$$
$$y_4 + z_2 = y_1 + z_3,$$
$$y_4 + z_1 = y_2 + z_3.$$

The solution to this system can be expressed as follows.

$$y_3 = 2y_2 - y_1,$$
$$z_2 = y_1 - y_2 + z_1,$$
$$z_3 = -y_2 + y_4 + z_1.$$

Choosing $y_1 = 4$, $y_2 = 2$, $y_4 = 5$, and $z_1 = 0$ gives us the following L-canon.

0	1	3.61	4.72	5.72	6.72	8.53	9.72	10.79
4	·	·	2	·	0	5	·	·
·	6	4	2	7	·	·	·	·
·	·	·	·	7	·	5	3	8

$$(4.6)$$

Since y_3 was a basic variable (not free!), we did not have the freedom to choose any melody that we desired, as we were able to in the case of tiling canons. This is the price that we often pay for allowing overlapping between the parts. The following example is a more extreme case, which admits no non-trivial solution.

0	2	3	4	6	8	12
x	x	·	·	x	·	·
x	·	·	x	·	·	x
·	x	·	x	·	x	·
·	·	x	x	x	·	·
·	·	·	·	x	x	x

$$(4.7)$$

Solving the same type of monophony-enforcing linear system that we did with the previous example yields $y_1 = y_2 = y_3$ and $z_1 = z_2 = z_3 = z_4 = z_5$. If we are determined to melodize this canon, we will need to take a more relaxed approach to melodic imitation. One possible approach is given in the following definition.

Definition 4. We will refer to an order $\preceq$ on $\{1, 2, \ldots, n\}$ with reflexivity, transitivity, and totality (a total preorder), as a contour.

We will say that a sequence of real numbers $\{u_i\}_{i=1}^n$ has contour $\preceq$ provided that $u_j \leq u_k$ if and only if $j \preceq k$. If each part of a canon is of the same contour, then we will say that it exhibits imitation of contour (as oppose to strict melodic imitation).

Suppose that we wish to melodize (4.7) with a contour defined by $1 \prec 3 \prec 2$, meaning that the first note should be lowest, the third should be higher, and the second should be highest. Let $u_{i,j}$ represent the jth note of part i. The following inequalities enforce our chosen contour.

$$u_{1,1} \leq u_{1,3} - 1 \qquad\qquad u_{1,3} \leq u_{1,2} - 1$$

$$u_{2,1} \leq u_{2,3} - 1 \qquad\qquad u_{2,3} \leq u_{2,2} - 1$$

$$\vdots \qquad\qquad\qquad\qquad \vdots$$

The following equations enforce monophony.

$$u_{1,1} = u_{2,1}$$

$$u_{1,2} = u_{3,1}$$

$$u_{2,2} = u_{3,2}$$

$$u_{3,2} = u_{4,2}$$

$$\vdots$$

Linear programming can be used to search for a feasible solution to this system of equations and inequalities. In this case, a solution does exist. The following gives one of the possible solutions.

0	2	3	4	6	8	12
0	2	.	.	1	.	.
0	.	.	4	.	.	2
.	2	.	4	.	3	.
.	.	0	4	1	.	.
.	.	.	.	1	3	2

Problem 5. Given a rhythmic L-canon, what is the most efficient method for determining the contours with which it may be melodized?

Problem 6. For some rhythmic L-canons, any melody (or contour) may be applied. We will say that these rhythmic L-canons possess melodic freedom. What necessary and/or sufficient conditions can we give for a rhythmic L-canon to possess melodic freedom? This question can be asked in the context of strict or contour imitation.

Problem 7. For fixed $|X|, |A| \in \mathbb{N}$, what is the shortest rhythmic L-canon with melodic freedom (in terms of either strict or contour imitation)?

Problem 8. Given a specific melody (or contour) and fixed $|A| \in \mathbb{N}$, what is the shortest L-canon based on that melody (or contour)?

4.4.   S-canons and G-canons

In this section we will look at two special types of L-canons with melodic freedom. The first type, which will call S-canon, is constructed using Sidon sets.

Definition 5. A Sidon set is a set of integers with distinct pairwise sums, disregarding those sums that are necessarily equal due to the commutative property of addition.

The following is the addition table of the set $\{0, 2, 3, 7\}$.

+	0	2	3	7
0	0	2	3	7
2	2	4	5	9
3	3	5	6	10
7	7	9	10	14

The commutative property of addition is responsible for the symmetry over the main diagonal, but the sums in the upper right triangle of the table are distinct. So $\{0, 2, 3, 7\}$ is a Sidon set. The rows of this addition table determine the parts of the following rhythmic S-canon.

0	2	3	4	5	6	7	9	10	14
x	x	x	.	.	.	x	.	.	.
.	x	.	x	x	.	.	x	.	.
.	.	x	.	x	x	.	.	x	.
.	.	.	.	.	.	x	x	x	x

$$(4.8)$$

A rhythmic S-canon can be melodized by choosing a melody (any melody) in the form of a sequence of numbers, and then determining the melody of each part from the addition table of that sequence. For example, let us choose the melody $\{5, 6, 1, 8\}$, whose addition table is as follows.

$+$	5	6	1	8
5	10	11	6	13
6	11	12	7	14
1	6	7	2	9
8	13	14	9	16

Applying the rows of this addition table to the parts of (4.8) yields the following S-canon.

0	2	3	4	5	6	7	9	10	14
10	11	6	.	.	.	13	.	.	.
.	11	.	12	7	.	.	14	.	.
.	.	6	.	7	2	.	.	9	.
.	.	.	.	.	.	13	14	9	16

Note that each part overlaps each of the other parts exactly once (in unison), and the notes of part 1 determine the entries of parts 2–4 in terms of their pitch and time. Interestingly, another part, based on the main diagonal of each addition table, can be joined to this S-canon. This extra part imitates the first part by augmentation of its rhythm and melody by a factor of 2.

0	2	3	4	5	6	7	9	10	14
10	11	6	·	·	·	13	·	·	·
·	11	·	12	7	·	·	14	·	·
·	·	6	·	7	2	·	·	9	·
·	·	·	·	·	·	13	14	9	16
10	·	·	12	·	2	·	·	·	16

It is left for the reader to verify that the construction exemplified above works (i.e., all of the noted properties hold) for any Sidon set and any melody.

Now let us turn to the construction of G-canons, which are based on sets of numbers called Golomb rulers.

Definition 6. A Golomb ruler is a set of integers with distinct pairwise differences, disregarding those differences that are necessarily zero.

It is a well-known result and easy exercise to show that a finite set of integers is a Sidon set if and only if it is a Golomb ruler. For example, the set $\{0, 2, 3, 7\}$, which we used as an example of a Sidon set, is also a Golomb ruler. We can construct a G-canon in the same way that we constructed the S-canon above, merely by replacing addition with subtraction. The subtraction tables

$-$	0	2	3	7
0	0	2	3	7
2	-2	0	1	5
3	-3	-1	0	4
7	-7	-5	-4	0

and

$-$	5	6	1	8
5	0	1	-4	3
6	-1	0	-5	2
1	4	5	0	7
8	-3	-2	-7	0

yield

-7	-5	-4	-3	-2	-1	0	1	2	3	4	5	7
.	.	.	.	.	.	0	.	1	-4	.	.	3
.	.	.	.	-1	.	0	-5	.	.	.	2	.
.	.	.	4	.	5	0	.	.	.	7	.	.
-3	-2	-7	.	.	.	0	.	.	.	.	.	.

Note that all of the parts overlap (in unison) on beat 0 and are otherwise disjoint. The entries of the parts are determined, in terms of their pitch and time, by a retrograde-inversion of the theme. It is left for the reader to verify that these properties hold for any G-canon constructed by this method, regardless of what Golomb ruler and melody are chosen.

So far, G-canons and S-canons have been defined in terms of Golomb rulers and Sidon sets, which are sets of integers. Let us extend our definitions to the real numbers as follows.

Definition 7. A rhythmic S-canon is a rhythmic L-canon (X, A) where $X \subset \mathbb{R}$ and $A = \{1\} \times X$. A rhythmic G-canon is a rhythmic L-canon (X, A) where $X \subset \mathbb{R}$ and $A = \{1\} \times (-X)$.

The following problems are posed in terms of S-canons. There are, of course, analogous problems concerning G-canons.

Problem 9. For a fixed order, there are infinitely many rhythmic S-canons. However, we can define similarity among rhythmic S-canons just as we did with rhythmic L-canons. How many similarity classes of rhythmic S-canons exist for each order? What is the most efficient algorithm for generating all of them? For higher orders, in which case generating a complete list is intractable, can we at least give bounds on the number of similarity classes.

Problem 10. Suppose that we are given a note order. Are there any useful necessary and/or sufficient conditions for the existence of a corresponding rhythmic S-canon?

Problem 11. Researchers have expended much time and effort searching for shortest (integer-valued) Sidon set of each order [3]. Is there anything to be gained by expanding the search to real-valued rhythmic S-canons, or are the best solutions always integral? In the former case, what are the shortest rhythmic S-canons of each order? In the latter case, can we prove it?

4.5. Some Extensions of L-, S-, and G-canons

In this section, I will describe some extensions of the most basic types of L-, S-, and G-canons described above. For the sake of simplicity, we have only considered cases in which A is finite. We could extend our investigation to canons which endlessly repeat (just as conventional tiling canons do) by considering A of the form $\{(b, c + pk) : (b, c) \in A', k \in \mathbb{Z}\}$ where A' is a finite subset of $\mathbb{R}$ and $p \in \mathbb{R}$ is the period of repetition. In this context, the optimization problems that we have discuss would be reframed in terms of minimizing the period. The following is an example in which $X \approx \{0, 0.249, 0.579, 1\}$, $A' \approx \{(10.24, 0), (9.55, 3.55), (8.67, 8.08), (8.03, 12.1)\}$, and $p = 15.2$. For clarity, the initial note of each part is given in boldface.

0	1.55	2.55	3.55	4.93	5.93	8.08	9.08	10.24	12.1	13.1	14.1	15.2
12	.	10	.	.	5	.	.	11	.	.	.	:||
.	.	.	**7**	.	5	.	0	.	.	6	.	:||
.	**12**	12	.	.	.	**13**	.	11	.	6	.	:||
.	12	.	.	18	.	.	.	.	**19**	.	17	:||

In order to have more freedom in melodization, we could decide to allow octaves between overlapping notes as in the following canon.

0	2	3	4	5	6	9	10	11	14	15	16	17
6	7	.	.	12	2	.	.	.	.	.	.	:||
.	.	.	**13**	.	14	19	9	.	.	.	.	:||
.	.	.	.	.	.	**7**	.	8	13	3	.	:||
.	**19**	9	.	.	.	.	.	.	13	.	14	:||

Additionally, there is variety of other imitative techniques that could be employed within L-canons. For example, the following L-canon involves inversion and retrogradation (in addition to the other techniques we have used so far).

0	1	2	3	4	5	6	7	9
5	·	·	8	0	·	·	·	·
·	·	16	8	·	·	11	·	·
·	·	·	·	·	3	11	·	8
·	3	·	·	·	·	·	0	8

In [5] I explored how two or more rhythmic S-canons can be used to tile $\mathbb{Z}_n$, resulting in multi-S-canons. In the following example, notice that parts 1–4 form an S-canon, and that parts 5–8 form a second S-canon, which happens to be a retrograde-inversion of the first. Also, note the high note-to-length ratio of $\frac{32}{20} = 1.6$.

0	1	2	3	4	5	6	7	8	9	10	11	12	13	14	15	16	17	18	19	20
6	·	10	·	·	·	3	·	·	4	·	·	·	·	·	·	·	·	·	·	‖
·	·	10	·	14	·	·	·	7	·	·	8	·	·	·	·	·	·	·	·	‖
·	·	·	·	·	·	3	·	7	·	·	·	0	·	·	1	·	·	·	·	‖
·	·	·	·	·	·	·	·	·	4	·	8	·	·	·	1	·	·	2	·	‖
·	·	·	·	·	·	·	·	17	·	·	18	·	·	·	11	·	15	·	·	‖
·	·	·	·	·	·	·	·	·	18	·	·	19	·	·	·	12	·	16	·	‖
·	5	·	9	·	·	·	·	·	·	·	·	·	11	·	·	12	·	·	·	‖
·	·	·	9	·	13	·	·	·	·	·	·	·	·	·	15	·	·	16	·	‖

Bibliography

[1] M. Andreattu and C. Agon, Special issue: Tiling problems in music guest editor's forward, *J. Math. Music* **3**(2) (2009), http://dx.doi.org/10.1080/17459730903086140.

[2] C. Argon and M. Andreatta, Modeling and implementing tiling rhythmic canons in the Open Music visual programming language, *Perspect. New Music* **49**(2) (2011), http://www.perspectivesofnewmusic.org/TOC492.pdf.

[3] Distributed.net, (2017), http://www.distributed.net/Main_Page (Accessed: 11/1/2017).

[4] F. Jedrzejewski, Enumeration of vuza canons, preprint (2013), arXiv:1304.6609 [math.HO].

[5] J. Kastine and M. Montiel, The s-canon and the multi-s-canon: an introduction, *J. Math. Music* **10**(3) (2016), 211–225, http://dx.doi.org/10.1080/17459737.2016.1168492.

Chapter 5

The Fibonacci Sequence as Metric Suspension in Luigi Nono's *Il Canto Sospeso*

Jon Kochavi

*Department of Music and Dance,
Swarthmore College,
500 College Avenue,
Swarthmore, PA 19081, USA
jkochav1@swarthmore.edu*

5.1. The Fibonacci Sequence in Music

In recent years, there has been a proliferation of books addressing the intersections between the fields of mathematics and music, frequently aimed at general audiences. One of the topics that often arises is the application of the Fibonacci sequence and the golden mean to musical organization and structure, though the treatment in these texts tends to be general and observational.[a] In depth examples of these types of connections also abound and can be found scattered the scholarly literature of both fields.

---

[a]See [4, 5, 11] for example. Other authors have summarily dismissed the potential connections, as in [6, 23]. In the introduction to the latter text, author Wright explicitly rejects Fibonacci as a mechanism useful for viewing or understanding music, explaining, "Connections that seem (to the author) to be distant or arguable, such as the golden ratio, are omitted or given only perfunctory mention [in the book]" (p. xi).

Perhaps the best known examples originate in Hungarian theorist Ernő Lendvai's exhaustive and contentious analyses of works by Béla Bartók, first undertaken in the 1950s, with widespread distribution of his writings in the early 1970s as they began appearing in English [10]. Lendvai strove towards a unifying theory explaining Bartók's compositional process and its results, and he looked to the Fibonacci sequence and the golden mean for the theoretical foundation of his system. In Lendvai's vision, the Fibonacci sequence and golden mean govern both durational aspects of Bartók's works and formal arrangement of pitches and pitch structures in his music. His examples are numerous, but perhaps his most familiar and compelling come from Bartók's *Music for Strings, Percussion and Celesta*, in which Lendvai maps out the architecture of the individual movements and finds that important moments of musical articulation divide up the piece based on approximate golden ratios, specifically referencing integers appearing in the Fibonacci sequence (Figs. 5.1 and 5.2). Many of Lendvai's analyses press well beyond reasonable analytical limits; this and his universal claims proved to be fatal flaws, encouraging the reader to dismiss his overly broad claims along with the balance of his work. Nevertheless, some of the individual analyses such as these have withstood the test of close scholarly review and are generally understood to be one useful lens through which to understand Bartók's careful control of large scale proportion.

A different line of inquiry revolves around durational proportions in sonata form movements from the Classical era. At the 1994 Joint Mathematics Meetings in Cincinnati, mathematician John Putz presented his findings testing the hypothesis that movements from

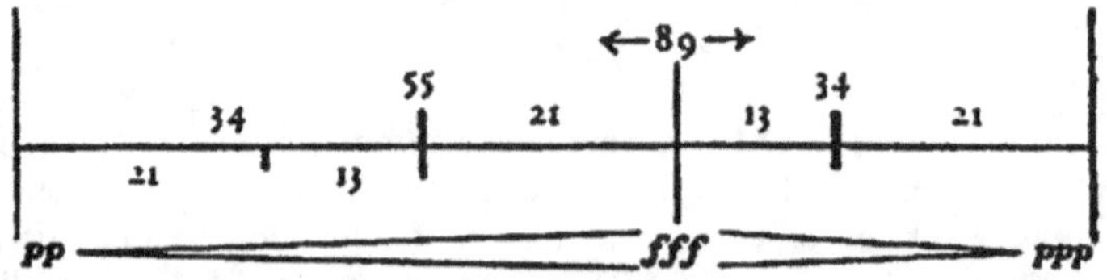

Fig. 5.1. Lendvai's analysis of Bartók, *Music for Strings, Percussion, and Celesta*, first movement.

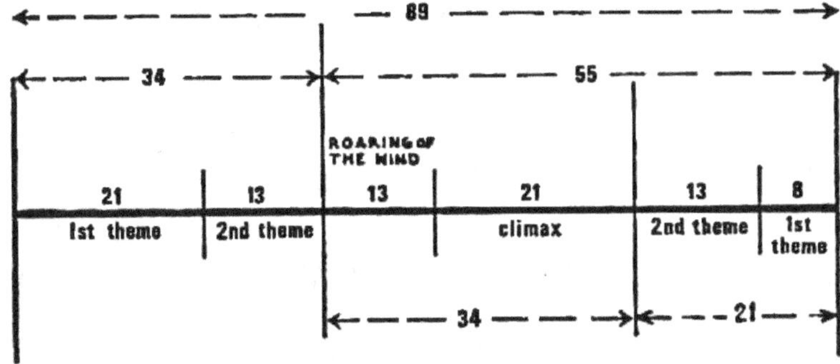

Fig. 5.2. Lendvai's analysis of Bartók, *Music for Strings, Percussion, and Celesta,* third movement.

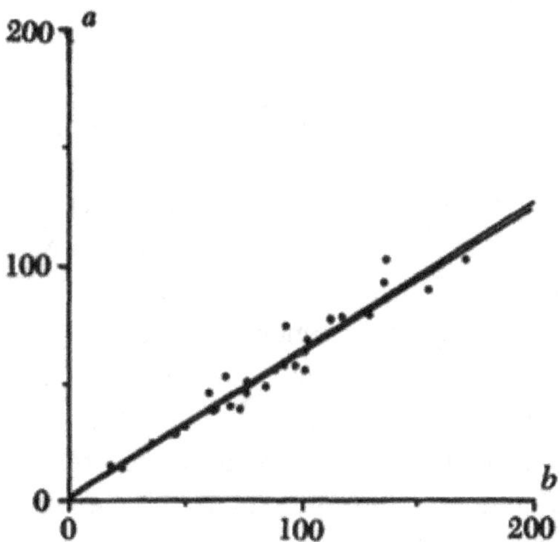

Fig. 5.3. Figure 8 from Putz's "The golden section and the Piano Sonatas of Mozart".

Mozart's piano sonatas in sonata form exhibit durational proportions approximating the golden ratio when comparing the lengths of the expositions to the development and recapitulations (codas, if present, were omitted from the calculations). This work, in which Putz rather elegantly showed the conformity of the ratios was both compelling and highly suspect, was subsequently published in *Mathematics Magazine* [15]. Figure 5.3 shows Putz's Figure 8, with the durational ratios of the two sections of each of the 29 movements plotted with

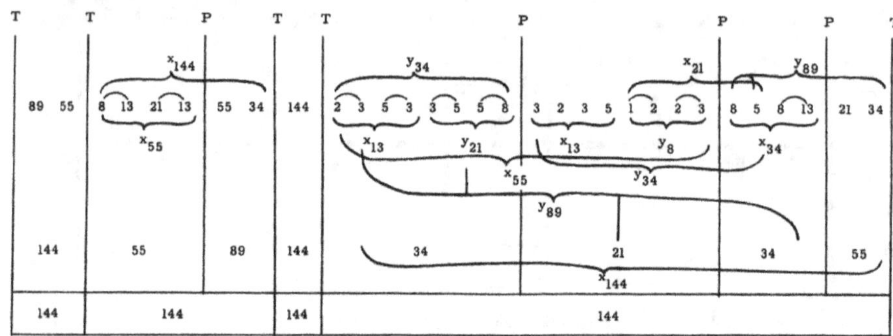

Fig. 5.4. Example 5 from Kramer's "The Fibonacci series in twentieth-century music".

the regression line as well as the line $y = \phi x$ (lying very slightly below the regression line).

A small treasure trove of examples of various types of appearances of the Fibonacci sequence and golden ratio in music can be found in Jonathan Kramer's 1973 *Journal of Music Theory* article [8] addressing the topic generally. Figure 5.4 is a reproduction of Kramer's Example 5, a formal diagram representing durations of salient musical cells in Stockhausen's *Adieu* for wind quintet, which all fall into Fibonacci patterns. Kramer's chapter also discusses the appearance of Fibonacci numbers in music of another Darmstadt school composer: Luigi Nono. It is Kramer's discussion of Nono's *Il canto sospeso* that serves as a jumping off point for the current investigation. This chapter will begin by fleshing out and clarifying Kramer's analytical observations. From there, the discussion will explore the relationship between the choice of the Fibonacci sequence as a compositional determinant and the expressive aims that Nono is striving for in his setting of the second movement of the piece.

5.2.　Nono and *Il Canto*

Nono was a leading member of the Darmstadt School, teaching there from 1954 to 1960. He championed the 12-tone techniques drawn directly from Schoenberg and Webern, filtered through his elder Italian colleague Luigi Dallapiccola, embracing the extended application of these techniques which characterizing the serialist compositional style. Nono viewed his approach to serialism on a continuum, writing in 1958: "There is an absolute historical and logical continuity of development between the beginnings of 12-tone music and its present state. No sort of 'break' exists. We just needed a certain amount of time to become aware of this historical development".

Il canto sospeso is a prime example of Nono's application of integral serialism to control multiple dimensions of musical content and structure. One of the extraordinary aspects of the piece is its emotional immediacy, despite its strict compositional procedure. Italian musicologist Massimo Mila observed shortly after *Il canto sospeso*'s premiere: "...invention and technical command serve to communicate a message whose fully perceptible depth is what counts. The emotional response which *Il canto sospeso* produces even in non-musicians is...the elimination of that much-deprecated chasm which divides modern art from the common man, an elimination obtained without the shadow of a concession and without compromising strict adherence to idiomatic originality".[b] But what if this depth of expression arises from the structural scaffolding itself, not in spite of it?

The text for the nine-movement work comes from excerpts from a 1954 collection of letters from anti-fascist prisoners about to be executed for their work in the resistance. The second movement — containing the first text heard in the piece — is scored for chorus alone, divided into eight parts, with text from a letter written by Anton Popov, a Bulgarian teacher, writer, and resistance fighter executed in Sofia in 1942. The idealistic defiance expressed here is common to all of the texts Nono sets in the cycle:

> *...io per un mondo che splender con luce tanto forte*
> *con tale belaezza che il mio stesso sacrificio non*
> *nulla. Per esso sono morti millioni di uomini sulle*

---

[b]From [12, p. 311], as quoted and translated in [14, pp. 83–84].

> *barricate e in guerra. Muoio per la giustizia. Le nostre*
> *idee vinceranno.*
>
> ...I am dying for a world which will shine with light
> of such strength and beauty that my own sacrifice
> is nothing. Millions of men have died for this on the
> barricades and in war. I am dying for justice. Our
> ideas will triumph.

5.3. Fibonacci in *Il Canto*: The Mechanics

As noted by Kramer [8], in order to follow Nono's use of Fibonacci
numbers, we need to track *contextual* durations, as measured in
multiples of four fundamental durational units. These beat divisions
themselves are highly irregular by our usual Western standards of
hierarchical meter:

1. 8ths (duple division of the quarter "beat", i.e., beat division by
   2; approximately 500 ms in length),
2. triplet 8ths (beat division by 3; approximately 333 ms),
3. 16ths (beat division by 4; approximately 250 ms),
4. quintuplet 16ths (beat division by 5; approximately 200 ms).

So, each note in the score is assigned an integer based on
its durational multiple of the fundamental durational unit that it
references. Figure 5.5 demonstrates these integers for the opening
notes of the movement. On the downbeat, the four upper vocal parts
sing a dotted 8th note (equaling a duration of three 16ths), a quarter-
note triplet 8th (or two triplet 8ths), a quarter note (five quintuplet
16ths), and an 8th note (one 8th), giving contextual durational
multiples 3, 2, 5, and 1. The next attack comes in the Alto 2, spanning
a total of four quarter beats (eight 8ths), and then the Bass 1's C
spanning four quarters and a triplet 8th (13 triplet 8ths). Next is the
Soprano 1's $F\sharp$ spanning 13 16ths (doubled in Tenor 2), etc.[c]

------

[c]As it turns out, there are a couple of oddities in these opening three measures.
The Soprano 1 and Tenor 2 unison is the only instance of vocal parts doubling a
line in the movement. Furthermore, the tie across the second barline in the Bass

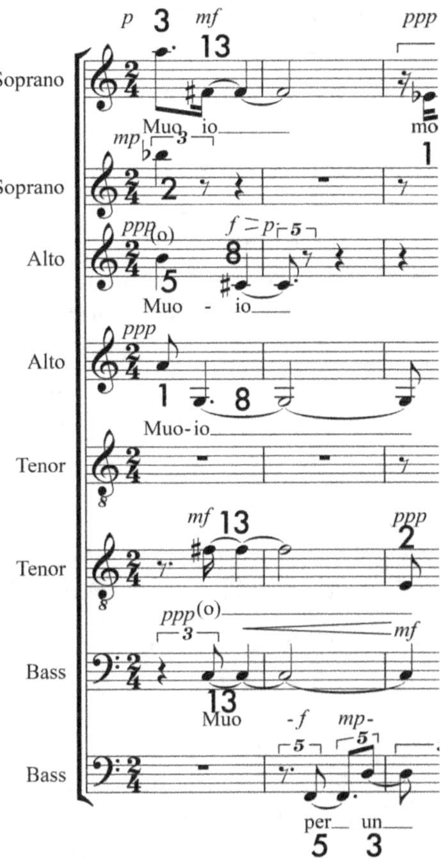

Fig. 5.5. Opening measures of *Il canto sospeso*, second movement with relative durational multiples.

The upshot here is that although there are eight vocal parts, there are only four musical lines, each one restricted to durations based on one of the four fundamental durational units. These lines are distributed freely and fluidly among the eight vocal parts. If we isolated any one of these lines, we would find that the melody proceeds without pause, snaking through the individual vocal parts, so that at any given timepoint in the music, there are exactly four

---

2 part should maintain a quintuplet rhythmic grouping rather than shifting to a triplet — most likely a printing error in the score.

vocal parts singing, one to each of the four lines defined by the reference durational unit.[d] It is a striking technique that supports the idea of *sospeso* here as in "floating" or "ethereal", a timelessness with no beginnings, no ends. As we will see, the use of Fibonacci durations further adds to this sense of "suspension".

In order to uncover the Fibonacci pattern of durations, we must track the *chronological* progression of each relative durational value (i.e., the integers produced by each note in time order, regardless of the note's underlying durational reference unit — its "line" — or the voice it appears in — its "vocal part"). Once we do, the straightforward serial structure of these rhythmic durations emerges. An analytical animation of the full score, annotated with the relative durational integers associated with each note, together with a synced performance, can be found at

https://www.swarthmore.edu/nono-animation;
https://vimeo.com/kochavi/nonoanimation.[e]

The numbers representing those relative durational values in each of the four lines are color-coded as follows:

red = 8ths
purple = triplet 8ths
blue = 16ths
green = quintuplet 16ths

In the animation, the relative durations appear below the score in the order they appear chronologically in the music. The durational integers are broken up in the analysis into groups of twelve, for reasons that will become clear.

As mentioned, the four lines are sung without pause, at least until the closing section. The analysis here and in [8] show that each line

---

[d]The exceptions to this rule come in the two anomalies detailed in the previous footnote, and then in the "coda" of the movement, which will be discussed.

[e]I wish to thank the ITS team at Swarthmore College for helping implement this animation: Jeremy Polk, Donna Fournier, Michael Kappeler, Kyungchang Min, Julian Turner, and Xintong Tian. The recording is from 1993, with the Rundfunkchor Berlin conducted by Claudio Abbado.

Fig. 5.6. Tone row in *Il canto sospeso*.

projects a serialized pattern of relative durational integers consisting of twelve Fibonacci numbers, beginning with

$$1\text{-}2\text{-}3\text{-}5\text{-}8\text{-}13\text{-}13\text{-}8\text{-}5\text{-}3\text{-}2\text{-}1,$$

and then systematically rotated, so that the second iteration begins on 2, circling back to the original 1:

$$2\text{-}3\text{-}5\text{-}8\text{-}13\text{-}13\text{-}8\text{-}5\text{-}3\text{-}2\text{-}1\text{-}1,$$

etc. After fifteen rotations of the row,[f] the final portion of the movement changes the pattern somewhat, organizing the durations chronologically within each individual line rather than composing patterns of durations drawn from mixed durational units. Although this chapter does not focus pitch structure, the 12-note durational pattern set with a fixed all-interval tone row, a wedge row radiating out from the initial *A* (Fig. 5.6).

5.4. Beat Class Distribution and Metric Hierarchy

What are the implications of using the Fibonacci sequence as an organizing principle in determining durational units in music and how does this choice support Nono's expressive aims? We will explore this question abstractly in order to work towards a structural and conceptual understanding of Nono's use of this process here.

Western rhythmic notation itself is biased towards hierarchical understanding. The "beat" or "tactus" can be broken up into two

---

[f]There are a few anomalies occurring in m. 135, beginning with the penultimate note of the 12th iteration of the row (marked with a red underline in the animation). The author suspects a typo in the score, with the dot on the Alto 2's E mistakenly displaced to the E♭ that follows. This would restore the expected durational patterns, with the caveat that the 13th iteration of the row is actually the result of a "double" rotation, skipping over the expected durational pattern that would have duplicated the initial durational row of the movement, beginning instead with 2-3-5-⋯.

or three equal parts (simple or compound meter, respectively), and these beats are typically grouped into twos, threes, or fours (duple, triple, or quadruple meter, respectively). Once this pattern is established, it is generally maintained, and if it shifts, then an adjustment is made to the notation in the form of a drastic time signature change or use of duplets or triplets to indicate alternate beat divisions. Any grouping or division that falls outside these normative parameters must be notated awkwardly via unusual meters and/or specialized "tuplets". The presence of barlines, of course, immediately implies beat hierarchy, another bias.

Suppose we were presented with a score with the barlines removed (along with possible notational adjustments for ties across these barlines), and that we wanted to reinsert these barlines accurately. Furthermore, suppose that we wish to do using only rhythmic information (ignoring the hierarchical implications of melody, harmony, articulation, dynamics, register, etc.). One mechanism for doing so would be to examine the placement of the note onsets relative to an assumed underlying "beat" (now considering a beat to be a reference duration, not necessarily the tactus). Do these onsets tend to appear at particular recurring rhythmic time intervals? Of course, we would not expect every beat at a given time interval to be attacked — some might be skipped.

One way to capture this analysis mathematically would be to look at the placement of the attacks at various moduli of the underlying reference duration. As an example, consider the opening of the final variation from Bach's *Goldberg Variations* (see Fig. 5.7). Consider the initial 48 16th note pulses, beginning with the left hand D at pulse point 0. The chart in Fig. 5.8 tallies the attack point appearances modulo 16, showing that the note onsets tend to occur

Fig. 5.7. Opening of Bach, *Goldberg Variations*, Variation 30.

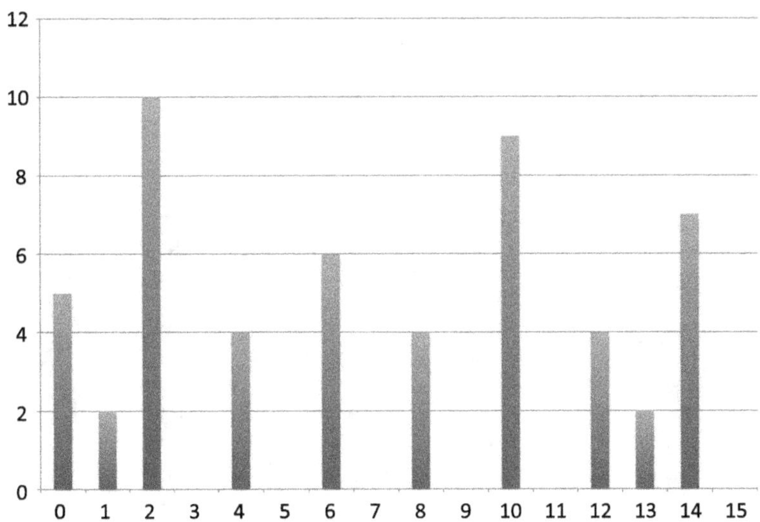

Fig. 5.8. Total 16th note beat class attack counts (mod 16), through m. 3.

on the even beat classes, with spikes at evenly spaced beat classes 2, 6, 10, and 14, just as we would expect in a 4/4 context with an 8th note upbeat. (Note the higher prevalence of attacks on bcs 2 and 10, suggesting the further hierarchization implied by the meter.) If we attempt a different modular organization, say mod 7, we find an essentially undifferentiated array of bcs, as shown in the chart in Fig. 5.9. The slightly counterintuitive reading of the graph: the more stable the bc multiplicities, the more chaotic or random the music will seem assuming that particular modular interpretation.

But suppose a composer wanted to create music via a compositional technique based on the traditional serial model but that systematically determined "chaotic" patterns no matter which beat modulus was chosen? The music would have to avoid the usual pattern of repeated durational values: $n, n, n, \ldots$ of course. Successions of durational intervals involving regular multiples of a base duration (i.e., $n, 2n, 3n, \ldots$) and those that involve successive powers (i.e., $n, n^2, n^3, \ldots$) are the natural choices with our familiar notational system, but neither will produce a system that ambiguates the underlying pulse.

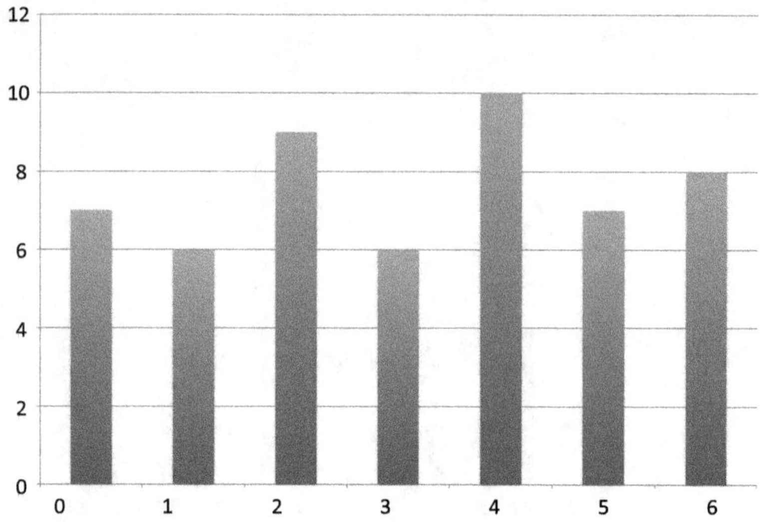

Fig. 5.9. Total 16th note beat class attack counts (mod 7), through m. 3.

It turns out that the ordered Fibonacci sequence provides an excellent basis for this system, a claim that is explored below. (Recall the notation for Fibonacci numbers is F_m, where $F_0 = 0$ and $F_1 = 1$. For the purpose of this discussion, we will consider the "first" element in the Fibonacci sequence to be $F_1 = 1$.)

5.5.   Fibonacci Durations and (Absence of) Metric Hierarchy

Let us consider the Fibonacci numbers to represent integer multiples of a base rhythmic duration. Note that if two consecutive durations share a non-trivial multiple of this base duration as a factor, the composer would risk implying a durational hierarchy on the time divisions used in the piece. Using the Fibonacci sequence precludes this possibility, as guaranteed by the following well-known property (a proof is included for the interested reader).

Theorem 1. *Any two consecutive Fibonacci numbers are relatively prime.*

Proof. Suppose not. Then there exists an integer $n > 1$ such that n divides two consecutive Fibonacci numbers F_i and F_{i+1}. Suppose that n does not divide F_{i-1}. Then $(F_{i-1} + F_i)$ cannot be a multiple of n, but since $(F_{i-1} + F_i) = F_{i+1}$, this is a contradiction. Therefore, F_{i-1} is divided by n. By analogous argument, n divides every element of the sequence prior to F_i. Since the sequence begins with 1, this is a contradiction. $\qquad\square$

Of course, this property only gets us so far. To really understand the metric implications of using the Fibonacci sequence as a durational structural determinant in music, we need to know something about the sequence mod n, for various integers n. To do so, we will pivot from the bc interpretation of mod n integer classes used in the Bach observations to one that views the integer classes as modular durations, or ordered bc intervals.

First, let us consider two interpretations of the Fibonacci durations mod n:

(1) Values of 0 mod n indicate multiples of a (non-trivial) base durational unit, the modulus. A large prevalence of 0 mod n values will imply normative rhythmic groupings, at least locally, suggesting an underlying meter.

(2) Values of k mod n for a fixed k indicate durations of a particular initial value "tied" to multiples of the base durational unit. A prevalence of these values will also imply a metric structure, though more weakly.

Note also that in the context of *Il canto sospeso*, any duration k implies an initial attack at timepoint x followed by a second attack at timepoint $x + k$. Therefore, prevalence of k mod n values for a fixed k could produce a pattern of recognizable attack point intervals implying a meter organized around the modulus n. In this case, the value of k in its absolute difference from n gives a kind of upbeat duration in the implied metrical structure.

To create a floating, suspended effect lacking metrical hierarchy, therefore, we are looking for a mod n sequence without a prevalence of zeros, and a wide distribution of other values. Importantly, we are looking for these properties for **all** values of n.

There are some known fact on this topic that will be useful to us here. It is not difficult to show Theorem 2, another standard result from the literature.

Theorem 2. *Taking the numbers of the Fibonacci sequence mod n will always result in a repeated pattern of modular integers at some period.*

Proof. Take two consecutive numbers in the Fibonacci sequence, A and B. Let $a = A$ mod n and $b = B$ mod n, where a and b are modular residues expressed in reduced form. Then $a + b = A + B$ mod n. Therefore, looking at the terms of the sequence occurring after the values A and B, the Fibonacci sequence with elements reduced mod n is completely determined by the ordered pair (a, b). So any duplicate appearance of this pair in the sequence will result in a replication of the sequence from that point. Since there are only n^2 possible ordered pairs under this modulus, there must always be a repeated pattern with period $\pi(n) \leq n^2$. $\square$

This has given rise to the following definition.

Definition 1. The number of integers classes in the repeated pattern of the Fibonacci sequence mod n is called the **Pisano period**, or $\boldsymbol{\pi(n)}$.

Note from the proof of Theorem 2 that since a given modular pair will always generate the rest of the repeated pattern, the initial two modular integer classes will both determine the pattern itself and begin the second (and subsequent) iterations of the pattern. In other words, when the initial ordered pair of integer classes — which will always be $(1, 1)$ — returns, it signals the (prior) conclusion of the pattern, and gives the Pisano period. Figure 5.10 gives the first 25 numbers of the Fibonacci sequence for moduli 2 to 7.

From Fig. 5.10, we see that the Pisano periods for $n = 2$ to 7 are:

$$\begin{aligned}
\pi(2) &= 3 & \pi(5) &= 20 \\
\pi(3) &= 6 & \pi(6) &= 24 \\
\pi(4) &= 8 & \pi(7) &= 16
\end{aligned}$$

	2	3	4	5	6	7
1	1	1	1	1	1	1
1	1	1	1	1	1	1
2	0	2	2	2	2	2
3	1	0	3	3	3	3
5	1	2	1	0	5	5
8	0	2	0	3	2	1
13	1	1	1	3	1	6
21	1	0	1	1	3	0
34	0	1	2	4	4	6
55	1	1	3	0	1	6
89	1	2	1	4	5	5
144	0	0	0	4	0	4
233	1	2	1	3	5	2
377	1	2	1	2	5	6
610	0	1	2	0	4	1
987	1	0	3	2	3	0
1597	1	1	1	2	1	1
2584	0	1	0	4	4	1
4181	1	2	1	1	5	2
6765	1	0	1	0	3	3
10946	0	2	2	1	2	5
17711	1	2	3	1	5	1
28657	1	1	1	2	1	6
46368	0	0	0	3	0	0
75025	1	1	1	0	1	6

Fig. 5.10. Modular values of the first 25 Fibonacci numbers. The Fibonacci sequence is listed down the first column, with subsequent columns giving the sequence under the modulus listed at the column head, $n = 2$ to 7. The first complete cycle is highlighted in each column.

Given these facts, we can easily prove the next theorem.

Theorem 3. *Given any integer n, there are an infinite number of integers in the Fibonacci sequence that are divisible by n.*

Proof. This fact can be restated as follows: given the Fibonacci sequence reduced modulo n, the repeated periodic pattern must always include a 0. Since the (modular) pattern always begins with the integer classes $(1, 1)$, there are an infinite number of subsequences of the form $(\ldots, m, 1, 1, \ldots)$ in the sequence. So, $m + 1 = 1 \bmod n$, so therefore m must be 0 mod n. $\qquad\square$

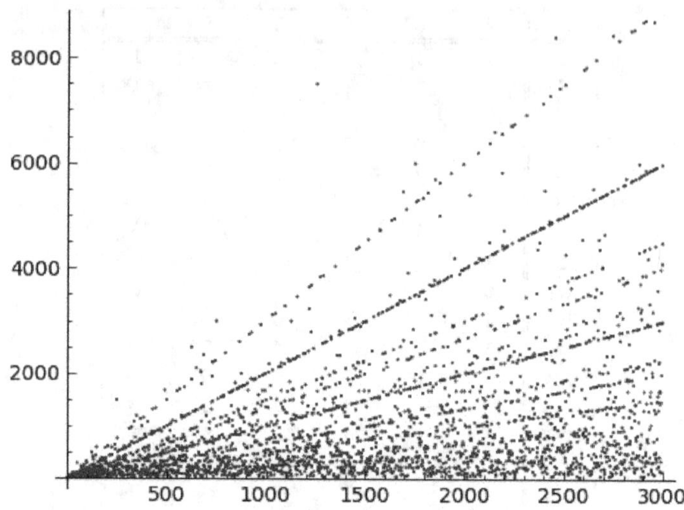

Fig. 5.11. Pisano periods (along the y-axis) for the Fibonacci sequence mod n (values of n along the x-axis. From [18].

The natural question that arises is: What is the relationship between the modulus n and the Pisano period for that modulus, $\pi(n)$? In particular, longer periods would mean a greater variety of attack point patterns at that modulus. Perhaps surprisingly, the precise relationship between the modulus and the period is still an open question, though progress has been made on it over the past few decades based on the prime factorization of n. Figure 5.11 shows a graph created by Marc Renault which gives the Pisano periods for n up to 3000 (from [18]).

You will notice that many of these relationships are linear, with varying slopes, but a significant number of values fall off of these lines as well.

For us, it is enough to notice that the Pisano periods are generally quite long relative to n, though there are exceptions.[g] To give some sense of just how long some are (along with accounting for some

---

[g]A lower bound on the Pisano period has been established by [2] based on the modulus and the ordinal position of the highest Lucas sequence number that is less than or equal to that modulus.

of the linear relationships), Theorem 4 gives the Pisano periods for certain classes of cardinalities which have been established.[h]

Theorem 4.

(a) *Suppose $n = 2^k$. Then $\pi(n) = 3 \times 2^{k-1}$.*
(b) *Suppose $n = 5^k$. Then $\pi(n) = 4 \times 5^k$.*

To go further, it would be useful to understand the distributions of the modular values within a single iteration of the pattern to see how varied they are. A counterpart to Theorem 3 is that these 0 mod n values will be rather widely distributed in the sequence. In particular, we know that strings of identical integer classes essentially do not occur in mod n Fibonacci sequences, as stated in Theorem 5. This is important to Nono's goal since such strings would quickly begin to create a feeling of a regular pulse at that durational modulus (given the rest-less nature of the musical lines).

Theorem 5. *In a given mod n Fibonacci sequence, there will never be a string of more than two identical integer classes consecutively.*

Proof. Suppose not. Then there exists a string of three consecutive Fibonacci numbers, F_i, F_{i+1}, F_{i+2} equal to k mod n. There for $k + k = k$ mod n, which implies that $k = 0$ mod n, implying that the entire sequence reduces to a string of 0 mod n values, a contradiction for non-trivial n. $\square$

Specific distributions of residue classes of mod n Fibonacci sequences were by and large unstudied in the mathematics literature until about 30 years ago, and a full characterization is still an open problem. However, certain specific results are helpful. We can begin with several standard definitions (see [17]), the first of which concerns the string's length before coming back to zero.

---

[h]These results come from [22]. More generally, if $n = p^k$, then $\pi(n) = \pi(k) \times p^{k-1}$, given the non-existence of so-called Wall–Sun–Sun primes. Further results can be found in [16]. Generally, breaking down n into a product of prime factors raised to the appropriate powers, $\pi(n)$ is the least common multiple of the Pisano periods of each of the members of this product.

Definition 2. The **rank** of the Fibonacci sequence mod n, denoted $\alpha(n)$, is defined as the smallest positive integer k such that $F_k = 0$ (mod n).

Note, of course, that $\pi(n) \geq \alpha(n)$, since every period of a Fibonacci sequence mod n begins with $[1, 1]$ and therefore must end with a 0 (mod n) to set up the $[1, 1]$ that begins the next period.

Definition 3. The **order** of the Fibonacci sequence mod n is defined as the period divided by the rank: $\omega(n) = \pi(n)/\alpha(n)$.

The order is indeed always an integer, as given by the following theorems, proofs of which can be found in [21].

Theorem 6. *The zero values in the mod n Fibonacci sequence are equally spaced (and the rank of the sequence always divides its Pisano period).*

Theorem 7. *For the Fibonacci sequence mod n, $\omega(n)$ is always 1, 2, or 4.*

Theorem 6 tells us that the order represents the number of zeros contained in a single cycle, and Theorem 7 tells us that this number is limited and necessarily small. So, at the very least, we know that although there are an infinite number of Fibonacci sequence numbers divisible by a given integer n, the spacing of these numbers precludes excessive "bunching" that might lead to a quickly implied beat hierarchy. Furthermore, Theorem 7 suggests a dearth of such zero values.

More generally, one fact about integer class distributions in mod n Fibonacci sequences for specific values of n that has been known for some time (since the early 1970s) is given in Theorem 8.

Theorem 8 (Kuipers and Shiue [9]; Niederreiter). *The distribution of integer classes in a single iteration of a mod n Fibonacci sequence is strictly uniform iff $n = 5^k$, for $k \geq 1$. In these cases, each integer class will appear 4 times in the iteration.*

0	1	2	3	4	5	6	7	8	9	10	11	12	13	14	15	16	17	18	19	20	21	22	23	24	25	26	27	28	29	30	31
2	3	8	1	0	3	0	1	2	3	0	1	0	3	0	1	2	3	0	1	0	3	0	1	2	3	0	1	0	3	0	1

Fig. 5.12. Multiplicities of integer classes in a single cycle of a mod 2^k Fibonacci sequence (for $k \geq 5$, with integer classes themselves taken mod 32, listed in the top row). (Shown in [7].)

0	1	2	3	4	5	6	7	8	9	10	11	12	13	14	15	16	17	18	19	20	21	22	23	24	25	26
2	8	2	2	2	2	2	2	?	2	2	2	2	2	2	2	2	2	2	?	2	2	2	2	2	2	8

Fig. 5.13. Multiplicities of integer classes in a single cycle of a mod 3^k Fibonacci sequence (for $k \geq 3$, with integer classes themselves taken mod 27, listed in the top row). (Shown in [19].)

Other than this early result, the characterizations of these distributions are messy. Jacobson [7] determines the distribution of integer classes for values of n that are powers of 2 greater than 16. The results are summarized in Fig. 5.12. The string of integers here gives the multiplicities of occurrences for each of the corresponding integer classes themselves taken mod 32, so the string loops back to the beginning for integer classes of larger sizes (when applicable). Finally, Shiu and Chu [19] published work in 2005 characterizing the distribution of integer classes for values of n that are powers of 3 greater than 9, with results shown in Fig. 5.13. Multiplicities of larger integer classes are again found by looping this string back. (The question marks represent variable multiplicities, further explored in [1].)

5.6. Fibonacci in Nono: An Embrace of Collective Humanity

Although these results only characterize the distributions of integer classes within Fibonacci mod n cycles for limited values of n, the properties do suggest that the Fibonacci sequence proves a robust source for generating a very wide variety of integer class patterns at multiple modular levels. For a composer looking to strictly control durational structure and yet provide rhythmic profile that does not fit into any type of hierarchical array, the Fibonacci sequence is a readymade resource. The fact that Nono applies these durational

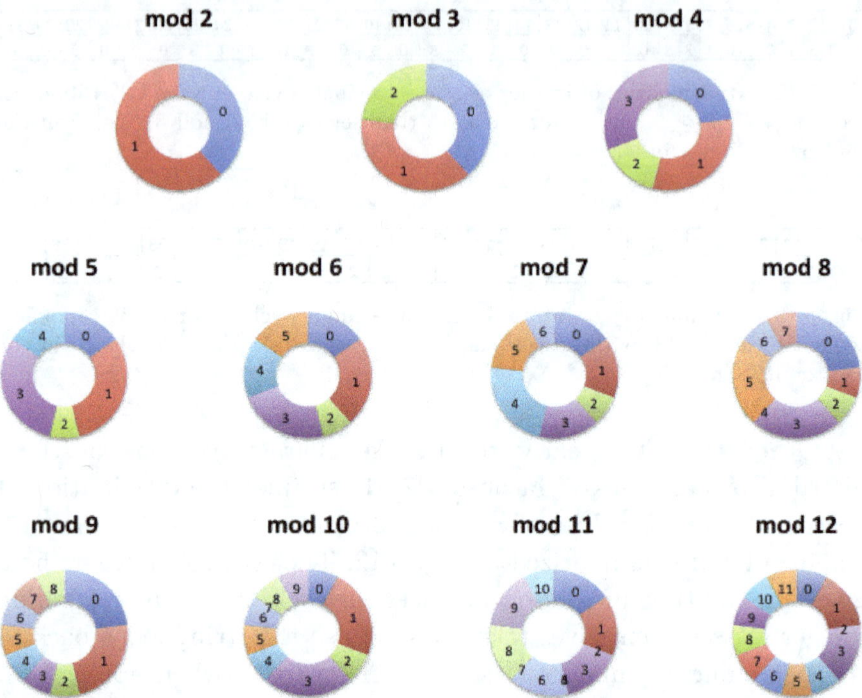

Fig. 5.14. Distributions of mod n values of the note attack timepoints of a single iteration of Nono's row (using durations 1-2-3-5-8-13-13-8-5-3-2-1).

multiples to four nearly incommensurable base durational units only serves to further distance the rhythms from any hierarchical status.

In Nono's actual realization, of course, he is only using the initial few terms of the Fibonacci sequence, and they are ordered in ascent and descent. In Fig. 5.14, we return to the type of analysis we looked at in the Bach excerpt: we now consider not isolated note duration in modulus, but modular values of the attack points in Nono's music — the durations as they accumulate, so to speak. Analyzed here are the distributions of the mod n values of the attack point positions in a single iteration of the row (assuming a 13th attack after the final note of the row). (Note they we are assuming a fixed base duration, as occurs in the final section of the movement.) You will notice a strikingly varied array of values at each modulus through $n = 12$, confirming the non-hierarchical rhythmic patterns that we experience while listening to the piece.

The poems of *Il canto sospeso* are moving and actually quite jarring in their juxtaposition of expressions of resolute defiance in the face of certain death. Corporeal imprisonment is juxtaposed with emotional and political freedom expressed in the most direct and irrefutable way. With his compositional choices, Nono is attempting to create this feeling of freedom in music that actually is tightly and almost obsessively controlled: he is organizing the music using systematized (and relatively simple) techniques that nevertheless produce unexpected sonic realizations from the perspective of Western art music. In this way, Nono's work is the musical embodiment of the fundamental contradiction of the poetry itself.

The Italian word *sospeso* can have various English translations: "suspended", "floating", "interrupted", or even "unsung" in this context. Nono's choice of this title plays on the word's ambiguity, and the "Song Unsung" is not simply a lament for what could have been, but statement about the role of artistic creation as resistance, or *resistenza*.[i]

Stockhausen's famous take on the piece was that Nono's fragmentation of the text among voices eliminates the meaning of the text. This observation was skewed by the pointillistic Webernian lens through which many in the Darmstadt school viewed the piece at the time. Stockhausen wrote:

> Nono composed the text as if to withdraw its meaning
> from the public eye, where it has no place. . . . he con-
> ceals [the text] in a musical form which is so uncom-
> promisingly strict and dense that when listening, they
> are hardly comprehensible. Why then have text at

---

[i]Nielinger [14] discusses the origin of Nono's title for the piece, discovered fairly recently: the Italian translation of a 1953 poem written by Ethel Rosenberg, the American accused along with her husband of spying for the Soviets. The poem, entitled "If We Die", was addressed to the couple's two young sons. Written a few months before the couple's execution, the verse has strong resonance in tone and content with the texts Nono sets in *Il canto sospeso*.

> all, and this text in particular? One can understand it as follows: in specially setting those passages from the letters which make one most ashamed that they ever had to be written, the musician acts purely as a composer....He does not interpret, he does not comment[20].

Nono disagreed strongly with Stockhausen's interpretation (which led to a serious rift between the two Darmstadt composers). His artistic stance was bound inextricably to his fierce political leanings, expressed directly by the *resistenza*. The goals of the *resistenza* and of the artist were the same: the banding together of peoples against oppression, towards freedom. Nono makes this clear in his response to Stockhausen:

> The pointillistic view [of the piece] is totally alien to me....it has always been clear that a human being can realize himself only in his relations with other people and society? [Rather than] gratuitously destroying the texts...in fact, I was after something quite different. I wanted a horizontal melodic construction encompassing all registers; floating from sound to sound, from syllable to syllable: a line which sometimes consists of a succession of individual tones or pitches, and sometimes thickens into chords.

Nono is urging a hearing of the piece that relies on a holistic view. The individual lines are tightly controlled by the Fibonacci durational patterns in a way to ensure their democratization in the overall context. The participation of each of the eight vocal parts is crucial to every one of the four lines: if one of the eight were to become "unsung", the entire texture would fall part — not one of the four lines would retain its meaning and coherence.

His approach here is reflective of the larger post-War artistic vision in Italy. "Beyond *Guernica*", a manifesto (the complete title was "Manifesto of Realism for Painters and Sculptors") published 1946, was developed and signed by ten Italian artists, including

Emilio Vedova, who would become a close friend and collaborator of Nono's. Relevant to our current study, a section of the manifesto reads:

> Painting and sculpting is for us an act of participation in the total reality of mankind, in a specific time and place, a reality that is contemporaneity and, in it continuity, history. We therefore consider the positive function of individualism to be exhausted, and we reject those aspects in which it has become corrupted (escape, sensibility, intuition).... Realism [means]... the concretized reality of one person, when this participates in, coincides with, and is equivalent to the reality of others; when it becomes in short, a common measurement of reality itself.[j]

Individualism is rejected in favor of interrelatedness, and the idea of art as a form of escapism is scorned. With this philosophical backdrop, Nono's vigorous refutation of Stockhausen's assessment of *Il canto* is not surprising.

Figures 5.15 and 5.16 show two of Vedova's paintings from the period. *Image of Time (Barrier)* (Fig. 5.15) was completed in 1951. The interplay of formalism and a feeling of impetuosity produce an emotionally charged painting. Overall the impression of is one of chaos, but the composition and individual elements are highly controlled: consider the shape, angle, and color here. *Concentration Camp (People and Barbed Wire)* from 1950 (Fig. 5.16) is clearly a more conservative work, but it still pushes in the direction of expressive freedom with a visual grammar that remains firmly within the confines of geometry and form. Here, the elements are distilled, with background and foreground merged (or blurred) and color obviously used symbolically (and shockingly). No "barrier" between image and expression exists here — they cannot be separated. This

---

[j]Translation from [3, p. 48].

Fig. 5.15.   Emilio Vedova, *Image of Time* (*Barrier*), 1951.

painting was later used in the production of Nono's one-act opera *Intolleranza 1960* in 1961.

Nono's reliance on the Fibonacci sequence as a compositional determinant is not merely a technique to generate the rhythmic structure of *Il canto*. The powerful emotions evoked by the texts of the letters that he sets come from the striking juxtaposition of the textual content and their context. Nono chooses a musical depiction that does not attempt to reconcile the individual tragedy and collective defiance of these texts. Instead, his setting frames this seeming dichotomy, tweaking and developing the tools in his expressive musical arsenal to do so. Just as in the Vedova images, the artistic methodology — the Fibonacci sequence — *is* the expression. Nono's process here does not attempt to capture the emotion directly, but instead mirrors musically the aspect of the text that makes it so powerful in the first place, and by doing so, produces an artistic response to the horrors of the war that is deeply political and profoundly moving, while avoiding the mawkish.

Fig. 5.16. Emilio Vedova, *Concentration Camp (People and Barbed Wire)*, 1951.

Acknowledgments

The author would like to thank Swarthmore College, which provided institutional support for this project via the Special Projects for Educational Exploration and Development (SPEED) program, coordinated by Information Technology Services and Swarthmore College Libraries. Particular thanks go to Jeremy Polk and Donna Fournier, whose involvement in this project extended well beyond the original parameters of the SPEED program.

Bibliography

[1] P. Bundschuh and R. Bundschuh, Distribution of Fibonacci and Lucas numbers modulo 3^k, *Fibonacci Quart.* **49**(3) (2011) 201–210.
[2] P. Catlin, A lower bound for the period of the Fibonacci series modulo M, *Fibonacci Quart.* **12**(4) (1974) 349–351.

[3] A. Duran, *Painting, Politics, and the New Front of Cold War Italy* (Ashgate Publishing Company, Burlington, VT, 2014).

[4] J. Fauvel, R. Flood and R. J. Wilson, *Music and Mathematics: From Pythagoras to Fractals* (Oxford University Press, 2003).

[5] T. H. Garland and C. V. Kahn, *Math and Music: Harmonious Connections* (Dale Seymour Publications, Palo Alto, CA, 1995).

[6] L. Harkleroad, *The Math Behind the Music* (Cambridge University Press, 2006).

[7] E. Jacobson, Distribution of the Fibonacci numbers mod 2^k, *Fibonacci Quart.* **30**(3) (1992) 211–215.

[8] J. Kramer, The Fibonacci series in twentieth-century music, *J. Music Theory* **17**(1) (1973) 110–148.

[9] L. Kuipers and J.-S. Shiue, A distribution property of the sequence of Fibonacci numbers, *Fibonacci Quart.* **10**(4) (1972) 375–376, 392.

[10] E. Lendvai, *Béla Bartók: An Analysis of His Music* (Kahn & Averill, London, 1971).

[11] G. Loy, *Musimathics: The Mathematical Foundations of Music*, Vol. 1 (MIT Press, 2006).

[12] M. Mila, La linea Nono, *La Rassegna Musicale* **30** (1960) 297–311.

[13] H. Niederreiter, Distribution of the Fibonacci numbers mod 5^k, *Fibonacci Quart.* **10**(4) (1972) 373–374.

[14] C. Nielinger, The Song Unsung: Luigi Nono's *Il canto sospeso*, *J. Roy. Musical Assoc.* **131**(1) (2006) 83–150.

[15] J. F. Putz, The golden section and the Piano Sonatas of Mozart, *Math. Mag.* **68**(4) (1995) 275–282.

[16] M. Renault, Properties of the Fibonacci sequence under various moduli, Master's thesis, Wake Forest University (May 1996).

[17] M. Renault, The period, rank, and order of the (a, b)-Fibonacci sequence mod m, *Math. Mag.* **86** (2013) 372–380.

[18] M. Renault, The Fibonacci sequence modulo m, Marc Renault. Web (N.d.) (accessed April 26, 2017). http://webspace.ship.edu/msrenault/fibonacci/fib.htm.

[19] W.-C. Shiu and C. I. Chu, Distribution of the Fibonacci numbers modulo 3^k, *Fibonacci Quart.* **43**(1) (2005) 22–28.

[20] K. Stockhausen, Musik und sprache, *Die Reihe* **6** (1958) 36–58. English translation adapted from Ruth Koenig (1964).

[21] J. Vinson, The relation of the period modulo m to the rank of apparition of m in the Fibonacci sequence, *Fibonacci Quart.* **1**(2) (1963) 37–45.

[22] D.D. Wall, Fibonacci series modulo m, *Amer. Math. Monthly* **67**(6) (1960) 525–532.

[23] D. Wright, *Mathematics and Music*, (AMS Mathematical World Series, American Mathematical Society, 2009).

Chapter 6

One Note Samba: Navigating Notes and Their Meanings Within Modes and Exo-modes

Thomas Noll

Escola Superior de Música de Catalunya,
Barcelona, Spain
thomas.mamuth@gmail.com

6.1. Orientation

This text ties in with an earlier joint paper of Mariana Montiel and the author [12] about the music-theoretical application of some known mathematical results about Special Sturmian morphisms and associated lattice path transformations, which had been achieved by Valérie Berthé, and Aldo de Luca, and Christophe Reutenauer [3] as well as by Pierre Arnoux and Shunji Ito [2]. The main focus of the present work is on the elaboration of the underlying music-theoretical concepts, which had only been sketched in the earlier contribution. The explanations are presented in one flow without a division in definitions, propositions and corollaries, because the mathematical facts are already established. A distinctive feature of the concrete music-theoretical interpretation is the *passive* interpretation of the lattice path transformations.

6.2. Regener's Note Interval System Revisited

Singers of the late medieval time memorized the predecessor of the modern note C_4, — the "middle C" — under the name *c-sol-fa-ut*, which combines the Latin note name with the three different positions of this note within the hard, soft and natural hexachords, respectively. The notes of each of the hexachords are associated with the syllables *ut, re, mi, fa, sol, la*. The natural and hard hexachords cover the white note diatonic system and contain *c-ut* and *c-fa*, respectively. The soft and natural hexachords cover the system with a one-flat-signature and contain *c-sol* and *c-ut*, respectively. In the diatonic nomenclature with the seven syllables *do, re, mi, fa, so, la, ti* only two characters are associated with the note C_4, namely C_4/do in the white note system and C_4/fa in the one-flat system.

The first section of the present chapter takes the cue from these combined names and explores a family of *modal signature morphisms* mediating between notes and modal meanings. The latter are expressed as pairs of syllables and scale degrees, such as $(fa, \hat{1})$, $(do, \hat{3})$, $(mi, \hat{3})$, and generalizations thereof. The present approach also seizes suggestions by Julian Hook [6], Steven Rings [14] and Eytan Agmon [1] about the parameterization of musical tones, and refines them by adding the width dimension. In this regard the chapter also presents continued work into the direction of [4].

6.2.1. *Notes and width/height-degrees*

Traditional musical pitch notation is a remarkably rich source for the conceptual understanding of the diatonic modes. The interplay of staff lines, clef, signature and note heads reveals valuable insights into the modal semantics of musical tones. Eric Regener's [13] seminal model of traditional musical pitch notation, especially, may demonstrate this and therefore it provides a suitable starting point for the present investigation. This subsection recapitulates and slightly extends its main ingredients. Regener's model is a fully valid forerunner and — a posteriori — a paradigmatic instance of a *Generalized Interval System* (*GIS*) in the sense of David Lewin (see [8, Definition 2.3.1, p. 26]). To be more precise, its core is an injective morphism

$$\sigma = (\sigma_S, \sigma_G) : \mathcal{N} = (S_\mathcal{N}, G_\mathcal{N}, \mathrm{int}_\mathcal{N}) \hookrightarrow \mathcal{D} = (S_\mathcal{D}, G_\mathcal{D}, \mathrm{int}_\mathcal{D})$$

of two GISes, embedding a *note space* $\mathcal{N}$ into a *degree space* $\mathcal{D}$ by virtue of a *modal signature morphism* σ.[a] This means that the interval functions $\mathrm{int}_\mathcal{N}$ and $\mathrm{int}_\mathcal{D}$ and the two components $\sigma_S : S_\mathcal{N} \hookrightarrow S_\mathcal{D}$ and $\sigma_G : G_\mathcal{N} \hookrightarrow G_\mathcal{D}$ of the modal signature morphism σ satisfy the equation $\sigma_G \circ \mathrm{int}_\mathcal{N} = \mathrm{int}_\mathcal{D} \circ (\sigma_S \times \sigma_S)$.

In the course of this section it shall be demonstrated that each modal signature morphism assigns a unique modal interpretation to the notes and note intervals. A few mathematical preliminaries need to be sketched in advance.

The two interval groups $G_\mathcal{N}$ (of note intervals) and $G_\mathcal{D}$ (of degree intervals) are both abstractly isomorphic to the free commutative group $\mathbb{Z} \oplus \mathbb{Z}$ of rank 2. But, as will be seen in the subsequent explanation, the concrete group homomorphism $\sigma_G : G_\mathcal{N} \to G_\mathcal{D}$ is not an isomorphism. The cardinality of the factor group $G_\mathcal{D}/\sigma(G_\mathcal{N})$ coincides with the cardinality 7 of the diatonic scale.

The group of note intervals can be freely generated by many pairs of intervals. The present investigations highlights two such pairs, namely (1) the perfect fifth $P5$ and fourth $P4$ and (2) the major and minor seconds $M2$ and $m2$. In both cases we obtain corresponding coordinate representations in terms of integer pairs. A linear coordinate transformation mediates between these coordinate representations (see Fig. 6.1):

$$M : \mathbb{Z} \cdot P5 \oplus \mathbb{Z} \cdot P4 \cong \mathbb{Z}^2$$

$$\to \mathbb{Z}^2 \cong \mathbb{Z} \cdot M2 \oplus \mathbb{Z} \cdot m2 \quad \text{with } M(v) = \begin{pmatrix} 3 & 2 \\ 1 & 1 \end{pmatrix} \cdot v.$$

Remark 6.2.1. Although Regener considers several other bases and base transformations, it is crucial for the present approach to consider precisely these two bases $(P5, P4)$ and $(M2, m2)$ together

---

[a]The terminology of the present chapter slightly differs from Regener's.

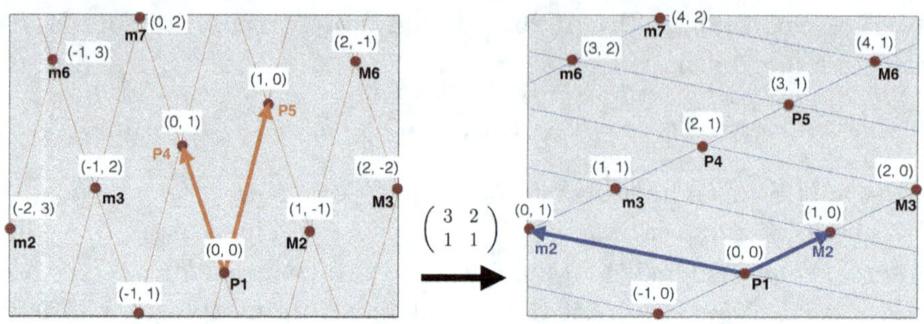

Fig. 6.1. Two bases: $(P5, P4)$ and $(M2, m2)$, generating the group $G_{\mathcal{N}}$ of note intervals. The 2×2-matrix denotes the linear transformation mediating between the respective coordinate representations. The concrete placement of the coordinate axes has no particular meaning at this stage. It anticipates the embedding σ_G (cf. Fig. 6.2).

with the transformation $M = \begin{pmatrix} 3 & 2 \\ 1 & 1 \end{pmatrix}$. We call it the *diatonic Regener transformation*. The choice of other matrices $M = \begin{pmatrix} a & b \\ c & d \end{pmatrix} \in \mathrm{SL}_2(\mathbb{N})$ with natural number entries a, b, c, d and determinant $ad - bc = 1$ provides an appropriate generalization of this approach for other tone systems. One instance is the *chromatic Regener transformation* with the matrix $\begin{pmatrix} 3 & 2 \\ 4 & 3 \end{pmatrix}$. It mediates between note intervals in $(P5, P4)$-coordinates and note intervals in $(A1, m2)$-coordinates. We discuss this case later in Sec. 6.5.2.

The space $S_{\mathcal{N}}$ of notes can be represented in terms of the conventional note names of the form Ls_r, each consisting of a Latin letter $L \in \{A, B, C, D, E, F, G\}$, an accidental $s \in \{\ldots, \flat\flat, \flat, \epsilon, \sharp, \sharp\sharp, \ldots\}$ and a subindex $r \in \mathbb{Z}$ for the octave register. Examples are $\ldots, C_2, C_3, C_4, \ldots$ and $\ldots, D\flat\flat_2, D\flat_2, D_2, D\sharp_2, D\sharp\sharp_2, \ldots$. A spatial representation of $S_{\mathcal{N}}$ can be inherited from $G_{\mathcal{N}}$ through the identification of an arbitrarily chosen note $X \in S_{\mathcal{N}}$ with the perfect prime $P1$ and the identification of all the other notes Y with the corresponding intervals $\mathrm{int}_{\mathcal{N}}(X, Y) \in G_{\mathcal{N}}$. It is convenient to choose the "middle C" for that identification, i.e., $X = C_4$ (see also the right side of Fig. 6.2). It is also sometimes convenient to use integer coordinates for notes: $Y = (y_1, y_2)$ if and only if $\mathrm{int}_{\mathcal{N}}(C_4, Y) = (y_1, y_2)$. In this regard is also convenient to notate the action $target_\mathcal{N} : S_\mathcal{N} \times G_\mathcal{N} \to S_\mathcal{N}$ of note intervals on notes as an addition of coordinate pairs.[b]

The degree space $\mathcal{D} = (S_\mathcal{D}, G_\mathcal{D}, \mathrm{int}_\mathcal{D})$ is a two-dimensional extension of the traditional one-dimensional arrangement of the staff lines[c] and demonstrates Regener's original insight most effectively. Its interval group $\mathrm{Int}(\mathcal{D})$ is freely generated by two elements w and h, which shall be examined now: The element h denotes the "vertical" *height degree* step-interval. It literally represents the step interval from staff lines to the interspace above them and from interspaces to the staff lines above them. The element w denotes an analogous "horizontal" *width degree* step-interval.[d] With w and h as basis intervals one may represent every degree interval in $G_\mathcal{D}$ as an integer pair $(r, s) \in \mathbb{Z} \oplus \mathbb{Z}$.

6.2.2. *Signature morphisms*

The motivation for the two-dimensional construction of the degree space $\mathcal{D}$ becomes manifest under the inspection of the linear embedding σ_G. It is insightful to first consider the pure height degree intervals $(0, s)$ and pure width degree intervals $(r, 0)$: Among the note intervals only perfect primes $P1$, perfect octaves $P8$ (and their multiples $-P8$, $P15$ etc.), as well as their altered variants $A1$, $d1 = -A1$, $A8$, $d8$, etc. have the property, that they can be identified in staff notation without the specification of a clef and a signature. In particular, any pair of note heads without accidentals, which are separated by precisely seven height degrees represent a perfect octave $P8$, independent of clef and signature. Hence, the map σ_G sends the note interval $P8 \in G_\mathcal{N}$ to the pure height degree interval $\sigma_G(P8) = 7h = (0, 7) \in G_\mathcal{D}$. Analogously, the note interval of the augmented prime $A1 \in G_\mathcal{N}$ is being sent to the pure width degree interval $\sigma_G(A1) = 7w = (7, 0) \in G_\mathcal{D}$.

---

[b]In the sense we may treat $\mathcal{N}$ as a *canonical GIS* in the sense of Oren Kolman (see [7]).

[c]It is silently assumed that the number of staff lines is unlimited.

[d]Regener calls the height degree step-interval *diatone* and the width degree step-interval *quint* (w), respectively.

The basis intervals $P5$ and $P4$ are mapped to the width/height degree intervals $\sigma_G(P5) = w + 4h = (1, 4)$ and $\sigma_G(P4) = -w + 3h = (-1, 3)$, respectively. The members of the step interval basis $M2$ and $m2$ are mapped to the degree intervals $\sigma_G(M2) = 2w + h = (2, 1)$ and $\sigma_G(m2) = -5w + h = (-5, 1)$, respectively. The sum of $P5$ and $P4$ is the octave $P8$, and — as to be expected — $\sigma_G(P5) + \sigma_G(P4) = (1, 4) + (-1, 3) = (0, 7) = \sigma_G(P8)$. Likewise, the difference between $M2$ and $m2$ is the augmented prime $A1$, and — as to be expected — one obtains $\sigma_G(M2) - \sigma_G(m2) = (2, 1) - (-5, 1) = (7, 0) = \sigma_G(A1)$. The left side of Fig. 6.2 displays the group embedding $\sigma_G : G_\mathcal{N} \hookrightarrow G_\mathcal{D}$. This particular group embedding σ_G models the diatonic peculiarities of traditional notation and shall be fixed for the most part of this chapter.

The set of GIS morphisms $\sigma = (\sigma_S, \sigma_G)$ sharing our fixed group embedding σ_G is precisely parameterized by the width/height-degree space $S_\mathcal{D}$ itself through the free choice of the image $\sigma_S(C_4) \in S_\mathcal{D}$. All other images are determined by the properties of the GIS-morphism: To any note Y we find a unique width/height-degree $\sigma_S(Y)$, satisfying $\sigma_G(\mathrm{int}_\mathcal{N}(C_4, Y)) = \mathrm{int}_\mathcal{D}(\sigma_S(C_4), \sigma_S(Y))$.

An equivalent — and musically more intuitive — characterization of a given modal signature morphism σ can be provided by the choice of the uniquely determined note T, whose image has the form $\sigma_S(T) = (m, 0)$: Suppose $\sigma_S(C_4) = (r, s)$. Then the integer $r - 2s$ can be uniquely decomposed into an integral multiple of 7 and a residue $0 \leq m < 7$, i.e., $r - 2s = 7k + m$. The desired note T can then be chosen such that the interval between C_4 and T satisfies $\mathrm{int}_\mathcal{N}(C_4, T) = -s \cdot M2 - k \cdot A1$. This particular choice of T yields $\sigma_S(T) = (r, s) - s(2, 1) - k(7, 0) = (r - 2s - 7k, 0) = (m, 0)$. The said note T shall be called the *tonic* note of σ. The pair (T, m) unambiguously identifies the modal signature morphism σ. In consideration of the long tradition of relative solmization, each of the seven numbers $m = 0, 1, \ldots, 6$ can be associated with a solmization syllable $\breve{m}$ indicating the modal character attributed to the tonic note under σ (see Table 6.1).

Two alternative notations may be useful for the identification of a modal signature morphism σ:

Table 6.1. Width degrees $0 \le m < 7$ and syllables $\breve{m}$.

0	1	2	3	4	5	6
fa	do	so	re	la	mi	ti

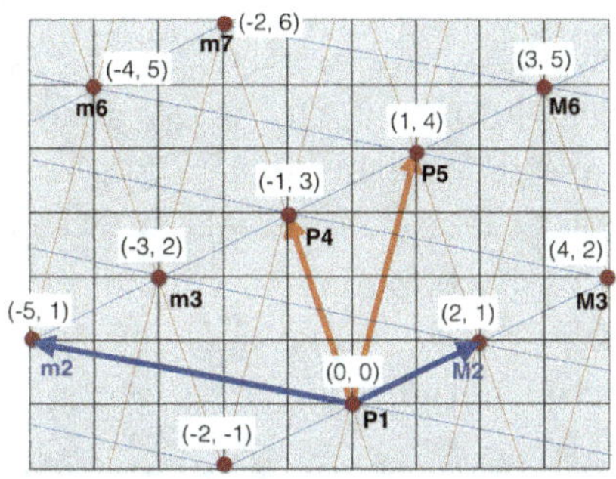
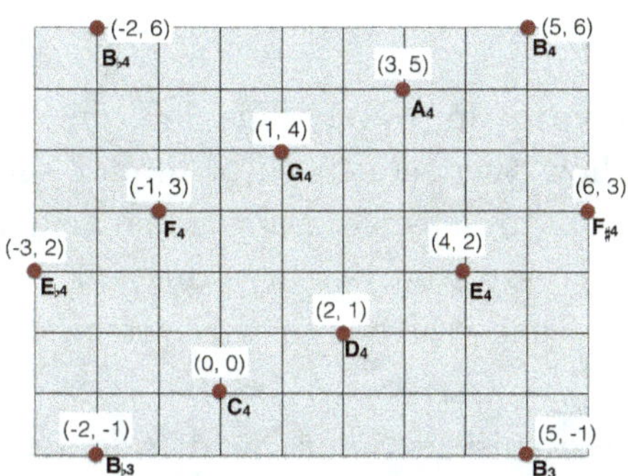

Fig. 6.2. The left subfigure displays the embedding $\sigma_G : G_{\mathcal{N}} \hookrightarrow G_{\mathcal{D}}$ (compare with Fig. 6.1). Only every seventh point of the width/height-degree lattice is inhabited by the image of a note interval. The right subfigure displays an example for an embedding $\sigma_S^{C_4/fa} = \sigma_S^{(0,0)} : S_{\mathcal{N}} \hookrightarrow S_{\mathcal{D}}$ of the note space into the degree space. The choice of any other width/height-degree (r, s) for the note C_4 defines another embedding $\sigma_S^{(r,s)}$.

(1) $\sigma_S^{T/\breve{m}}$ denotes the morphism with $\sigma_S^{T/\breve{m}}(T) = (m, 0)$,

(2) $\sigma^{(r,s)}$ denotes the morphism with $\sigma_S^{(r,s)}(C_4) = (r, s)$.

The right side of Fig. 6.2 displays the example of the map σ_S, where $\sigma_S(C_4) = (0, 0)$, i.e., where $T = C_4$ is the tonic note and where the residue $m = 0$ is associated with the modal character $\breve{0} = fa$. Hence, we may write σ as $\sigma^{C_4/fa} = \sigma^{(0,0)}$.

The consideration of two or more modal signature morphisms in combination provides considerable descriptive flexibility. Recall that to any note $X \in \mathcal{N}$ and any width/height degree $(r, s) \in \mathcal{D}$ there is a unique modal signature morphism σ, such that $\sigma_S(X) = (r, s)$. Hence, for any two notes $X, Y \in \mathcal{N}$, any width/height degree $(r, s) \in \mathcal{D}$ and any degree interval $(i, j) \in G_{\mathcal{D}}$ there are unique modal signature morphisms σ and σ', such that $\sigma_S(X) = (r, s)$ and $\mathrm{int}_{\mathcal{D}}(\sigma_S(X), \sigma'_S(Y)) = (i, j)$. Whenever two notes X and Y are mapped to the same degree (r, s) this means, that they share the same modal meaning under the respective modal signature morphisms. The group $G_{\mathcal{D}}$ of width/height-intervals parameterizes all possible variations in the modal meanings.

Example 6.2.1. The left and right columns of the array below show two sequences of modal signature morphisms in the application to selected notes. The width/height-degree intervals between successive entries are constant within each column. On the left side the successive intervals are pure height-degree steps such as $\mathrm{int}_{\mathcal{D}}$ $(\sigma_S^{C_4/so}(F_4), \sigma_S^{C_4/fa}(G_4)) = h = (0, 1)$. On the right side the successive intervals are pure width-degree steps such as $\mathrm{int}_{\mathcal{D}}(\sigma_S^{C_4/la}(B\flat_4),$ $\sigma_S^{C_4/re}(B\flat_4)) = -w = (-1, 0)$.

$$
\begin{array}{ll}
\sigma_S^{C\flat_4/do}(C\flat_4) = (1, 0) & \sigma_S^{C\flat_4/do}(B\flat_4) = (6, 6) \\[4pt]
\sigma_S^{C_4/ti}(D\flat_4) = (1, 1) & \sigma_S^{C\flat_4/fa}(B\flat_4) = (5, 6) \\[4pt]
\sigma_S^{C_4/la}(E\flat_4) = (1, 2) & \sigma_S^{C_4/ti}(B\flat_4) = (4, 6) \\[4pt]
\sigma_S^{C_4/so}(F_4) = (1, 3) & \sigma_S^{C_4/mi}(B\flat_4) = (3, 6) \\[4pt]
\sigma_S^{C_4/fa}(G_4) = (1, 4) & \sigma_S^{C_4/la}(B\flat_4) = (2, 6) \\[4pt]
\sigma_S^{C_4/mi}(A_4) = (1, 5) & \sigma_S^{C_4/re}(B\flat_4) = (1, 6) \\[4pt]
\sigma_S^{C_4/re}(B_4) = (1, 6) & \sigma_S^{C_4/so}(B\flat_4) = (0, 6) \\[4pt]
\sigma_S^{C\sharp_4/do}(C\sharp_5) = (1, 7) & \sigma_S^{C_4/do}(B\flat_4) = (-1, 6)
\end{array}
$$

As a consequence of the combined signature morphisms the note interval $\mathrm{int}_{\mathcal{N}}(C\flat_4, C\sharp_5) = AA8$ of the doubly-augmented octave between the first and last notes in the left column (upper staff of Fig. 6.3) corresponds to a pure height-degree interval $\mathrm{int}_{\mathcal{D}}(\sigma_S^{C\flat_4/do}(C\flat_4), \sigma_S^{C\sharp_4/do}(C\sharp_5)) = 7h = (0, 7)$, which under σ_G is the

Fig. 6.3. The upper and lower staff represent the examples from the left and right columns in musical notation, respectively. The information about the tonic notes is not contained, but the key signatures are in concordance with their syllables.

image of the perfect octave. The zero-width results in this case from the shared *do*-character of the two notes $C\flat_4$ and $C\sharp_5$. The perfect prime $int_{\mathcal{N}}(B\flat_4, B\flat_4) = P1$ between the first and last notes in the right column (lower staff of Fig. 6.3) corresponds to a pure width-degree interval $int_{\mathcal{D}}(\sigma_S^{C\flat_4/do}(B\flat_4), \sigma_S^{C_4/do}(B\flat_4) = -7w = (0, -7)$, which under σ_G is the image of the descending augmented prime $-A1$. The zero-height results from the fact that in both cases $B\flat_4$ is seven degrees above the respective tonics $C\flat_4$ and C_4.

6.3.   Notes and Modes

The choice of the seven solmization syllables as differentiators for the signature morphisms anticipates their close connection to the diatonic modes. This connection shall be made explicit in this section.

6.3.1.   *The principal diatonic domain*

The augmented prime $A1$ and the perfect octave $P8$ together with their images $\sigma_G(A1) = 7w$ and $\sigma_G(P8) = 7h$ serve as framing intervals for the diatonic modes. To that end we consider the subgroups $G_{\mathcal{N}}^0 = \langle A1, P8 \rangle \subset G_{\mathcal{N}}$ and $G_{\mathcal{D}}^0 = \langle 7w, 7h \rangle \subset G_{\mathcal{D}}$. The factor group $G_{\mathcal{N}}/G_{\mathcal{N}}^0$ is cyclic of order 7 and the factor group $G_{\mathcal{D}}/G_{\mathcal{D}}^0$ is isomorphic to $\mathbb{Z}_7 \oplus \mathbb{Z}_7$.

For the non-transitive action $target_{\mathcal{D}} : S_{\mathcal{D}} \times G_{\mathcal{D}}^0 \to S_{\mathcal{D}}$ we choose a privileged fundamental domain

$$Dia = \{(r, s) \in \mathbb{Z}^2 \,|\, 0 \leq r, s < 7\} \subset S_{\mathcal{D}}$$

which shall be called the *principal diatonic domain*. For every modal signature morphism σ the preimage $\mu_\sigma = \sigma_S^{-1}(Dia) \subset S_{\mathcal{N}}$ forms

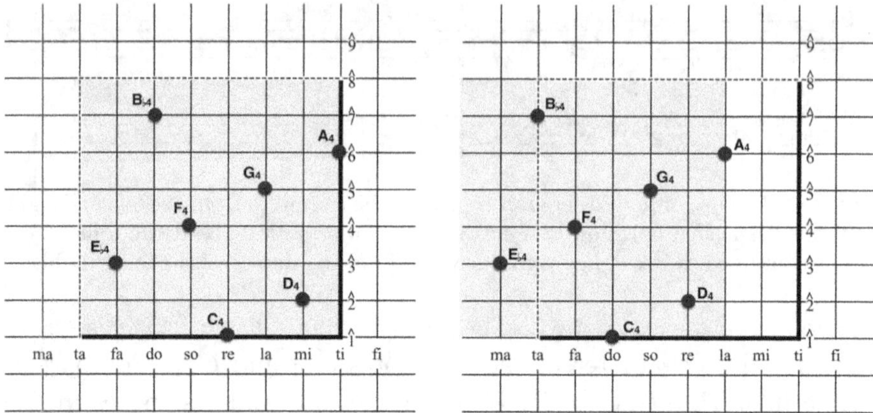

Fig. 6.4. The notes of the C_4-Dorian species of the octave as a principal *re*-mode (left figure) and an accidental *do*-mode (right figure). The altered width degrees are designated by the altered syllables from relative solmization.

a particular fundamental domain for the action $target_{\mathcal{N}} : S_{\mathcal{N}} \times G^0_{\mathcal{N}} \to S_{\mathcal{N}}$. It consists of seven notes and forms the carrier set for a *diatonic mode*. To be more precise, the restriction $\sigma_{S|\mu_\sigma} : \mu_\sigma \to Dia$ of σ_S to the set μ_σ is called a *principal diatonic mode*. The set μ_σ itself is called a *modal note set*. The restriction of a different modal signature morphism $\sigma'_{S|\mu_\sigma} : \mu_\sigma \to S_{\mathcal{D}}$ to the same underlying modal set μ_σ is called an *accidental diatonic mode*. The distinction between principal and accidental modes can be illustrated in terms of the example in Fig. 6.4.

In modulating tonal pieces passages in a major key may interact with passages both in relative and parallel minor. In solfège practice this is sometimes reflected in a preference for *la*-based minor for the passages in relative and *do*-based minor for the passages in parallel minor. The distinction between principal and accidental modes provides an adequate theoretical framework for this.

6.3.2. *Essential and accidental modes*

Each note can be associated with any of the $49 = 7 \cdot 7$ height/width-degrees in the principal diatonic domain. In each case this determines a unique principal mode. Figure 6.5 displays those seven, wherein

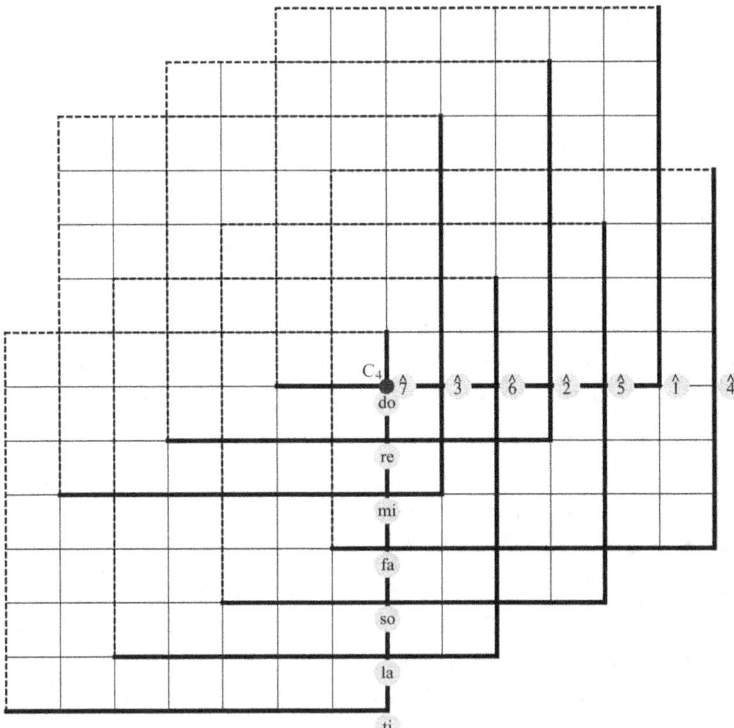

Fig. 6.5. Seven copies of the principal diatonic domain are arranged around the note C_4 in order to illustrate the variation of the Ionian meanings of that same note.

the note C_4 belongs to a Ionian species of the octave, namely $\sigma_S^{C_4/do}(C_4) = (do, \hat{1})$ and

$$\sigma_S^{B\flat_3/do}(C_4) = (re, \hat{2}), \ \sigma_S^{A\flat_3/do}(C_4) = (mi, \hat{3}), \ \sigma_S^{G_3/do}(C_4) = (fa, \hat{4}),$$
$$\sigma_S^{F_3/do}(C_4) = (so, \hat{5}), \ \sigma_S^{E\flat_3/do}(C_4) = (la, \hat{6}), \ \sigma_S^{D\flat_3/do}(C_4) = (ti, \hat{7}),$$

The arrangement of the seven copies of the principal diatonic domain in Fig. 6.5 is reminiscent of the shape of a single Ionian mode, which is flipped horizontally as well as vertically. The higher the height degree of C_4, the lower the associated domain, The sharper ($=$ more rightward) a width degree of C_4, the flatter ($=$ more leftward) the associated domain. This arrangement anticipates a transformational description of these modal meanings in Sec. 6.5.

6.3.3.　*Hook's signature transformations revisited*

The term *signature morphism* stands for a single interpretation of the note system in terms of width/height degrees. For frequent applications of this concept it could be convenient to omit the term *morphism* and to simply speak of *signatures*. Julian Hook [6] investigated a concept of *signature transformation* which addresses a higher transformational level. A signature transformation acts on a set of signatures as well as on diatonic score constructs, which are notated within these signatures. The present approach offers suitable means to recast Hook's concept. The present subsection sketches this very briefly in anticipation of a more detailed elaboration.

We consider the (right) action $target_{\mathcal{D}} : S_{\mathcal{D}} \times G_{\mathcal{D}} \to S_{\mathcal{D}}$ of the interval group on the degree space.[e] By bringing the trivial action of $G_{\mathcal{D}}$ on itself into play, we may interpret the action $target_{\mathcal{D}}$ as a group of GIS-endomorphisms of the degree space $\mathcal{D}$. For each degree interval $u \in G_{\mathcal{D}}$ we write $\tau_u = target_{\mathcal{D}}(\_,u)$. By concatenating signature morphisms σ with target-morphisms τ_u from the right we obtain again signature morphisms $\sigma \circ \tau_u$. Thus, if SIG denotes the set of all signature morphisms $\sigma : \mathcal{N} \to \mathcal{D}$, we obtain a right action $J : SIG \times G_{\mathcal{D}} \to SIG$ of *signature transformations*. We write synonymously $J_u(\sigma) = J(\sigma, u)$. The width/height degree interval $(i, j) \in G_{\mathcal{D}}$ induces the signature transformation with

$$J_{(i,j)}((\sigma_S^{(r,s)}, \sigma_G^{(r,s)})) = (\sigma_S^{(r+i,s+j)}, \sigma_G^{(r,s)}).$$

The interval component $\sigma_G^{(r,s)}$ of $\sigma^{(r,s)}$ remains unchanged under $J_{(i,j)}$. To write down its action on the degree-space $\sigma_S^{(r,s)}$ it is straightforward to use the C_4-based designation. For the tonic-based designation one requires a more complicated case distinction. This can be illustrated with an example. Hook's signature transformation s_1 — the appending of one sharp — corresponds to the

---

[e]Recall that Lewin introduces the group $STRANS$ by converting the right action of the interval group $IVLS$ ($G_{\mathcal{D}}$ in our case) into a left action of the opposite group $STRANS$. In the commutative case it is one and the same group. Nevertheless it is appropriate to stick to the concept of a right action here.

Table 6.2. Action of the signature transformation $J_{(-1,0)}$ on signature morphisms in tonic-based designation.

$\sigma^{X/fa}$	$\sigma^{X/do}$	$\sigma^{X/so}$	$\sigma^{X/re}$	$\sigma^{X/la}$	$\sigma^{X/mi}$	$\sigma^{X/ti}$
$\sigma^{X\sharp/ti}$	$\sigma^{X/fa}$	$\sigma^{X/do}$	$\sigma^{X/so}$	$\sigma^{X/re}$	$\sigma^{X/la}$	$\sigma^{X/mi}$

transformation $J_{(-1,0)}$, which is induced by the elementary width degree interval $(-1,0) \in G_{\mathcal{D}}$. Its action on a given signature $\sigma^{X/\check{m}}$ in tonic-based designation is listed in Table 6.2

The elementary height interval $(0,-1)$ induces the transformation $J_{(0,-1)}$, which corresponds to Hook's diatonic transposition operator t_1.[f] Figure 6.6 revisits a subnetwork of Fig. 6.4 in [6]. It connects three signature morphisms through three signature transformations as follows:

$$J_{(-1,0)} : \sigma^{A_4/do} = \sigma^{(-2,-5)} \mapsto \sigma^{A_4/fa} = \sigma^{(-3,-5)},$$
$$J_{(0,-3)} : \sigma^{A_4/do} = \sigma^{(-2,-5)} \mapsto \sigma^{D_5/fa} = \sigma^{(-2,-8)},$$
$$J_{(-1,3)} : \sigma^{D_5/fa} = \sigma^{(-2,-8)} \mapsto \sigma^{A_4/fa} = \sigma^{(-3,-5)}.$$

The maps σ_S are defined on notes, but they can be naturally extended to sets of notes, sequences of notes and other score-related constructs. The same holds for other maps in the following consideration.[g] In Hook's example we have three four-note sequences $A, B, C \in S_{\mathcal{N}}^4$ each of which is mapped into the principal diatonic domain by one of the three signature morphisms. We show how the signature transformations induce transformations on such score constructs. For each signature morphism σ let $\pi_\sigma : Dia \cap \sigma_S(S_{\mathcal{N}}) \to \{\hat{1}, \hat{2}, \hat{3}, \hat{4}, \hat{5}, \hat{6}, \hat{7}\}$ denote the (bijective) projection from the image of σ within the principal diatonic domain to the height degree coordinates and consider the height-degree parameterization of its carrier set,

---

[f]At this stage of modeling without enharmonic identification, t_1 is independent of s_1.

[g]A formalization of this claim can be made in the framework of Guerino Mazzola's Denotator calculus [9]. For score-based denotators see Mariana Montiel's thesis [10].

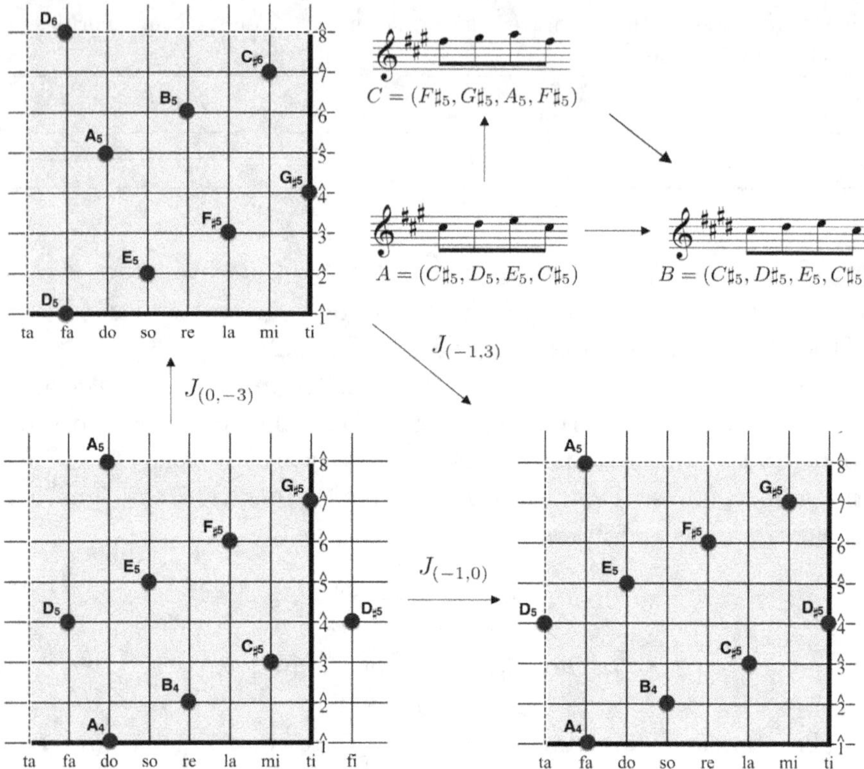

Fig. 6.6. Network of three signature morphisms and three signature transformations referring to Fig. 6.4 in [6]. Each of the three four-note sequences A, B, C in the upper left corner is embedded by the corresponding signature morphism into the principal diatonic domain. All three share the same pattern of height degrees, namely $\hat{3}$-$\hat{4}$-$\hat{5}$-$\hat{3}$.

namely

$$\mathrm{carr}_\sigma := (\sigma_S)^{-1} \circ (\pi_\sigma)^{-1} : \{\hat{1}, \hat{2}, \hat{3}, \hat{4}, \hat{5}, \hat{6}, \hat{7}\} \to S_\mathcal{N}.$$

Now consider a signature morphism σ and a signature transformation J_u and some score construct K built from notes in $S_\mathcal{N}$. Then the signature transformation J_u induces the score construct $K' = \mathrm{carr}_{J_u(\sigma)}(\pi_\sigma(\sigma_S(K)))$. It satisfies $\pi_{J_u(\sigma)}(J_u(\sigma)(K')) = \pi_\sigma(\sigma_S(K))$ which means that K and K' are mapped onto the same height-degree pattern.

In this sense the four-note sequences $B = (C\sharp_5, D\sharp_5, E_5, C\sharp_5)$ and $C = (F\sharp_5, G\sharp_5, A_5, F\sharp_5)$ are induced from the four-note sequence $A = (C\sharp_5, D_5, E_5, C\sharp_5)$ by the signature transformations $J_{(-1,0)}$ and $J_{(0,-3)}$, respectively.

6.4.   Modes as Transformations

There are two main transformational aspects connected to the present approach. The signature morphisms and signature transformations are transformations at two different levels, but they are both manifestations of a semantics, where notes and note intervals gain musical meaning in terms of width/height-degrees and their intervals. The diatonic Regener transformation affects a different level of description, namely a change of perspective *within* the group of note intervals. In the last three subsections it is the second aspect, which shall be further developed and refined. In the first subsection, however, we proceed with our method to catch modal note sets and affiliated structures with the help of signature transformations.

6.4.1.   *Anchored note intervals and lattice paths*

A pair $(X, i) \in S_{\mathcal{N}} \times G_{\mathcal{N}}$ of a note X and a note interval i shall be called an *anchored note interval*. For any pair of generators $\{i, j\}$ of the interval group $G_{\mathcal{N}}$ let $\mathcal{B}[i, j] = \{(X, i) \mid X \in \mathcal{N}\} \sqcup \{(X, j) \mid X \in \mathcal{N}\}$ denote the set of all anchored instances of i and j. Let further $\mathcal{F}[i, j]$ denote the real vector space of all finite linear combinations $\sum_{k=0}^{n-1} r_k(X_k, i_k)$ of anchored note intervals $(X_k, i_k) \in \mathcal{B}[i, j]$ with $n \in \mathbb{N}$, $i_k \in \{i, j\}$ and $r_k \in \mathbb{R}$. Elements of this space, whose non-vanishing coefficients satisfy $r_k = 1$ throughout, are called *geometric elements*. And finally, a geometric element $\sum_{k=0}^{n-1}(X_k, i_k)$ is called a *lattice path*, if its anchored intervals form a contiguous chain, i.e., if the target note $target_{\mathcal{N}}(X_k, i_k) = X_k + i_k$ of each anchored interval coincides with the anchor note X_{k+1} of a successor for $k = 0, \ldots, n - 2$.

Now we associate two lattice paths with each modal note set as follows: Consider the seven notes of a modal note set $\mu_\sigma \subset S_{\mathbb{N}}$.

We access (and designate) them in two ways: For $k = 0, \ldots, 6$ let $Y_k \in \mu_\sigma$ designate the note whose height degree is k, i.e., $\sigma_S(Y_k) = (\psi_\sigma(k), k)$ for some width degree $\psi_\sigma(k)$. Similarly, for $k = 0, \ldots, 6$ let $X_k \in \mu_\sigma$ designate the note whose width degree is $6 - k$, i.e., $\sigma_S(X_k) = (6 - k, \phi_\sigma(k))$ for some height degree $\phi_\sigma(k)$. The intervals $j_k = \mathrm{int}_\mathcal{N}(Y_k, Y_{k+1})$ for $k = 0, \ldots, 5$ are either major or minor seconds. Define $j_6 := P8 - \sum_{k=0}^{5} j_k \in \{M2, m2\}$ and $Y_7 = \mathrm{target}_\mathcal{N}(Y_6, j_6)$, such that $\mathrm{int}_\mathcal{N}(Y_0, Y_7) = P8$. Y_7 is a boundary note for μ_σ, but not an element. Similarly, the intervals $i_k = \mathrm{int}_\mathcal{N}(X_k, X_{k+1})$ for $k = 0, \ldots, 5$ are either descending perfect fifths or ascending perfect fourths. Define $i_6 := -A1 - \sum_{k=0}^{5} i_k \in \{-P5, P4\}$ and $X_7 = \mathrm{target}_\mathcal{N}(X_6, i_6)$, such that $\mathrm{int}_\mathcal{N}(X_0, X_7) = -A1$. X_7 is a boundary note for but not an element of μ_σ, for which we have: $\mu_\sigma = \{X_0, \ldots, X_6\} = \{Y_0, \ldots, Y_6\}$. The *fifth/fourth-folding path* associated with μ_σ is then defined as $\Phi_\sigma := \sum_{k=0}^{6}(X_k, i_k) \in \mathcal{F}[-P5, P4]$. The *step-interval path* associated with μ_σ is the defined as $\Psi_\sigma := \sum_{k=0}^{6}(Y_k, j_k) \in \mathcal{F}[-P5, P4]$. In the case of the modal signature morphism $\sigma^{C4/do}$ the corresponding fifth/fourth-folding path is $Phi_{\sigma^{C4/do}} = (B4, -P5) + (E4, P4) + (A4, -P5) + (E4, A4) + (A4, -P5) + (D4, P4) + (G4, -P5) + (C4, P4) + (F4, P4)$ and the step-interval path is $Psi_{\sigma^{C4/do}} = (C_4, M2) + (D_4, M2) + (E_4, m2) + (F_4, M2) + (G_4, M2) + (A_4, M2) + (B_4, m2)$.

6.4.2. *Lattice path transformations: Motivation*

To this point we introduced the modal note sets and the associated lattice paths with the help of the signature transformations and the principal diatonic domain. The remainder of this section is dedicated to an alternative method to obtain them. In contrast to the approach taken above, it does not involve a recurrence to the degree space $\mathcal{D}$. It is entirely based upon refinements of the diatonic Regener transformations within the note interval space $\mathcal{N}$. Traditionally a mode μ_σ is called an *authentic mode*, if its boundary interval of the perfect octave is decomposed into a perfect fifth and perfect fourth and if the step interval fillings are inherited from this division. The authentic division of an anchored octave

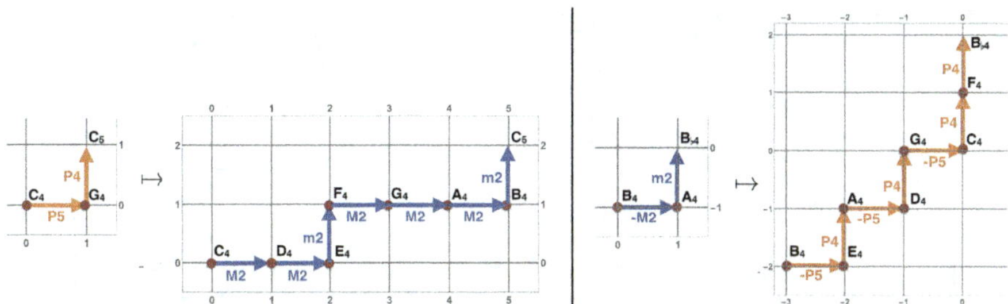

Fig. 6.7. The left subfigure illustrates the desired lattice path transformation on the basis of a single example. It sends the lattice path $(C_4, P5) + (G_4, P4) \in \mathcal{F}[P5, P4]$ to $\Psi_{\sigma C_4/do} \in \mathcal{F}[M2, m2]$. The right subfigure shows an instance of another lattice path transformation, which is closely related to the transformation on the left side. It sends the step interval folding path $(B4, -M2) + (A4, m2) \in \mathcal{F}[-M2, m2]$ to the fifth/fourth-folding path $\Phi_{\sigma C_4/do} \in \mathcal{F}[-P5, P4]$.

$(X, P8)$ can be described in terms of the lattice path $(X, P5) + (target_{\mathcal{N}}(X, P5), P4) \in \mathcal{F}[P5, P4]$. In order to transformationally describe its step interval-filling we are constructing a so-called *lattice path transformation*, i.e., a suitable linear map from $\mathcal{F}[P5, P4]$ to $\mathcal{F}[M2, m2]$. In the case of our example $\Psi_{\sigma C_4/do}$ — as-yet captured with the help of the signature transformations $\sigma^{C_4/do}$ — it sends the lattice path $(C_4, P5) + (G_4, P4) \in \mathcal{F}[P5, P4]$ to $\Psi_{\sigma C_4/do} \in \mathcal{F}[M2, m2]$ (see left subfigure in Fig. 6.7).

The idea of the authentic division of the perfect octave has a dual counterpart in terms of a division of the "flat" augmented prime $-A1$ into a descending major second $-M2$ and an ascending minor second $m2$. The fifth/fourth-folding path $\Phi_{\sigma C_4/do} \in \mathcal{F}[-P5, P4]$ then appears as the image of the step interval folding path $(B4, -M2) + (A4, m2) \in \mathcal{F}[-M2, m2]$ under a suitable lattice path transformation. The signature transformation $\sigma^{C_4/do}$ can be used in order to transport the lattice paths into the width-height-degree space $\mathcal{D}$. Thereby every summand (X, i) is mapped to the pair $(\sigma_S^{C_4/do}(X), \sigma_G^{C_4/do}(i)) \in S_{\mathcal{D}} \times G_{\mathcal{D}}$. Figure 6.8 shows the result of this transfer. It is more suggestive with respect to the duality between the two lattice paths. The mathematical definition of the lattice path transformations involves two stages, which are the respective subjects of the subsequent subsections.

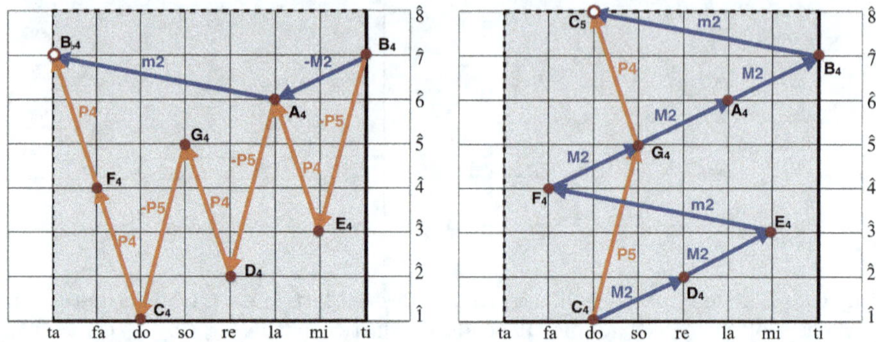

Fig. 6.8. The lattice paths, by definition, inhabit the note interval space. The figure shows, however, that these paths can be transported in to the degree space.

6.4.3. *Interlude: Special sturmian morphisms*

The contiguity of the lattice paths allows us to consider just the underlying sequences of intervals (disregarding their anchor notes) as an intermediate stage in the transformational setup. So, for $\sigma^{C_4/do}$ and any other $\sigma^{X/do}$ the interval sequence $(P5, P4)$ is mapped to the sequence $(M2, M2, m2, M2, M2, M2, m2)$. Both sequences have in common that their entries are drawn from the two intervals of a basis of the group $G_{\mathcal{N}}$. Disregarding the concrete base intervals, we may encode the pattern of their respective occurrences in a sequence in terms of a word in two letters x and y, namely with the word xy for $(P5, P4)$ and with the word $xxyxxxy$ for $(M2, M2, m2, M2, M2, M2, m2)$. These letters can be seen as two separate incarnations of the natural number 1, x for each record of the primary and y for each record of the secondary base interval. These words in the letters x and y on their part can be interpreted as (positive) elements of the free group F_2 with x and y as generators. The surjective group homomorphism $V : F_2 \to \mathbb{Z}^2$ maps the generator x to the pair $V(x) = \begin{pmatrix} 1 \\ 0 \end{pmatrix}$ and y to the pair $V(y) = \begin{pmatrix} 0 \\ 1 \end{pmatrix}$. Thus, just as coordinate pairs $\begin{pmatrix} r \\ s \end{pmatrix} \in \mathbb{Z}^2$ encode the formation of linear combinations "as such" (regardless of the concrete basis intervals to be combined), we may likewise say that words $w \in \{x, y\} \subset F_2$

encode the formation of sequences with entries drawn from a basis (but regardless of the concrete basis intervals). In other words: The two-letter words are refinements of number pairs and the homomorphism V "un-refines" them.

In expansion of this thought the construction of the respective transformation f_1 can be understood as a refinement of the diatonic Regener transformation c. In terms of the basis intervals, the perfect fifth $P5$ and perfect fourth $P4$ are mapped to the Ionian species $(M2, M2, m2, M2)$ of the fifth and $(M2, M2, m2)$ of the fourth, respectively. It is our intention, though, to understand this as a passive transformation, i.e., as a refinement of the coordinate transformation $\binom{1}{0} \mapsto \binom{3}{1}$, $\binom{0}{1} \mapsto \binom{2}{1}$. This means musically, that the species of the fifth and fourth are understood as manifestations of the "empty" intervals under a different perspective. But the fifth and the fourth as such are unchanged. In terms of refined coordinates, the transformation f_1 substitutes the letter x by the word $xxyx$ and the letter y to the word xxy. This substitution can be uniformly applied to any word $w \in \{x, y\}^*$. The pair $f_1(x, y) = (xxyx, xxy)$ is thus a refinement of the matrix $M = \left(\begin{smallmatrix} 3 & 2 \\ 1 & 1 \end{smallmatrix}\right)$, in the sense that the entries $3 = |xxyx|_x$ and $2 = |xxy|_x$ denote the multiplicities of the letter x in the words $f_1(x)$ and $f_1(y)$, and the entries $1 = |xxyx|_y$ and $1 = |xxy|_y$ the multiplicities of the letter y in these words, respectively. The matrix $M_{f_1} = \left(\begin{smallmatrix} |f_1(x)|_x & |f_1(y)|_x \\ |f_1(x)|_y & |f_1(y)|_y \end{smallmatrix}\right) = \left(\begin{smallmatrix} 3 & 2 \\ 1 & 1 \end{smallmatrix}\right)$ is called the incidence matrix of f_1.

Recall that the monoid $\mathrm{SL}_2(\mathbb{N})$ of special linear matrices with natural numbers as coefficients is (freely) generated by the two triangular matrices $R = \left(\begin{smallmatrix} 1 & 1 \\ 0 & 1 \end{smallmatrix}\right)$ and $L = \left(\begin{smallmatrix} 1 & 0 \\ 1 & 1 \end{smallmatrix}\right)$. The matrix of the diatonic Regener transformation has a decomposition $M = RRL$. In our passive interpretation of the transformations the decomposition goes along with a concatenation of three base changes in the interval group.

$$\mathbb{Z} \cdot P5 \oplus \mathbb{Z} \cdot P4 \to \mathbb{Z} \cdot M2 \oplus \mathbb{Z} \cdot P4 \to \mathbb{Z} \cdot M2 \oplus \mathbb{Z} \cdot m3$$

$$\to \mathbb{Z} \cdot M2 \oplus \mathbb{Z} \cdot m2.$$

In refinement, consider the four monoid endomorphisms $G, \tilde{G}, D, \tilde{D}$ on two-letter words $\{x, y\}^*$:

$$G(x) = x, \quad G(y) = xy, \quad \tilde{G}(x) = x, \quad \tilde{G}(y) = yx,$$
$$D(x) = yx, \quad D(y) = y, \quad \tilde{D}(x) = xy, \quad \tilde{D}(y) = y,$$

$R = M_G = M_{\tilde{G}}$ is the incidence matrix for both G and $\tilde{G}$, while $L = M_D = M_{\tilde{D}}$ is the incidence matrix for both D and $\tilde{D}$. The submonoid $St_0 = \langle G, \tilde{G}, D, \tilde{D} \rangle \subset \mathrm{Aut}(F_2)$ within the automorphism group, which is generated by these four maps leaves the monoid $\{x, y\}^* \subset F_2$ of *positive words*[h] invariant. With respect to the restricted action $St_0 \times \{x, y\}^* \to \{x, y\}^*$, on positive words, St_0 is known under the name *Special Sturmian monoid.*

There are six $\mathrm{Aut}(F_2)$-conjugates of f_1 within St_0 (including f_1). Applying them to the word xy (representing the authentic division of the octave) we obtain the six step interval patterns of the authentic diatonic modes, such as exemplified by the signature morphisms $\sigma^{C_4/fa}$, $\sigma^{C_4/do}$, $\sigma^{C_4/so}$, $\sigma^{C_4/re}$, $\sigma^{C_4/la}$, $\sigma^{C_4/mi}$ (see Fig. 6.9 left side and upper staves).

It is interesting to observe, that the interval patterns of the associated fifth/fourth-foldings are not only generated by another conjugacy class of six Sturmian morphisms (see right side of Fig. 6.9), but that each individual folding pattern is generated by a morphism f_i^* that is in an illuminating relation to the corresponding morphism f_i that generates the associated step interval pattern. The comparison of the right and left sides of Fig. 6.9 reveals that each morphism f_i^* is the image of f_i under the anti-automorphism of St_0, which exchanges D and $\tilde{D}$, and which leaves G and $\tilde{G}$ fixed: $D^* = \tilde{D}$, $\tilde{D}^* = D$, $G^* = G$, $\tilde{G}^* = \tilde{G}$. In each f_i^* the images of the four generators under $*$ are concatenated in the reverse order in comparison to their appearances in f_i. The anti-automorphism $* : St_0 \to St_0$ is called *Sturmian involution.*

---

[h]Positive words use only x and y as letters, and not their inverses x^{-1} and y^{-1}.

$$f_4 = GG\tilde{D} = (xxxy, xxy)$$

$$f_4^* = DGG = (yx, yxyxy)$$

$$f_1 = GGD = (xxyx, xxy)$$

$$f_1^* = \tilde{D}GG = (xy, xyxyy)$$

$$f_5 = G\tilde{G}\tilde{D} = (xxyx, xyx)$$

$$f_5^* = DG\tilde{G} = (yx, yxyyx)$$

$$f_2 = \tilde{G}GD = (xyxx, xyx)$$

$$f_2^* = \tilde{D}G\tilde{G} = (xy, xyyxy)$$

$$f_6 = \tilde{G}\tilde{G}\tilde{D} = (xyxx, yxx)$$

$$f_6^* = D\tilde{G}\tilde{G} = (yx, yyxyx)$$

$$f_3 = \tilde{G}\tilde{G}D = (yxxx, yxx)$$

$$f_3^* = \tilde{D}\tilde{G}\tilde{G} = (xy, yxyxy)$$

Fig. 6.9. The conjugated Sturmian morphisms $f_1, \ldots, f_6$ generate the species of the fifth and of the fourth for the six authentic diatonic modes. Here this family is represented by the tropes with the common finalis C_4. Under Sturmian involution the six morphisms are transformed into another family of conjugated Sturmian morphisms $f_1^*, \ldots, f_6^*$. They generate the associated folding-species of the flatward major second and minor second.

6.4.4. *Lattice path transformations: Continuation*

We access the interval group $G_{\mathcal{N}}$ through the choice of a basis, such as $(P5, P4)$, or $(M2, m2)$. The respective coordinate spaces $\mathbb{Z} \cdot P5 \oplus \mathbb{Z} \cdot P4$ and $\mathbb{Z} \cdot M2 \oplus \mathbb{Z} \cdot m2$ are representations of one and the same group $G_{\mathcal{N}}$. The situation is slightly different, though, for lattice paths and the linear spaces $\mathcal{F}[P5, P4]$, $\mathcal{F}[M2, m2]$, $\mathcal{F}[-P5, P4]$ and $\mathcal{F}[-M2, m2]$, containing them. Shall we regard the lattice path $(C_4, P5) + (G_4, P4) \in \mathcal{F}[P5, P4]$ describing the authentic division of the anchored octave $(C_4, P8)$ and the Ionian step interval path $\Psi_{\sigma C_4/do} = (C_4, M2) + (D_4, M2) + (E_4, m2) + (F_4, M2) + (G_4, M2) + (A_4, M2) + (B_4, m2)$ as two different "coordinate representations" of that same anchored interval $(C_4, P8)$? For the present we prefer to treat them as inhabitants of different spaces. In order to manage transformations between them we distinguish between neutral and interpreted lattice path transformations. Let $\mathcal{F}$ denote the real vector space of all finite linear combinations of the elements of the set $\mathcal{B} = \{(X, e_x) \mid X \in \mathbb{Z}^2\} \sqcup \{(X, e_y) \mid X \in \mathbb{Z}^2\}$, where $e_x = \binom{1}{0}, e_y = \binom{0}{1} \in \mathbb{Z}^2$. *Neutral lattice path transformations* are linear endomorphisms $E(f)$ of $\mathcal{F}$, which are induced by a special Sturmian morphism $f \in St_0$ as follows.[i] For given f it is sufficient to define the transformation $E(f)$ on the elements of $\mathcal{B}$, i.e., for the neutral anchored base intervals (W, e_z) for $W \in \mathbb{Z}^2$ and for $z = x$ or $z = y$. The image $E(f)(W, e_z)$ is determined by the word $w = f(z)$ and is a sum of elements from $\mathcal{B}$. More precisely, the base interval of the kth summand is determined by the kth letter $L_k(w)$ of the word w: it is e_x if $L_k(w) = x$ and it is e_y if $L_k(w) = y$. The associated lattice point of the kth summand is obtained as the image of the given lattice point W under an affine transformation: $M_f \cdot W + V(P_k(w))$. Its linear part M_f is the incidence matrix of the Sturmian morphism f and its translation part $V(P_k(w)) \in \mathbb{Z}^2$ is the commutative image of the prefix $P_k(w)$ of length $k - 1$ of the word w. Thus one has:

$$E(f)(W, e_z) := \sum_{k=1}^{|f(z)|} (M_f \cdot W + V(P_k(f(z))), e_{L_k(f(z))}).$$

For every basis (i, j) of the interval group $G_\mathcal{N}$ we have a linear isomorphism $\iota_{[i,j]} : \mathcal{F}[i,j] \to \mathcal{F}$, sending i and j to e_x and e_y, respectively. An *interpreted lattice path transformation* is a concatenation $\overline{E(f)} = \iota_{[k,l]}^{-1} \circ E(f) \circ \iota_{[i,j]} : \mathcal{F}[i,j] \to \mathcal{F}[k,l]$, such that the incidence matrix M_f of the Sturmian morphism $f \in St_0$ serves as the coordinate transformation $M_f : \mathbb{Z} \cdot i \oplus \mathbb{Z} \cdot j \to \mathbb{Z} \cdot k \oplus \mathbb{Z} \cdot l$ of the interval group $G_\mathcal{N}$.

---

[i]The same formula works more generally. Berthe *et al.* consider a monoid homomorphism $E : \mathrm{End}(\{x, y\}^*) \to \mathrm{End}(\mathcal{F})$. See [3, Theorem 5.1].

In our diatonic situation, we have six conjugated Sturmian morphisms

$$f_1 = GGD, \quad f_2 = \tilde{G}GD, \quad f_3 = \tilde{G}\tilde{G}D,$$
$$f_4 = GG\tilde{D}, \quad f_5 = G\tilde{G}\tilde{D}, \quad f_6 = \tilde{G}\tilde{G}\tilde{D},$$

sharing the diatonic Regener transformation $M = RRL = \left(\begin{smallmatrix} 3 & 2 \\ 1 & 1 \end{smallmatrix}\right)$ as their incidence matrix (see Fig. 6.9). The associated lattice path transformations $\overline{E(f_i)} : \mathcal{F}[P5, P4] \rightarrow \mathcal{F}[M2, m2]$ can be applied to paths in the fifth/fourth-lattice and yield paths in the diatonic step interval lattice. For example, applying the interpreted lattice path transformation of the Dorian morphism $f_2 = \tilde{G}GD = (xyxx, xyx)$ to the path $(C_4, P5) + (G_4, P4) \in \mathcal{F}[P5, P4]$ yields the C_4-Dorian lattice path $(C_4, M2) + (D_4, m2) + (E\flat_4, M2) + (F_4, M2) + (G_4, M2) + (A_4, m2) + (B\flat_4, M2) \in \mathcal{F}[M2, m2]$. Dually, we have the six conjugated Sturmian morphisms

$$f_1^* = \tilde{D}GG, \quad f_2^* = \tilde{D}G\tilde{G}, \quad f_3^* = \tilde{D}\tilde{G}\tilde{G},$$
$$f_4^* = DGG, \quad f_5^* = D\tilde{G}G, \quad f_6^* = D\tilde{G}\tilde{G},$$

sharing the diatonic Regener transformation $M^* = LRR = \left(\begin{smallmatrix} 1 & 2 \\ 1 & 3 \end{smallmatrix}\right)$ as their incidence matrix. The associated lattice path transformations $\overline{E(f_i^*)} : \mathcal{F}[-M2, m2] \rightarrow \mathcal{F}[-P5, P4]$ can be applied to step-interval folding paths in the twisted step-interval lattice and yield fifth/fourth-folding paths in the twisted fifth/fourth-lattice. For example, applying the Sturmian involution $f_2^* = \tilde{D}G\tilde{G} = (xy, xyyxy)$ of the Dorian morphism f_2 to the path $(A_4, -M2) + (G_4, m2) \in \mathcal{F}[-M2, m2]$ yields the lattice path $(A_4, -P5) + (D_4, P4) + (G_4, -P5) + (C_4, P4) + (F_4, P4) + (B\flat_4, -P5) + (E\flat_4, P4) \in \mathcal{F}[-P5, P4]$.

There are also other music-theoretically interesting lattice paths, which are images under the path transformation $\overline{E(f_2)}$. If it is applied to the plagal division $(G_3, P4) + (C_4, P5) \in \mathcal{F}[P5, P4]$ of the anchored octave $(G_3, P8)$ one obtains the Hypodorian step interval path $(G_3, M2) + (A_3, m2) + (G_3, M2) + (B\flat_4, M2) + (C_4, M2) + (D_4, m2) + (E\flat_4, M2) + (F_4, M2) \in \mathcal{F}[M2, m2]$. If one applies it to a path of ascending fourths $(D_3, P4) + (G_3, P4) + (C_4, P4) + (F_3, P4)$ one obtains an extended Dasian chain of Dorian species of the fourth:

$$\overline{E(f_2)}((D_3, P4) + (G_3, P4) + (C_4, P4) + (F_3, P4))$$
$$= (D_3, M2) + (E_3, m2) + (F_3, M2)$$
$$+ (G_3, M2) + (A_3, m2) + (B\flat_3, M2)$$
$$+ (C_4, M2) + (D_4, m2) + (E\flat_4, M2)$$
$$+ (F_4, M2) + (G_4, m2) + (A\flat_4, M2).$$

6.5. Dual Lattice Path Transformations and Exo-modes

At the end of Sec. 6.3.2 a first attempt was made to parameterize the different modal meanings of one and the same note. Figure 6.5 shows a superposition of seven copies of the principal diatonic domain in degree space relative to the different images of the note C_4 from the note space. This final section is again dedicated to the motto of this chapter. This time, however, the *one note samba* takes place solely in the space of anchored note intervals and the lattice paths built upon them. We will disassemble the construction plan of the transformation $\overline{E(f)} : \mathcal{F}[P5, P4] \to \mathcal{F}[M2, m2]$ in order to extract its inverse combinatorics, namely to list all the occurrences of a given anchored step interval within the modal fillings of all possible anchored fifths and fourths. This can be done most elegantly with the help of the dual transformation $\overline{E(f)}^* : \mathcal{F}[M2, m2]^* \to \mathcal{F}[P5, P4]^*$. As will be explicated below, the dual map sends the dual of any given anchored step interval to the sum of the duals of those anchored fifths and fourths, in the $\overline{E(f)}$-images of whose the given step interval occurs as a summand.

Following the procedure in Sec. 6.4.4 we first calculate the dual $E(f)^* : \mathcal{F}^* \to \mathcal{F}^*$ of a neutral lattice path transformation $E(f) : \mathcal{F} \to \mathcal{F}$ and then we concatenate it with the duals of the appropriate interpretation maps. We turn our attention to the duals of elements of $\mathcal{F}$, i.e., to linear functions $\phi : \mathcal{F} \to \mathbb{R}$. The evaluation of ϕ on a element $X \in \mathcal{F}$ we write in bracket notation as $\langle \phi | X \rangle \in \mathbb{R}$.

Each anchored neutral base interval $(W, e_z) \in \mathcal{B}$ has a dual $(W, e_z)^*$, satisfying

$$\langle (W, e_z)^* | (W', e_{z'}) \rangle = \begin{cases} 1 & \text{if } W = W' \text{ and } z = z', \\ 0 & \text{otherwise.} \end{cases}$$

We write these dual elements also in the form (W, e_z^*) and call them anchored neutral base co-intervals, accordingly. Let $\mathcal{B}^*$ denote the set of all anchored neutral base co-intervals and $\mathcal{F}^*$ the linear space, generated by $\mathcal{B}^*$. Note that this is only a subspace within the full dual space of $\mathcal{F}$, but for our purpose it is sufficient to work with $\mathcal{F}^*$.

For a given Sturmian morphism $f \in St_0$ and $(W, e_z^*) \in \mathcal{B}^*$ we start from the definition of the dual map and obtain through equivalent conversion:

$$\langle E(f)^*(W, e_z^*) | (W', e_z') \rangle = \langle (W, e_z^*) | E(f)(W', e_z') \rangle$$

$$= \left\langle (W, e_z^*) \left| \sum_{k=1}^{|f(z')|} (M_f \cdot W' + V(P_k(f(z'))), e_{L_k(f(z'))}) \right. \right\rangle$$

$$= \sum_{k=1}^{|f(z')|} \langle (W, e_z^*) | (M_f \cdot W' + V(P_k(f(z'))), e_{L_k(f(z'))}) \rangle$$

$$= \sum_{k=1}^{|f(z')|} \langle (M_f^{-1} \cdot (W - V(P_k(f(z')))), e_z^*) | (W', e_{L_k(f(z'))}) \rangle$$

$$= \begin{cases} \displaystyle\sum_{L_k(f(x))=z} \langle (M_f^{-1} \cdot (W - V(P_k(f(z')))), e_z^*) | (W', e_x) \rangle \\ \qquad\qquad \text{if } z' = x, \\[1em] \displaystyle\sum_{L_k(f(y))=z} \langle (M_f^{-1} \cdot (W - V(P_k(f(z')))), e_z^*) | (W', e_y) \rangle \\ \qquad\qquad \text{if } z' = y. \end{cases}$$

Each summand in line 3 equals either 1 or 0. And it is 1 if and only if $W = M_f \cdot W' + V(P_k(f(z')))$ and $z = L_k(f(z'))$. Line 4 follows from the fact that $W = M_f \cdot W' + V(P_k(f(z')))$ if and only if $M_f^{-1} \cdot (W - V(P_k(f(z')))) = W'$. Finally, we have to determine, when the condition $z = L_k(f(z'))$ holds. To that end we have to look for all instances of the letter z in the word $f(z')$. Thus we have to distinguish the case $z' = x$ from $z' = y$ and have sum over all instances where the letter z occurs in the word $f(z')$.

In the light of this calculation, the image of an anchored neutral base co-interval (W, e_z^*) under the dual lattice path transformation $E(f)^*$ can be written as a sum of two sums, comprising the distinction of the two cases above:

$$E(f)^*(W, e_z)^* = \sum_{L_j(f(x))=z} (M_f^{-1}(W - V(P_j(f(x)))), e_x)^*$$

$$+ \sum_{L_j(f(y))=z} (M_f^{-1}(W - V(P_j(f(y)))), e_y)^*.$$

In order to obtain the dual $\overline{E(f)}^*$ of the interpreted lattice path transformation

$$\overline{E(f)} = \iota_{[k,l]}^{-1} \circ E(f) \circ \iota_{[i,j]} : \mathcal{F}[i,j] \to \mathcal{F}[k,l]$$

the duals of the three maps have to concatenated in the reverse order:

$$\overline{E(f)}^* = \iota_{[i,j]}^* \circ E(f)^* \circ (\iota_{[k,l]}^{-1})^* : \mathcal{F}[k,l]^* \to \mathcal{F}[i,j]^*.$$

In addition to the algebraic level it is also interesting to consider a geometric interpretation of the co-intervals, which we inherit from [2]. Each note X is the lower left corner of a unit square in $\mathbb{Z}^2$. The co-vector (X, e_x^*) shall be represented by the line segment $[X + e_x, X + e_x + e_y] = \{X + (1,t) \,|\, 0 \le t \le 1\}$ and the co-vector (X, e_y^*) by the line segment $[X + e_y, X + e_y + e_x] = \{X + (t,1) \,|\, 0 \le t \le 1\}$, respectively. These line segments live in the continuous extension of the note space, but they should not be conflated with note intervals. These segments rather form continuous inter-note-spans. In the case of the $P5/P4$-basis the co-interval of an anchored fifth $(X, P5)$ is the continuous inter-note span of its octave complement $[X + P5, X + P8]$. Likewise, the co-interval of an anchored fourth $(X, P4)$ can be geometrically represented by the continuous inter-note span of its

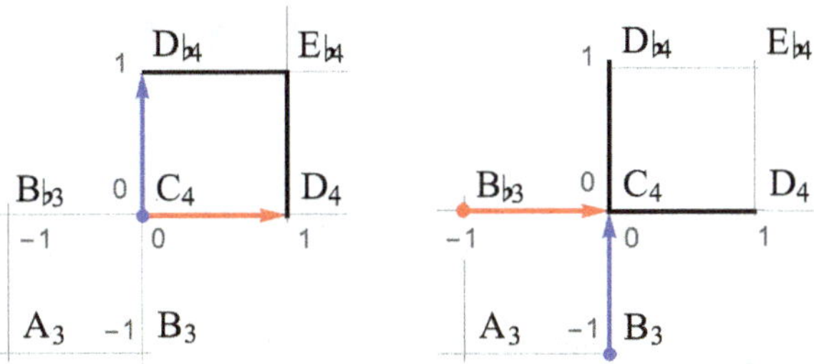

Fig. 6.10. Geometric representation of the dual lattice paths $\underline{P} = (C_4, M2^*) + (C_4, m2^*) \in \mathcal{F}[M2, m2]^*$ (left) and $\overline{P} = (B\flat_3, M2^*) + (B_3, m2^*) \in \mathcal{F}[M2, m2]^*$ (right) in the coordinate space $\mathbb{Z} \cdot M2 \oplus \mathbb{Z} \cdot m2$. The colored arrows represent the anchored note intervals, while the black line segments represent their duals.

octave complement $[X + P4, X + P8]$. In the case of the $M2/m2$-basis the co-intervals of the step intervals are represented by the inter-note-spans of their minor-third complements.

6.5.1. *Diatonic exo-modes for the note C_4*

The note C_4 is the anchor note of the two (anchored) step intervals $(C_4, M2)$ and $(C_4, m2)$. The associated co-intervals can be combined in form of the element $\underline{P} = (C_4, M2^*) + (C_4, m2^*) \in \mathcal{F}[M2, m2]^*$. Likewise, C_4 is the target note of the two anchored step intervals $(B\flat_3, M2)$ and $(B_3, m2)$ and their duals together form the element $\overline{P} = (B\flat_3, M2^*) + (B_3, m2^*) \in \mathcal{F}[M2, m2^*]$. Figure 6.10 displays the dual lattice paths $\underline{P}$ and $\overline{P}$.

We explore the dual lattice path transformations in the case of two (out of the six) special Sturmian morphisms f_4 (Lydian) and f_3 (Phrygian), by applying the associated dual lattice transformations $\overline{E(f_4)}^*$ and $\overline{E(f_3)}^*$ to both of the dual lattice paths $\underline{P}$ and $\overline{P}$. We begin with the case of the Lydian morphism $f_4 = GG\tilde{D} = (xxxy, xxy)$. Figure 6.11 illustrates the derivation of $\overline{E(f_4)}^*(\underline{P})$ and $\overline{E(f_4)}^*(\overline{P})$ in terms of the involved interval species in musical staff notation.

The musical staff above the left subfigure displays all Lydian species of the fourth and the fifth, containing the step intervals $(C_4, M2)$ or $(C_4, m2)$. These occurrences are encircled in red and blue, respectively. The respective Lydian width and height degrees

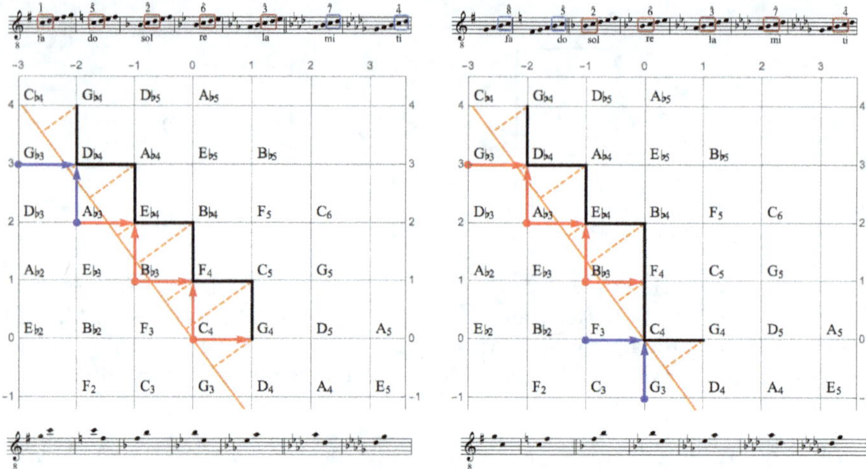

Fig. 6.11. The left and right figures display the Lydian exo-modes $\overline{E(f_4)}^*(\overline{P})$ and $\underline{E(f_4)}^*(\overline{P})$, respectively. Details are explained in the text.

of the note C_4 are annotated. The octave complements of the fifths and fourths are shown in the musical staff below the graphics. In the graphics itself all the anchored fifths and fourths from the upper staff are shown as colored arrows. The respective color indicates whether the interval contains the major second $(C_4, M2)$ (red) or the minor second $(C_4, m2)$ (blue). In other words, the colors identify the images $\overline{E(f_4)}^*(C_4, M2^*)$ and $\overline{E(f_4)}^*(C_4, m2^*)$ separately. The octave complements are shown as black line segments. It is important to observe that the seven line segments form a contiguous polygon. Reading it from bottom-right to top-left, we observe that the polygon is divided into one part of length 5, spanning the ascending minor second $[G_4, A\flat_4]$ and a another part of length 2 spanning the descending major second $[A\flat_4, G\flat_4]$ (in accordance with the coloring of the underlying intervals). In the right subfigure the same procedure is applied to the derivation of the dual lattice path $\overline{E(f_4)}^*(\overline{P})$. Here the subdivision of the long polygon starts with a part of length 2, spanning the descending major second $[G_4, F_4]$ and a another part of length 5 spanning the ascending minor second $[F_4, G\flat_4]$. The form of this divided polygon is reminiscent of the flatward folding pattern $(xy, yxyxy)$ of the authentic Phrygian mode (cf. Fig. 6.9). This

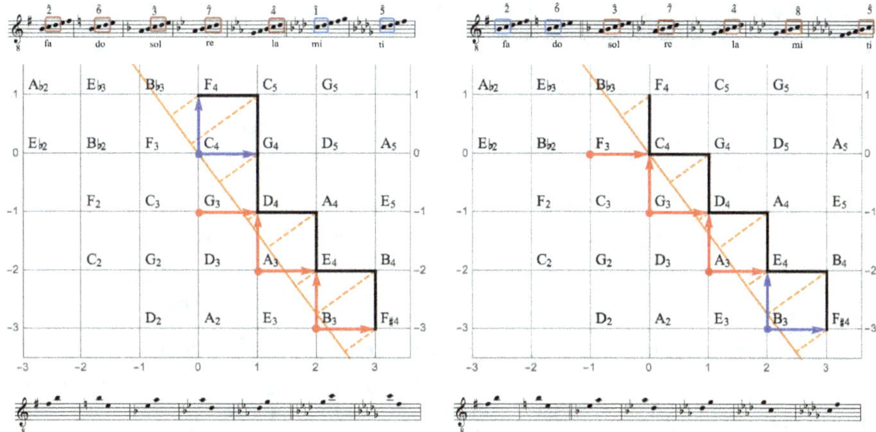

Fig. 6.12. The left and right figures display the Phrygian exo-modes $\overline{E(f_3)}^*(\underline{P})$ and $\overline{E(f_3)}^*(\overline{P})$, respectively.

observation motivates the term *exo-mode* for the dual lattice paths.[j] Figure 6.12 shows the analogous derivation of the dual lattice paths $\overline{E(f_3)}^*(\underline{P})$ and $\overline{E(f_3)}^*(\overline{P})$, in the Phrygian case.

Remark 6.5.1. In all four subfigures of Figs. 6.11 and 6.12 the zigzag-lines of the dual lattice paths are approximating a diagonal line which is shown in orange color. It is the eigenspace of the diatonic Regener transformation with respect to its smaller eigenvalue. From each participating lattice point an orthogonally oriented line is leading to this eigenspace, indicating the respective projections. The orthogonal lines are oriented in the direction of the eigenspace of the dual diatonic Regener transformation $\left(\begin{smallmatrix} 3 & 1 \\ 2 & 1 \end{smallmatrix}\right)$ with respect to its larger eigenvalue. For the musical interpretation of these spaces see [11].

6.5.2. *The lydo-phrygian and standard chromatic modes*

A different—and unconventional—interpretation of the note interval and degree interval system is associated with the chromatic 12-note

---

[j]The study of the correspondences between foldings paths and exo-modes is an interesting question in its own right and relates to [3, Theorem 5.2].

scale. What differs in the basic GIS setup $\sigma = (\sigma_S, \sigma_G) : \mathcal{N} \hookrightarrow \mathcal{D}$ is the signature morphism. We will mark the chromatic signature morphism with an attached tilde: $\tilde{\sigma} = (\tilde{\sigma}_S, \tilde{\sigma}_G)$. The image $\tilde{\sigma}_G(P8)$ of the perfect octave is then the pure height degree interval $(0, 12)$. The role of a pure width interval falls to the diminished second $d2$, i.e., $\tilde{\sigma}_G(-d2) = \tilde{\sigma}_G(\text{int}_{\mathcal{N}}(G\flat_4, F\sharp_4)) = (12, 0)$. The principal chromatic domain is the square $Chrom = \{(r, s) \in \mathbb{Z}^2 \,|\, 0 \leq r, s < 12\} \subset S_{\mathcal{D}}$. It can be annotated with height degrees $0, \ldots, 11$ and 12 syllables $su, ru, lo, ma, ta, fa, do, so, re, la, mi, ti$ for the width degrees. As already indicated in Remark 6.2.1, the chromatic scale constitutes another relevant music-theoretical source for a Regener transformation $M' : \mathbb{Z} \cdot P5 \oplus \mathbb{Z} \cdot P4 \cong \mathbb{Z}^2 \to \mathbb{Z}^2 \cong \mathbb{Z} \cdot A1 \oplus \mathbb{Z} \cdot m2$ with $M'(v) = \left(\begin{smallmatrix} 3 & 2 \\ 4 & 3 \end{smallmatrix}\right) \cdot v$. The corresponding conjugacy class of Sturmian morphisms has eleven elements $g_1 = DGGD$, $g_2 = \tilde{D}GGD$, $g_3 = D\tilde{G}GD, \ldots, g_{11} = \tilde{D}\tilde{G}\tilde{G}\tilde{D}$ and they correspond to eleven authentic chromatic modes.[k] Just as in the case of the diatonic modes with six authentically divided species of the octave (represented by the word pairs $(f_k(x), f_k(y))$ with $k = 1, \ldots, 6$) and the exceptional "badly divided" locrian species $(m2, M2, M2, m2, M2, M2, M2)$ of the octave with a division into a diminished fifth $2\,M2 + 2\,m2 = d5$ and an augmented fourth $3\,M3 + 0\,m2 = A4$, there are eleven authentically divided chromatic species (represented by the word pairs $(g_k(x), g_k(y))$ with $k = 1, \ldots, 11$) and one "badly divided" chromatic species $(m2, m2, A1, m2, A1, m2, m2, A1, m2, A1, m2, A1)$ of the octave with a division into a diminished sixth $5\,m2 + 2\,A1 = d6$ and an augmented third $2\,m2 + 3\,A1 = A3$.

We examine two out of the eleven authentic modes. They are called the *Lydo-Phrygian* and the *Standard* chromatic mode, respectively. Their step interval patterns and fifth/fourth foldings are shown in Fig. 6.13.

---

[k]See [5] for an investigation into the matching orderings of the chromatic modes under conjugation, on the one hand, and the morphisms $g_1, \ldots, g_{11}$ with respect to the lexicographic orderings of words in the four generators $G, \tilde{G}\ D$, and $\tilde{D}$ of the Special Sturmian monoid St_0, on the other.

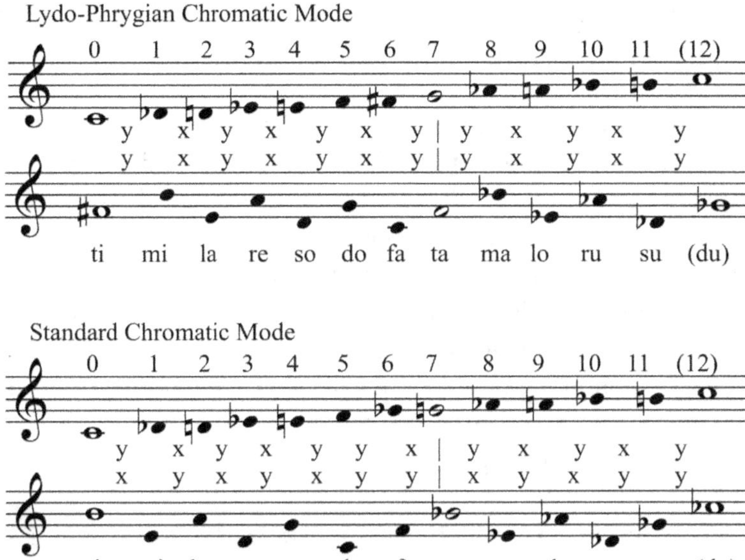

Fig. 6.13. Upper half: Ascending Lydo-Phrygian chromatic C_4-mode and its flatward fifth/fourth folding. The numbers $0, \ldots, 11, (12)$ and the syllables $ti, \ldots, su, (du)$ serve as labels for the height and width degrees in the principal chromatic domain in the degree space. Lower half: Ascending Standard chromatic C_4-mode and its flatward fifth/fourth folding. The modes differ only in the chromatic scale degree 6: $F\sharp_4$ versus $G\flat_4$.

A side aspect of the consideration of two (or more) modes is that we need to deliberate about the designation of the tonic width degrees within the principal chromatic domain. As in the diatonic case we distinguish between principal and accidental chromatic modes. Our designation of the principal width degrees is determined by the choice of the subsequence $su, ru, lo, ma, ta, fa, do, so, re, la, mi, ti$ of length 12 out of a (conventionalized) sequence of syllables, such as $\ldots, fu, du, su, ru, lo, ma, ta, fa, do, so, re, la, mi, ti, fi,$ $di, si, ri, li, \ldots$. Exactly one width degree $0 \leq d \leq 11$ can be designated with the syllable *do*. And this means: Exactly one out of the twelve principal chromatic modes has a tonic which is designated by the syllable *do*. This privilege is bestowed upon the chromatic *Standard mode* (see also Fig. 6.13). The Lydo-Phrygian mode becomes the chromatic *fa*-mode then. The deliberation of this choice is postponed to the musical example towards the end of this subsection.

Remark 6.5.2. The term *Lydo-Phrygian chromatic mode* is my own idiosyncratic nomenclature in the context of this chapter and is particularly motivated by the connection of this mode to the content of the previous Sec. 6.5.1. The term *Standard mode*, in contrast, is inherited by established mathematical nomenclature in the field of Algebraic Combinatorics on words. A Special Standard morphism is a special Sturmian morphism $f \in \langle G, D \rangle$, i.e., it is a concatenation of a finite combination of instances of the two generators D and G. This is the case for the Ionian diatonic morphism $f_1 = GGD$ and for the chromatic morphism $g_1 = DGGD$.

The study of the *Lydo-Phrygian chromatic mode* and its exo-modes suggests itself in consideration of a complete family of authentic diatonic modes with a shared tonic (or finalis): If one unifies the notes from all the six authentic C_4-modes in Fig. 6.8 one obtains the Lydo-Phrygian chromatic mode. This is true both for the stepwise ordered modes as well as for their fifth/fourth foldings. Thereby the Lydian and the Phrygian modes already cover the entire Lydo-Phrygian one, hence the name. The chromatic step interval pattern is generated by the morphism $g_6 = \tilde{D}G\tilde{G}D$ and the fifth/fourth-folding pattern is generated by the same morphism $g_6^* = \tilde{D}G\tilde{G}D$, because g_6 is invariant under Sturmian involution. This means that the step interval pattern and the fifth/fourth-folding pattern coincide, abstractly.

$$(g_6(x), g_6(y)) = \tilde{D}G\tilde{G}D(x, y) = \tilde{D}(yxxx, yxx) = (yxyxyxy, yxyxy)$$
$$(g_6^*(x), g_6^*(y)) = \tilde{D}\tilde{G}\tilde{G}D(x, y) = \qquad \ldots \qquad = (yxyxyxy, yxyxy).$$

The Sturmian morphism g_6 has two representations in terms of the four generators G, $\tilde{G}$, D, $\tilde{D}$: When the Lydian Sturmian morphism $f_4 = GG\tilde{D}$ is concatenated from the left with D, one obtains the representation $DGG\tilde{D}$ of g_6 The same morphism and the same generated mode can be obtained by concatenating the Phrygian

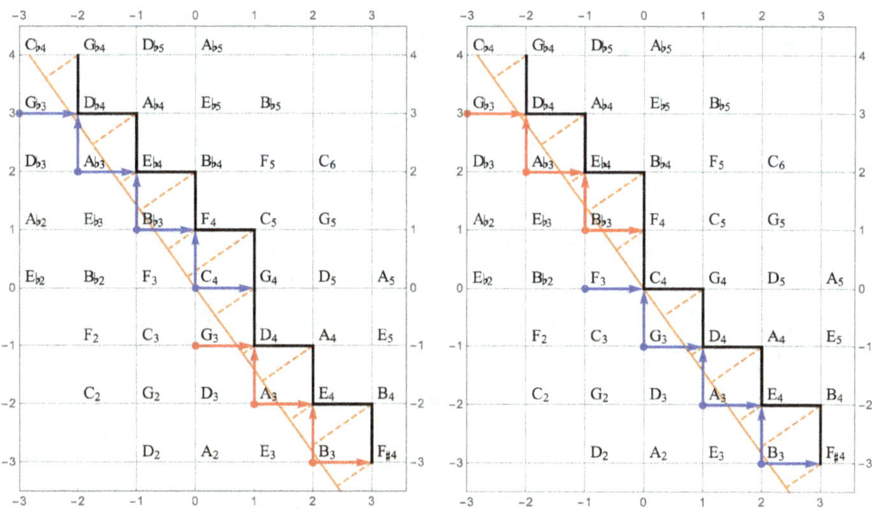

Fig. 6.14. The left and right figures display the Lydo-Phrygian chromatic exo-modes $\overline{E(g)}^*(\underline{P'})$ and $\overline{E(g)}^*(\overline{P'})$, respectively.

Sturmian morphism $f_3 = \tilde{G}\tilde{G}D$ from the left with $\tilde{D}$: $\tilde{D}\tilde{G}\tilde{G}D(x,y) = \tilde{D}(yxxx, yxx) = (yxyxyxy, yxyxy)$.[1]

The same observation can be made with the dual Lydo-Phrygian Lattice paths. To that end we consider the two dual lattice paths $\underline{P'} = (C_4, A1^*) + (C_4, m2^*)$ and $\overline{P'} = (C\flat_4, A1^*) + (B_3, m2^*) \in \mathcal{F}[A1, m2]^*$. To both of them we apply the dual lattice path transformation $\overline{E(g_6)}^*$. Figure 6.14 displays the resulting dual lattice paths. The zigzag polygons are the unions of the associated lydian and phrygian polygons from Figs. 6.11 and 6.12. In fact, these two are the remote poles in the family of all six diatomic modes and the zigzag polygons are covered by all six diatonic dual lattice paths.

Remark 6.5.3. The slopes of the orange lines and orthogonal projections are here determined by the eigenvectors of the chromatic Regener transformation, which is self-dual.

---

[1]See again [5] for a detailed combinatorial investigation into the representation of Sturmian morphisms in terms of the generators G, $\tilde{G}$, D, $\tilde{D}$.

Fig. 6.15. The opening four measures of *OneNoteSamba* by Antonio Carlos Jobim with common chord symbols.

A special musical example may shed some light upon the combinatorics of the exo-modes. Songs like *Ein Ton* by Peter Cornelius (1854), *It don't mean a thing* by Duke Ellington (1931) or *One note samba* by Antonio Carlos Jobim (1959) contain passages, where a constantly repeated tone systematically changes its meaning. Here we only will have a brief look at the note F_4 in the opening four measures of *One note samba*. From a diatonic bird's eye view of the tune, the repeated tone occupies the width/height degree $\sigma_S^{B\flat 3/do}(F_4) = (so, \hat{5})$ of the key of $B\flat$. In a second instance the changing meanings of the note F_4 would be of harmonic nature, in so far as it occupies changing roles as chord-tones or non-chord tones with respect to the given chords (see Fig. 6.15).

As Jazz theory emphasizes the connection between chords and modes, it provides us also with a certain motivation for the application of the present approach to these tone meanings. Following the first chord symbol Dm^7, we could interpret the first measure in terms of an accidental *do*-based dorian mode on D_4 and arrive at the interpretation $\sigma_S^{D_4/do}(F_4) = (ma, \hat{3})$. Following the chord-symbol Db^7 in measure 2 literally, would suggests a *do*-based mixolydian mode on Db_4 with the interpretation $\sigma_S^{Db_4/do}(F_4) = (mi, \hat{3})$. As a "confessing Ramist", however, I tend to assume a fundament G_3 with the provisional mixolydian interpretation $\sigma_S^{G_3/do}(F_4) = (ta, \hat{7})$. However, the diminished fifth above this fundament, the real bass Db_4 does not fit into the G_3-mixolydian mode. Before tackling this problem we go on with the other two chords. The interpretation of the third measure in the context of a *do*-based dorian mode on C_4 leads to $\sigma_S^{C_4/do}(F_4) = (fa, \hat{4})$. Eventually, in measure 4, it is worthwhile to envisage also the enharmonically related chord notation $Cb^{7\sharp 4}$. On the one hand, of course, the real bass Cb fits better with the Ramist assumption of a fundament F and with the melody note F_4

interpreted as $\sigma_S^{F_3/do}(F_4) = (do, \hat{8})$. On the other hand, however, even from a linear point of view it seems to be more attractive to assume $C\flat^{7\sharp4}$ with the interpretation $\sigma_S^{C\flat 4/do}(F_4) = (fi, \hat{4})$ rather than the common notation $B^{7\flat5}$ with the interpretation $\sigma_S^{B_3/do}(F_4) = (su, \hat{5})$. The resulting progression $(ma, \hat{3}) \rightsquigarrow (mi, \hat{3}) \rightsquigarrow (fa, \hat{4}) \rightsquigarrow (fi, \hat{4})$ of degree-meanings of the note F_4 over a linearly descending bass seems to be more consequential than $(ma, \hat{3}) \rightsquigarrow (mi, \hat{3}) \rightsquigarrow (fa, \hat{4}) \rightsquigarrow (su, \hat{5})$.

The "Ramist" interpretation with the fundament progression $D_4 \rightsquigarrow G_3 \rightsquigarrow C_4 \rightsquigarrow F_3$ leads to the following progression of width/height degrees for the note F_4: $(ma, \hat{3}) \rightsquigarrow (ta, \hat{7}) \rightsquigarrow (fa, \hat{4}) \rightsquigarrow (do, \hat{8})$. The falling fifths of the fundaments correspond to rising fifths in the meanings. Furthermore these fifths bring us closer to the logics of the dual lattice paths, which are contiguous in the $P5/P4$-space. But the diatonic modes lack flexibility for this purpose. Not only the lack of a diminished fifth in the mixolydian mode provides an obstacle. Even the change of the modal type from chord to chord prevents us from tracing the shifting modal roles of F_4 within a single dual lattice path. The chromatic modes are better suited for this purpose.

The Lydo-Phrygian mode comes to mind as it is the union of all the six common-finalis modes. It competes though, with the Standard chromatic mode (cf. Fig. 6.13 lower half), because the latter one contains the diminished fifth over the fundament. It is generated by the *Standard morphism* $g_1 = DGGD$ and its associated fifth/fourth-folding is generated by the Sturmian involution of g_1, namely $g_1^* = (DGGD)^* = \tilde{D}GG\tilde{D}$:

$$(g_1(x), g_1(y)) = (DGGD)(x, y) = D(xxyx, xxy) = (yxyxyyx, yxyxy),$$
$$(g_1^*(x), g_1^*(y)) = \tilde{D}GG\tilde{D}(x, y) = \tilde{D}(xxxy, xxy) = (xyxyxyy, xyxyy).$$

Figure 6.16 contrasts the analysis of the note F_4 within accidental dorian and mixolydian diatonic modes with its analysis within principal Standard chromatic modes. The choice of the syllable *do* for the designation of their tonic width degrees entails the incidence of the syllables *ma, ta, fa, do* in both readings. The chromatic signature

Fig. 6.16. Interpretation of the changing meanings of the Note F_4 in the first four measures of the song *One Note Samba*. The annotation to the upper staff summarizes my "Ramist" interpretation in terms of accidental diatonic *do*-modes relative to the fundament progression $D_4 \rightsquigarrow G_3 \rightsquigarrow C_4 \rightsquigarrow F_3$. The lower staff shows species of the perfect fifth $P5$ or the perfect fourth $P4$ within the associated chromatic Standard mode over these same fundaments. In each species the anchored semitone step interval $(E_4, m2)$ with target tone F_4 is marked. The associated chromatic width/height degrees in the principal chromatic domain are annotated to the note F_4. The choice of the syllable *do* for the tonic in the Standard modes implies that the syllables coincide in both annotations.

morphisms are applied to the note F_4 as follows:

$$\tilde{\sigma}_S^{D_4/do}(F_4) = (ma, 3) \rightsquigarrow \tilde{\sigma}_S^{G_3/do}(F_4) = (ta, 7)$$
$$\rightsquigarrow \tilde{\sigma}_S^{C_4/do}(F_4) = (fa, 5) \rightsquigarrow \tilde{\sigma}_S^{F_3/do}(F_4) = (do, 12).$$

The same signature morphisms can be applied to the real bass progression $D_4 \rightsquigarrow D\flat_4 \rightsquigarrow C_4 \rightsquigarrow C\flat_4$:

$$\tilde{\sigma}_S^{D_4/do}(D_4) = (do, 0) \rightsquigarrow \tilde{\sigma}_S^{G_3/do}(D\flat_4) = (su, 6)$$
$$\rightsquigarrow \tilde{\sigma}_S^{C_4/do}(C_4) = (do, 0) \rightsquigarrow \tilde{\sigma}_S^{F_3/do}(C\flat_4) = (su, 6).$$

The contiguity of the degrees $ma \rightsquigarrow ta \rightsquigarrow fa \rightsquigarrow do$ along the width degree axis within $\mathcal{D}$ is also reflected as a contiguity in the dual lattice path of the chromatic Standard mode, i.e., in the chromatic Standard exo-mode. The lower staff in Fig. 6.16 displays two chromatic Standard species of the fifth $m2A1m2A1m2m2A1$ and two chromatic Standard species of the fourth with the step interval pattern $m2A1m2A1m2$. Both of them are anchored in the two notes D_4 and C_4. In each of them there is an occurrence of the minor second $(E_4, m2)$ with F_4 as its target tone. The other chromatic interval $(F\flat, A1)$ with target tone F_4 does not occur within these four particular examples. Their octave complements can be assembled into a contiguous polygon of line segments, namely $([F_4, C_5], [C_5, G_4], [G_4, D_5], [D_5, A_4])$. Under the perspective of the above analysis and the geometric interpretation of co-intervals as octave complements we might dare to envisage the process of the changing meaning of the note F_4 as a continuous motion along this polygon. Such a speculation, however, would first require an advanced study of this geometrical representation.

Bibliography

[1] E. Agmon, *The Languages of Western Tonality* (Springer, Berlin, 2013).

[2] P. Arnoux and S. Ito, Pisot substitutions and rauzy fractals, *Bull. Belgian Math. Soc. Simon Stevin* **8** (2001) 181.

[3] V. Berthé, A. de Luca and C. Reutenauer, On an involution of christoffel words and sturmian morphisms, *European J. Combin.* **29** (2008) 535.

[4] D. Clampitt and T. Noll, Modes, the height-width duality, and handschin's tone character, *Music Theory Online* **17** (2011).

[5] D. Clampitt and T. Noll, *Matching Lexicographic and Conjugation Orders on the Conjugation Class of a Special Sturmian Morphism*, in *Combinatorics on Words: WORDS 2017 Proceedings*, ed. S. Brlek (Springer, 2017), pp. 85–96.

[6] J. Hook, Signature transformations, in *Music Theory and Mathematics: Chords, Collections, and Transformations*, eds. J. Douthett, M. M. Heyde and C. J. Smith (University of Rochester Press, Rochester, 2008), pp. 137–160.

[7] O. Kolman, Transfer principles for generalized interval systems, *Perspec. New Music* **42** (2004) 150.

[8] D. Lewin, *Generalized Musical Intervals and Transformations* (Yale University Press, New Haven, 1987).

[9] G. Mazzola, *The Topos of Music: Geometric Logic of Concepts, Theory, and Performance* (Birkhäuser, Basel, 2002).

[10] M. Montiel, *El Denotador: Su Estructura, construcción y Papel en la Teoría Matemática de la Musica* (UNAM, Mexico City, 1999).

[11] T. Noll, Handschins toncharakter: Plädoyer füer einen neuen anlauf, ausgehend von neueren musiktheoretischen und kognitionspsychologischen untersuchungen zu den tonalen qualia, *ZGMTH: Zeit. Gesellschaft Musiktheorie* **13** (2016).

[12] T. Noll and M. Montiel, Glarean's Dodecachordon revisited, in *Mathematics and Computation in Music*, eds. J. Yust, J. Wild and J. A. Burgoyne, Lecture Notes in Computer Science, Vol. 7937 (Springer, 2013), pp. 151–166.

[13] E. Regener, *Pitch Notation and Equal Temperament: A Formal Study* (University of California Press, Berkeley, 1973).

[14] S. Rings, *Tonality and Transformation* (Oxford University Press, 2011).

Chapter 7

Difference Sets and All-Directed-Interval Chords

Robert W. Peck

School of Music, Louisiana State University,
Baton Rouge, LA 70803, USA
rpeck@lsu.edu

7.1. Introduction

In this chapter, we examine sets of intervals among musical objects, where such intervals are defined as members of transformation groups following [10]. In particular, we are interested in sets of intervals that correspond in a general sense to the all-interval chords of traditional pitch-class set theory: the forty-eight members of set classes $[0, 1, 4, 6]_{12}$ and $[0, 1, 3, 7]_{12}$ that appear prominently in the music of the post-tonal era and that are discussed in the major post-tonal texts, including [3, 11, 12]. However, we impose here the stricter requirement that these sets possess a flat directed-interval distribution, which the canonical all-interval tetrachords do not. We call these structures all-*directed*-interval chords, and relate them to various types of difference sets from the field of combinatorics: planar and non-planar, cyclic, (non-cyclic) abelian, and non-abelian [5, 7]. Whereas the correspondence between all-directed-interval chords and difference sets was first introduced in [4], those authors explore only the connection to planar cyclic difference sets. This work furthers their investigation.

7.2.   All-Interval Chords

Before we begin our discussion of all-interval chords, we draw a distinction between *directed intervals* and *non-directed intervals* (or interval classes), and we introduce the concepts of interval vector and interval-class vector.

Definition 1. Let G be a group of order v with a regular (simply transitive) action on a set S. We call $g \in G$ a **directed interval** in S. Further, if $s \cdot g = t$ for some $g \in G$ and $s, t \in S$, then we say that the ordered pair $(s, t) \in S \times S$ is an **occurrence** of the directed interval g. Further, if $s \cdot g = t$ and $t \cdot g = s$ (i.e., if $g = g^{-1}$), then we call (s, t) and (t, s) **reciprocal** occurrences of g.

Definition 2. Let $\binom{S}{2}$ be the set of all 2-element subsets in S. An **interval class** (non-directed interval) is an equivalence class $[g]$ that includes both g and g^{-1}. (If $g^{-1} = g$, then $[g]$ is a singleton.) If g sends s to t (and g^{-1} sends t to s), as above, then we call the unordered pair $\{s, t\} \in \binom{S}{2}$ an **occurrence** of $[g]$.

An interval group G has v directed intervals, including the unison (identity) directed interval: $(s, s) \in S \times S$. To calculate the number c of interval classes for G, however, it is necessary first to know how many involutions are in G.[a] Let w be this number. Then, c is determined as follows.

$$c = \frac{w + v - 1}{2}. \tag{7.1}$$

Note: We do not consider the unison (identity) interval class, as $\{s, s\} \notin \binom{S}{2}$.

Theorem 1. $w + v - 1$ *is always divisible by* 2.

Proof. v may be either even or odd. If v is odd, then $w = 0$ by extension of Lagrange's theorem. Hence, w (zero) $+ v$ (odd) $- 1$ is

---

[a]Determining the number of involutions in a finite group is a classical result in character theory (see [6]).

divisible by 2. If v is even, then combine each $g \in G$ with g^{-1} for which $g \neq g^{-1}$. We see that this set of combinations involves an even number of elements of G. The remaining even number of elements of G includes the identity element and all the involutions. Removing the identity element leaves an odd number w of involutions; hence, w (odd) $+ v$ (even) $- 1$ is divisible by 2. $\qquad\qquad\square$

Definition 3. Let G be an interval group of order v with a regular action on a set S, as above, and let $D \subseteq S$. The **interval vector** of D, $\mathrm{IV}(D) = (i_1, i_2, \ldots, i_v)$, is a v-member array that tallies the occurrences in $D \times D$ of each directed interval in G (see [8]). In interval groups for which the constituent intervals are measurable (such as in a vector space), it is customary to list the coordinates of the interval vector in order of increasing intervallic size, beginning with the unison (identity) interval.

Definition 4. The **interval-class vector** of D, $\mathrm{IcV}(D) = [ic_1, ic_2, \ldots, ic_c]$, is a c-member array that tallies the occurrences in $\binom{D}{2}$ of each non-unison interval class in G (see [3]). Again, it is customary (when appropriate) to list the coordinates of the interval-class vector in order of increasing intervallic size, based on smallest directed interval in each interval class, and to omit the tally of occurrences of the unison (identity) interval.

Most generally, an all-interval chord D is a subset of a space S for which $\binom{D}{2}$ contains at least one occurrence of every interval class in the interval group G that acts on S. In other words, $\mathrm{IcV}(D)$ does not include any zeros. The pitch-class set $\{0, 1, 2, 4, 6\}_{12}$ is an example of an all-interval chord, as demonstrated by its interval-class vector: $[2, 3, 1, 2, 1, 1]$. In contrast, the pitch-class set $\{0, 1, 2, 3, 4\}_{12}$ is not all-interval, as its interval-class vector, $[4, 3, 2, 1, 0, 0]$, includes two coordinates with zeros. (Similarly, in the most general sense, an all-*directed*-interval chord D is one for which $D \times D$ contains at least one occurrence of every directed interval in G.)

Composers and music theorists since the early years of the twentieth century have been particularly interested in the all-interval tetrachords that occur in 12-tone pitch-class space—members of set classes $[0, 1, 4, 6]_{12}$ and $[0, 1, 3, 7]_{12}$. These forty-eight tetrachords are the only collections of cardinality $k \leq 4$ within this space to display the aforementioned all-interval property. A further remarkable characteristic of these tetrachords is that they possess one and only one occurrence of each interval class in the interval group $\mathbb{Z}_{12}$. Accordingly, the interval-class vector for this all-interval tetrachord (or for any all-interval tetrachord) is $[1, 1, 1, 1, 1, 1]$. It is largely this aspect of the all-interval tetrachords that has attracted the attention of composers and music theorists. Therefore, when we refer henceforth to all-interval chords of any size, we indicate chords with this particular property.

Definition 5. $D \subseteq S$ is an **all-interval chord** if $\binom{D}{2}$ possesses among its members exactly one occurrence of every (non-unison) interval class in the group G that acts on S.

Remark 1. If D is an all-interval chord of size k, then its interval group G must contain a triangular number $\Delta = \binom{k}{2}$ of interval classes.

The above remark follows naturally from the observation that a set D of size k has Δ two-element subsets. If (by Definition 5) each one of those subsets is an occurrence of a unique interval class in G, and these occurrences include all the interval classes in G, then G must have Δ interval classes.

In certain circumstances, we want to consider the natural action of a group G on itself under (right) multiplication. In this context, G acts as both interval group and space. As a result, we may describe an all-interval chord D as a subset of G. Put $g, h \in G$. The directed interval from g to h is the unique element $i = g^{-1}h$ that satisfies the equation $g \cdot i = h$ (recall that the action of a group on itself under right multiplication is regular). Further, $[i]$ is the interval class that includes i and i^{-1}. Then, $D \subseteq G$ is an all-interval chord if $\binom{D}{2}$ possesses among its members a single occurrence of every (non-unison) interval class in G.

We are particularly interested in the following special type of all-interval chord.

Definition 6. $D \subseteq S$ is an **all-directed-interval chord** if $D \times D$ possesses among its members exactly one occurrence of every non-identity directed interval in G.

We consider only non-identity directed intervals in the above definition because $D \times D$ always includes k occurrences of the identity directed interval. As a result, $\text{IV}(D)$ includes one coordinate (traditionally, the first) that is equal to k, which is equal to 1 only if D is a singleton set. Rather, it is the fact that each of the other, non-identity directed intervals in D has a singular occurrence in $D \times D$ that defines it as an all-directed-interval chord, as shown in the vector's remaining $v - 1$ coordinates. For example, consider the interval vector $(3, 1, 1, 1, 1, 1, 1)$ for the all-directed-interval chord $D = \{1, 2, 4\}_7$.

7.3.   Difference Sets

Definition 7. Let G be a finite (additively notated) group of order v, and let D be a k-member subset of G. D is a (v, k, λ) **difference set** in G if every non-identity element of G can be written as a difference $g - h$ of elements $g, h, \in D$ in exactly λ ways. (In multiplicatively notated groups, we use the product gh^{-1} as an analog for the difference $g - h$.) We call $n = k - \lambda$ the **order** of D.

Depending on the properties of the group in which it is situated, a difference set may be cyclic, non-cyclic abelian, or non-abelian. An example of a cyclic $(7, 4, 2)$ difference set is $D = \{0, 1, 2, 4\}_7$ in the group $\mathbb{Z}_7$. In this case, $\lambda = 2$, as each non-zero integer modulo 7 can be expressed two ways as a difference of elements in D.

$$
\begin{aligned}
1 - 0 &= 2 - 1 = 1 \ (\text{mod } 7) \\
2 - 0 &= 4 - 2 = 2 \\
4 - 1 &= 0 - 4 = 3 \\
4 - 0 &= 1 - 4 = 4 \\
2 - 4 &= 0 - 2 = 5 \\
1 - 2 &= 0 - 1 = 6.
\end{aligned}
\tag{7.2}
$$

Definition 8. Let D_1 be a difference set in a group G_1, and D_2 be a difference set in another group G_2. If D_1 and D_2 have the same (v, k, λ) parameters, we say that they are **isomorphic**. In this case, a group isomorphism may or may not exist from G_1 to G_2. If, however, a group isomorphism does exist from G_1 to G_2, then we say further that D_1 and D_2 are **equivalent**.

7.3.1. *Planar difference sets*

Definition 9. A **planar difference set** is one in which $\lambda = 1$.

The set $D = \{1, 2, 4\}_7$ is an example of a planar difference set, as every non-identity element of $\mathbb{Z}_7$ can be expressed in only one way as a difference of elements $g, h \in D$.

$$
\begin{aligned}
2 - 1 &= 1 \ (\mathrm{mod}\ 7) \\
4 - 2 &= 2 \\
4 - 1 &= 3 \\
1 - 4 &= 4 \\
2 - 4 &= 5 \\
1 - 2 &= 6.
\end{aligned}
\tag{7.3}
$$

Theorem 2. $\lambda = 1 \implies v = n^2 + n + 1$.

Proof. Let G be a group of order v, and let $G^{\#}$ be the set of non-identity elements of G. We note that, for a difference set D of size k in G, there exist exactly $2\binom{k}{2}$ possible differences $g - h$, where $g, h \in D$ and $g \neq h$. If $\lambda = 1$, then each non-identity element of G appears once and only once as a difference of elements in D; therefore, $G^{\#}$ is also of size $2\binom{k}{2}$. As $n = k - 1$ (using $n = k - \lambda$ and $\lambda = 1$) and $2\binom{k}{2} = k(k - 1)$, we observe that $k(k - 1) = (k - 1)^2 + (k - 1) = n^2 + n$. Adding back the identity element gives the order of G as $v = n^2 + n + 1$. $\qquad\square$

The following two classical results in the theory of difference sets (see [13]) are of particular significance to our later discussion. First, each isomorphism class of planar difference sets of order n

corresponds to the Desarguesian projective plane $\mathrm{PG}(2, n)$ of the same order, as suggested by Theorem 2. Second, for every cyclic planar difference set with $k \equiv 2$ (modulo 3), there exists an isomorphic non-abelian difference set.

7.4.   All-Directed-Interval Chords as Difference Sets

In additively notated (abelian) interval groups, the directed interval from g to h is reckoned in the same way as the difference $h - g$. Hence, the set of directed intervals in D is equivalent to the set of differences in D. In multiplicatively notated (non-abelian) groups, however, the directed interval from g to h is given as $g^{-1}h$, but the product that corresponds to the difference $h - g$ is hg^{-1}. Hence, we require an additional layer of structure in relating all-interval chords to non-abelian difference sets.

Definition 10. Let $(G, *)$ be a group. Then, $(G^{\mathrm{opp}}, *')$ is the **opposite group** for G if G^{opp}, as a set, is the same as G, and if $g *' h = h * g$ for all $g, h \in G$.

Theorem 3. *Put $\phi : G \to G$, where $\phi(g) = g^{-1}$ for all $g \in G$. Then, $\phi' : G \to G^{\mathrm{opp}}$, where $\phi'(g) = \phi(g)$, is an isomorphism.*

Proof. First, we show that ϕ is an anti-automorphism of G. As $\phi(g) * \phi(h) = g^{-1} * h^{-1}$ and $\phi(h * g) = g^{-1} * h^{-1}$, we note that $\phi(g) * \phi(h) = \phi(h * g)$ for all $g, h \in G$. Hence, ϕ is an anti-homomorphism. Then, by the group axiom that stipulates the existence of a unique inverse for every $g \in G$, ϕ is a bijection. Therefore, ϕ is an anti-automorphism.

Next, as

$$
\begin{aligned}
\phi'(g) *' \phi'(h) &= \phi(g) *' \phi(h) && (\text{by } \phi'(g) = \phi(g)) \\
&= \phi(h) * \phi(g) && (\text{by } g *' h = h * g) \\
&= \phi(g * h) && (\text{by definition of anti-automorphism}) \\
&= \phi'(g * h), && (\text{by } \phi'(g) = \phi(g))
\end{aligned}
$$

we observe that ϕ' is an isomorphism from G to G^{opp}. $\qquad\square$

Essentially, the set $\{g^{-1}h \mid g, h \in G\}$ of directed intervals in G and the set $\{hg^{-1} \mid g, h \in G\}$ of differences in G are opposite groups. However, because of the isomorphism ϕ', a subset $D \subseteq G$ is structurally the same as its image under ϕ', $D' \subseteq G^{\mathrm{opp}}$, and the properties in one set (such as its being an all-interval chord or a difference set) are inherited by the other. Consequently, we speak informally below of an equivalence between corresponding sets of intervals and differences without invoking ϕ': for our purposes, we treat intervals and differences interchangeably.

7.4.1. *All-directed-interval chords as planar difference sets*

Remark 2. An all-directed-interval chord D is a planar difference set.

The remark follows directly from the discussion immediately above. In a planar difference set D, every non-identity element of G can be written in only one way as $d_1 d_2^{-1}$ of elements $d_1, d_2 \in D$, whereas in an all-directed-interval chord D', every non-identity element of G can be written in only way as $d_2^{-1} d_1$. Hence, these sets are images of one another under ϕ', and we may consider them (informally) as being equivalent.

We note that every all-directed-interval chord is all-interval, but not every all-interval chord is all-directed interval. For instance, the pitch-class set $\{0, 1, 4, 6\}$ is an all-interval tetrachord in two different groups: $\mathbb{Z}_{12}$ and $\mathbb{Z}_{13}$. In both cases, it has the defining interval-class vector $[1, 1, 1, 1, 1, 1]$. However, the (directed) interval vector of the former is $(4, 1, 1, 1, 1, 1, 2, 1, 1, 1, 1, 1)$, which is not all-directed-interval, whereas that of the latter, $(4, 1, 1, 1, 1, 1, 1, 1, 1, 1, 1, 1, 1)$, is all-directed-interval. The reason for the difference is that $\{0, 1, 4, 6\}_{12}$ contains an involution: $-6 \equiv 6$ (modulo 12). In contrast, $\{0, 1, 4, 6\}_{13}$ has no involutions. The involution in $\{0, 1, 4, 6\}_{12}$ accounts for the 2 in its interval vector's seventh coordinate, as any interval class that contains an involution represents two occurrences of a directed interval.

Theorem 4. *The interval group of an all-directed-interval chord cannot contain an involution.*

Proof. We observe (in Remark 2) that an all-directed-interval chord is a planar difference set. Then, for the condition $\lambda = 1$ to hold, there must exist some $d_1, d_2 \in D$ for every non-identity element $g \in G$ such that $d_1 d_2^{-1} = g$, but for which

$$d_2 d_1^{-1} \neq g. \tag{7.4}$$

If, however, $g = g^{-1}$, then $d_2 d_1^{-1} = (d_1 d_2^{-1})^{-1} = g^{-1} = g$, contradicting the statement in (7.4). $\qquad\square$

Accordingly, the order, $v = n^2 + n + 1$, of an interval group for an all-directed-interval chord is always odd, and no group of odd order contains an involution (as implied by Lagrange's theorem).

We find an example of an all-directed-interval trichord in the head motive of the subject to the E major fugue, BWV 878, from Book 2 of Johann Sebastian Bach's *Das wohltemperierte Klavier* (see Fig. 7.1). If we model the E-major collection with $\mathbb{Z}_7$ (where 0 corresponds to the pitch class E, 1 corresponds to F$\sharp$, etc.), then the subset that comprises the first three pitches of the subject's head motive, $D = \{0, 1, 3\}$, is an all-directed-interval chord, as indicated by the following differences.

$$\begin{aligned}
1 - 0 &= 1 \ (\text{mod } 7) \\
3 - 1 &= 2 \\
3 - 0 &= 3 \\
0 - 3 &= 4 \\
1 - 3 &= 5 \\
0 - 1 &= 6
\end{aligned} \tag{7.5}$$

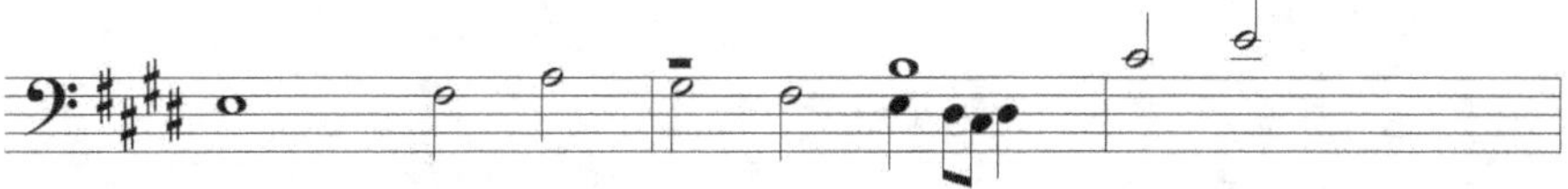

Fig. 7.1. Bach, E Major Fugue from *Das wohltemperierte Klavier*, Book 2, BWV 878, mm. 1–3.

A consequence of D's being an all-directed-interval chord is that translation of D by any element $g \in G^{\#}$ yields exactly one invariant member.

Theorem 5. *If D is an all-directed interval chord, then $|D \cap Dg| = 1$, for all $g \in G^{\#}$.*

Proof. By definition of an all-directed-interval chord, any $g \in G^{\#}$ can be expressed as some directed interval $d_1^{-1}d_2$ of elements d_1, $d_2 \in D$. Therefore, put $g = d_1^{-1}d_2$. As $d_1 \cdot g = d_1 \cdot d_1^{-1}d_2 = (d_1 d_1^{-1}) \cdot d_2 = d_2$, we note that $d_2 \in Dg$. Hence, $|D \cap Dg| \geq 1$. Furthermore, by definition of an all-directed-interval chord, $d_1^{-1}d_2$ is the only directed interval in D that yields g. Thus, d_2 is the only invariant between D and Dg. $\qquad\qquad\square$

Theorem 5 has certain implications regarding all-directed-interval chords as incidence structures and block designs.

Corollary 1. *Let D be an all-directed-interval chord for an interval group G. Each $d \in D$ is coincident to exactly $|D|$ G-translations of D.*

As a result, the set of all G-translations of an all-interval-chord form a Desarguesian projective plane of order n, $\mathrm{PG}(2, n)$. As such, it is conjectured that, for $\lambda = 1$, n must be the power of a prime (see [2], which has shown this conjecture to be true for all values of $n \leq 2,000,000$). For instance, Fig. 7.2 presents a reduction of mm. 1–3 of the fugue, consisting of three $\mathbb{Z}_7$-translations of $D = \{0, 1, 3\}$. The set of these translations is coincident in the pitch class E $= 0$ (shown in the reduction with white noteheads). As a similar incidence structure obtains for every $d \in D$, the full set of $\mathbb{Z}_7$-translations of $\{0, 1, 3\}$ yields a Fano plane, $\mathrm{PG}(2, 2)$.

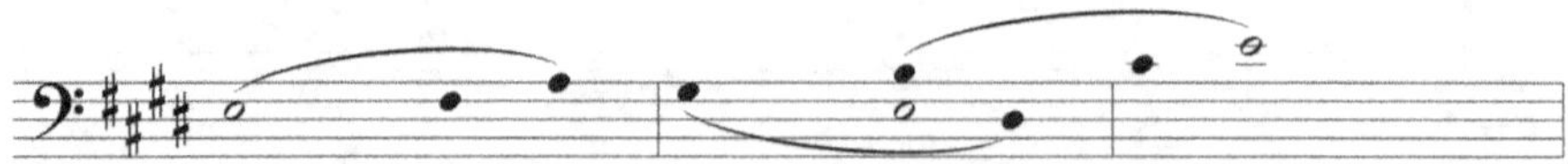

Fig. 7.2. Bach, E-major Fugue, mm. 1–3, Reduction to show co-incidence of $\mathbb{Z}_7$-translations of $\{0, 1, 3\}$ in the pitch class E $= 0$.

In addition to those found in cyclic groups, certain non-abelian interval groups also contain all-directed-interval chords.

Theorem 6. *Let D be a cyclic difference set with $\lambda = 1$ and $k \equiv 2$ (mod 3). Then there exists a non-abelian difference set with the same parameters. The associated projective planes are isomorphic* [7, *Theorem* 18.68].

The smallest non-abelian interval group in which we find an all-directed-interval chord occurs with the parameters $\lambda = 1$ and $k = 5$; hence, $v = 4^2 + 4 + 1 = 21$. Therefore, in addition to the all-directed-interval pentachords found in the cyclic group $\mathbb{Z}_{21}$, we find them also in the non-abelian group of order 21: $\mathbb{Z}_7 \rtimes \mathbb{Z}_3$.

Figure 7.3 shows the continuation (overlapping with Fig. 7.1) of the E-major fugue's exposition. An all-directed-interval pentachord occurs in the harmonic intervals of this passage, between the head motive of the subject's answer and its counterpoint. Let S be the set of two-element subsets of the E-major diatonic collection, modeled as $\binom{\mathbb{Z}_7}{2}$ with E $= 0$. We note that S is of size $\binom{7}{2} = 21$. Further, let $G = \langle\, x, y \mid x^7 = y^3 = 1;\ yx = x^2 y \,\rangle \cong \mathbb{Z}_7 \rtimes \mathbb{Z}_3$ be a non-abelian group with a regular action on S, such that, for all $\{s, t\} \in S$,

$$\{s, t\} \cdot x = \{s + 1, t + 1\} \ (\mathrm{mod}\ 7)$$

and

$$\{s, t\} \cdot y = \{2s, 2t\} \ (\mathrm{mod}\ 7).$$

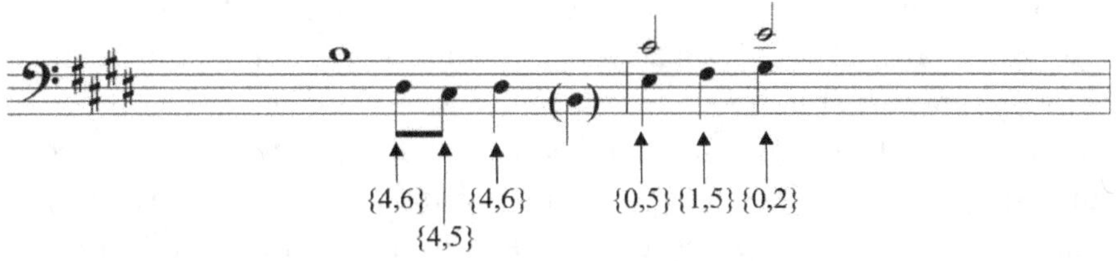

Fig. 7.3. Bach, E-major Fugue, mm. 2–3, showing the non-abelian all-directed-interval pentachord $D = \{\{0, 2\}, \{0, 5\}, \{1, 5\}, \{4, 5\}, \{4, 6\}\}$.

Table 7.1. $D = \{\{0,2\}, \{0,5\}, \{1,5\}, \{4,5\}, \{4,6\}\}$ as a non-abelian all-directed-interval pentachord.

Non-identity directed interval in G	occurrence in $D \times D$
$\{s,t\} \cdot y^1 x^1 = \{2s+1, 2t+1\} \pmod 7$	$(\{0,2\}, \{1,5\})$
$\{s,t\} \cdot y^1 x^2 = \{2s+2, 2t+2\} \pmod 7$	$(\{1,5\}, \{4,5\})$
$\{s,t\} \cdot y^1 x^3 = \{2s+3, 2t+3\} \pmod 7$	$(\{4,5\}, \{4,6\})$
$\{s,t\} \cdot y^1 x^4 = \{2s+4, 2t+4\} \pmod 7$	$(\{4,5\}, \{0,5\})$
$\{s,t\} \cdot y^1 x^5 = \{2s+5, 2t+5\} \pmod 7$	$(\{0,5\}, \{1,5\})$
$\{s,t\} \cdot y^1 x^6 = \{2s+6, 2t+6\} \pmod 7$	$(\{4,5\}, \{0,2\})$
$\{s,t\} \cdot y^1 x^7 = \{2s+0, 2t+0\} \pmod 7$	$(\{4,6\}, \{1,5\})$
$\{s,t\} \cdot y^2 x^1 = \{4s+1, 4t+1\} \pmod 7$	$(\{1,5\}, \{0,5\})$
$\{s,t\} \cdot y^2 x^2 = \{4s+2, 4t+2\} \pmod 7$	$(\{4,6\}, \{4,5\})$
$\{s,t\} \cdot y^2 x^3 = \{4s+3, 4t+3\} \pmod 7$	$(\{1,5\}, \{0,2\})$
$\{s,t\} \cdot y^2 x^4 = \{4s+4, 4t+4\} \pmod 7$	$(\{0,2\}, \{4,5\})$
$\{s,t\} \cdot y^2 x^5 = \{4s+5, 4t+5\} \pmod 7$	$(\{0,5\}, \{4,5\})$
$\{s,t\} \cdot y^2 x^6 = \{4s+6, 4t+6\} \pmod 7$	$(\{4,5\}, \{1,5\})$
$\{s,t\} \cdot y^2 x^7 = \{4s+0, 4t+0\} \pmod 7$	$(\{1,5\}, \{4,6\})$
$\{s,t\} \cdot y^3 x^1 = \{s+1, t+1\} \pmod 7$	$(\{4,6\}, \{0,5\})$
$\{s,t\} \cdot y^3 x^2 = \{s+2, t+2\} \pmod 7$	$(\{0,5\}, \{0,2\})$
$\{s,t\} \cdot y^3 x^3 = \{s+3, t+3\} \pmod 7$	$(\{4,6\}, \{0,2\})$
$\{s,t\} \cdot y^3 x^4 = \{s+4, t+4\} \pmod 7$	$(\{0,2\}, \{4,6\})$
$\{s,t\} \cdot y^3 x^5 = \{s+5, t+5\} \pmod 7$	$(\{0,2\}, \{0,5\})$
$\{s,t\} \cdot y^3 x^6 = \{s+6, t+6\} \pmod 7$	$(\{0,5\}, \{4,6\})$

Then, the set $D = \{\{0,2\}, \{0,5\}, \{1,5\}, \{4,5\}, \{4,6\}\}$, as shown in the figure, is an all-directed-interval pentachord.[b] Table 7.1 demonstrates the singular occurrence in $D \times D$ for each non-identity directed interval in G.

7.4.2. *All-interval chords and non-planar difference sets*

A non-planar difference set has $\lambda \geq 2$. Correspondingly, certain all-interval chords have a flat distribution of more than one occurrence of each directed interval in their respective interval groups. However, as we demonstrate below, the condition that an all-interval chord contains only one occurrence of each interval class in its interval group places a restriction on potential values for λ: in contrast to all-interval

---

[b]We do not consider the octave Bs at the end of m. 2, as $\{s,s\} \notin \binom{S}{2}$.

chords that are planar difference sets, in which no non-identity elements of the interval group are involutions, all-interval chords that are non-planar difference sets require every non-identity element of the group to be an involution; hence, $\lambda = 2$. Accordingly, we call such structures *twice*-all-directed-interval chords.

Definition 11. A **twice-all-directed-interval chord** is an all-interval chord with $\lambda = 2$.

Theorem 7. *If an all-interval chord D is a non-planar difference set, then G is an elementary abelian 2-group.*

Proof. D must meet two requirements. First, as a non-planar difference set, there need to be $\lambda \geq 2$ occurrences in $D \times D$ of each non-unison directed interval in G. Second, as an all-interval chord, there exists a single occurrence in $\binom{D}{2}$ of each interval class in G.

Put $s, t \in D$ and $g \in G$, such that $s \cdot g = t$. Then, $(s, t) \in D \times D$ is an occurrence of the directed interval g, and $\{s, t\} \in \binom{D}{2}$ is an occurrence of the interval class $[g]$. Now $\lambda \geq 2$ implies that there exists at least one other, distinct occurrence of g. Therefore, there must be some $(s', t') \in S \times S$, where $s' \cdot g = t'$, and for which $(s', t') \neq (s, t)$. However, $\{s', t'\}$ also represents an additional occurrence of $[g]$ unless $\{s', t'\} = \{s, t\}$, where $s' = t$, $t' = s$ and $t \cdot g = s$ (i.e., if $g = g^{-1}$).

To satisfy $\lambda \geq 2$, all the non-identity elements of G must then be involutions, which puts $\lambda = 2$, and G is an elementary abelian 2-group. $\qquad\square$

It is significant to note that not all difference sets with $\lambda = 2$ are twice-all-directed-interval chords. For instance, the $(7, 4, 2)$ difference set $D = \{0, 1, 2, 4\}_7$ is not a twice-all-directed-interval chord. Each of the interval classes in $\mathbb{Z}_7$ has two (not one) occurrences in $\binom{\mathbb{Z}_7}{2}$, as shown above in (7.2) by the equivalences among pairs of non-reciprocal differences (i.e., those with $s - t \neq t - s$). Further, not all elementary abelian 2-groups contain twice-all-directed-interval chords. Per Remark 1, G must contain a triangular number Δ of interval classes. As every non-identity element of $\mathbb{Z}_2^m$ is an involution,

Fig. 7.4. The set of tone rows used in Schoenberg, *Klavierstück*, Op. 33b, mm. 1–10.

G has a Mersenne number $|G^{\#}| = 2^m - 1$ of interval classes. Then, G can contain all-interval chords only if Δ is also a Mersenne number.

Other than $\mathbb{Z}_2$, for which $2 - 1$ is equal to the trivial triangular number $\binom{2}{2}$, the smallest group to contain a non-planar all-interval chord is the Klein 4-group, $\mathbb{Z}_2^2$, as $\binom{4}{2} = 2^2 - 1 = 3$. An example of a twice-all-directed-interval chord within this group is found in the opening of Arnold Schoenberg's *Klavierstück*, Op. 33b. The passage in mm. 1–10 contains four forms of a 12-tone row: P_{11} (the prime form of the row that begins on pitch-class 11), I_4 (a hexachordally combinatorial inversion of P_{11}), RP_{11} (the retrograde of P_{11}), and RI_4 (the retrograde of I_4) (see Fig. 7.4).[c] The group $G = \{T_0, I_3, RT_0, RI_3\}$ of pitch-class and order-position operators — which is isomorphic to $\mathbb{Z}_2^2$ — has a regular action on this set, forming a non-cyclic abelian GIS.

Schoenberg begins the passage with the work's principal theme (mm. 1–5), a single phrase that presents consecutively the P_{11} and the I_4 forms of the row (see Fig. 7.5). These two row forms are related to one another by the operator I_3, which itself assumes thematic significance and receives subsequent development in the work. The next phrase (mm. 5–10) begins a transition (Fig. 7.6). This phrase combines the retrogrades of the previous two row forms, but instead of placing them in succession, it presents them simultaneously (RI_4 in the right hand, and RP_{11} in the left). This arrangement develops the

---

[c]In fact, Op. 33b uses only these four row forms throughout.

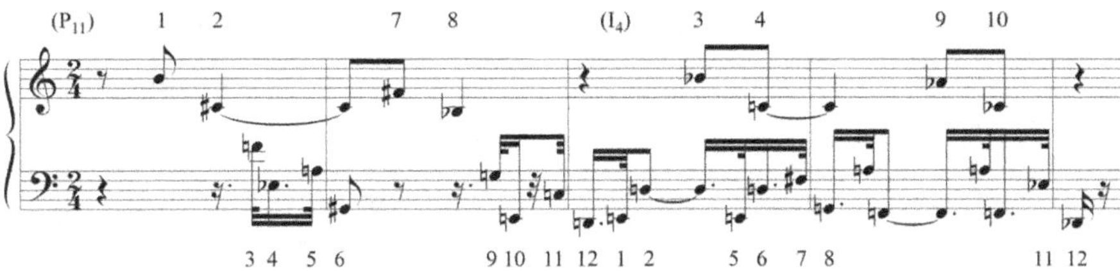

Fig. 7.5. Schoenberg, Op. 33b, mm. 1–5 (with row labels and order-position numbers).

Fig. 7.6. Schoenberg, Op. 33b, mm. 5–10 (with row labels and order-position numbers).

first phrase's inversional relationship in two ways. First, the inversion operator I_3 that relates the consecutive segments of the first phrase to one another *in time* is the same operator that relates simultaneous segments of the second phrase *in register*. Second, the *pitch-class* inversion between the two segments of the first phrase becomes an *order-number* inversion (retrograde) between the respective rows of the first and second phrases.

Figure 7.7 displays a graph of these row forms and the relationships among them. It demonstrates that the set $D = \{I_4, RP_{11}, RI_4\}$, consisting of the passage's three final row forms, is an all-interval trichord.[d] By virtue of their being involutions, everyone of the non-identity elements of G has two occurrences in D; hence, $\lambda = 2$, and D is a twice-all-directed-interval chord. Further, because the

---

[d]In fact, any three-element subset of the four row forms used in this passage constitutes an all-interval trichord. We select this particular trichord, as it contains the row forms that develop the initial I_3 relationship.

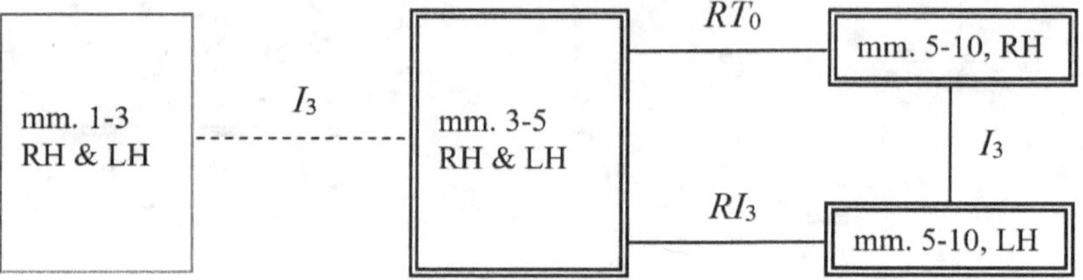

Fig. 7.7. Row form relations in Op. 33b, mm. 1–10.

Table 7.2. $D = \{I_4, RP_{11}, RI_4\}$ as a non-planar all-directed-interval trichord.

Non-identity directed interval in G	Reciprocal occurrences in $D \times D$
I_3	(RP_{11}, RI_4), (RI_4, RP_{11})
RT_0	(I_4, RI_4), (RI_4, I_4)
RI_3	(I_4, RI_{11}), (RI_{11}, I_4)

Table 7.3. Babbitt, instrumentation in *Composition for Four Instruments* ($0 =$ tacet, $1 =$ playing).

Section	1	2	3	4	5	6	7	8	9	10	11	12	13	14	15
Flute	0	1	0	1	1	0	1	0	1	0	0	1	1	0	1
Clarinet	1	0	1	0	1	0	0	1	1	0	0	1	0	1	1
Violin	0	1	0	1	1	0	0	1	0	1	1	0	0	1	1
Cello	0	1	1	0	0	1	1	0	1	0	1	0	0	1	1

occurrences in each pair are reciprocal, any such pair represents only one occurrence of an interval class in G. (See Table 7.2.)

A larger group that contains non-planar all-interval chords is $\mathbb{Z}_2^4$, which has $\binom{6}{2} = 2^4 - 1 = 15$ interval classes. We find examples of all-interval hexachords that incorporate this group in Milton Babbitt's *Composition for Four Instruments*. The piece is comprised of fifteen sections, each of which features a different combination of the four instruments, and every non-empty subset of the ensemble is used only once. Moreover, the sections are arranged so that the instrumentation of each odd-numbered section is the complement of the subsequent even-numbered one. Table 7.3 presents a model of this scheme, wherein 0 signifies that an instrument is tacet in the section, while 1 signifies that it plays.

Let S be the set of 4-tuples (s_1, s_2, s_3, s_4), $s_i \in \mathbb{Z}_2$ that represent the respective instrumentations of the sections. The first coordinate in any such 4-tuple indicates whether the flute is tacet or playing in the section; the second coordinate, the clarinet; the third, the violin; and the fourth, the cello. The group $G = \mathbb{Z}_2^4$ has a natural action on S. The interval in G between any two 4-tuples $(s_1, s_2, s_3, s_4), (t_1, t_2, t_3, t_4)$ is the element $g = (s_1 - t_1, s_2 - t_2, s_3 - t_3, s_4 - t_4)$ (mod 2), forming a GIS.[e] Then, the set $D_1 = \{(0, 1, 0, 0), (1, 0, 1, 1), (0, 1, 0, 1), (1, 1, 1, 0), (1, 0, 0, 1), (1, 1, 0, 1)\}$, consisting of the instrumentations of sections 1, 2, 3, 5, 7, and 9, is a non-planar all-interval hexachord (twice-all-directed-interval), as demonstrated by the equivalences of the following differences.

$$
\begin{aligned}
0101 - 1101 &= 1101 - 0101 = 1000 \\
1001 - 1101 &= 1101 - 1001 = 0100 \\
0101 - 1001 &= 1001 - 0101 = 1100 \\
1011 - 1001 &= 1001 - 1011 = 0010 \\
1011 - 1101 &= 1101 - 1011 = 0110 \\
0100 - 1110 &= 1110 - 0100 = 1010 \\
1011 - 0101 &= 0101 - 1011 = 1110 \\
0100 - 0101 &= 0101 - 0100 = 0001 \\
1110 - 1101 &= 1101 - 1110 = 0011 \\
1011 - 1110 &= 1110 - 1011 = 0101 \\
1110 - 1001 &= 1001 - 1110 = 0111 \\
0100 - 1101 &= 1101 - 0100 = 1001 \\
0101 - 1110 &= 1110 - 0101 = 1011 \\
0100 - 1001 &= 1001 - 0100 = 1101 \\
0100 - 1011 &= 1011 - 0100 = 1111.
\end{aligned}
\tag{7.6}
$$

To be all-interval, any such hexachord D_i needs to incorporate one pair of complementary instrumentations, as the interval class that includes the interval $(1, 1, 1, 1)$ must have one occurrence in $\binom{D_i}{2}$.

---

[e]Lewin explores this space and group as a GIS in [9].

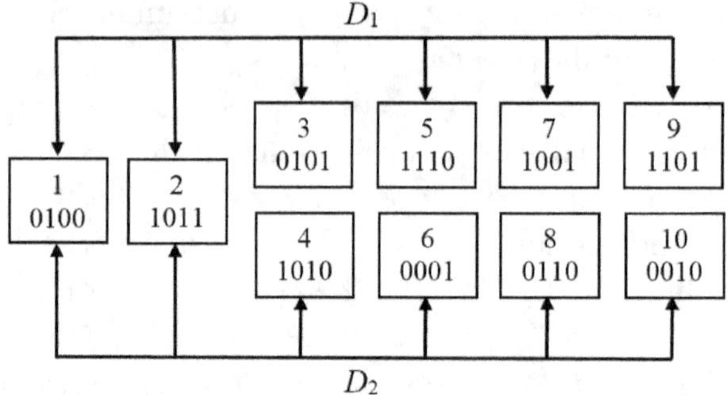

Fig. 7.8. All-interval hexachords with complementary instrumentations in *Composition for Four Instruments*.

Hence, we will find in D_i one odd-order-number section that is followed by the subsequent even-order-number section. However, D_i cannot contain more than one pair of sections with complementary instrumentations, as the interval class that contains $(1, 1, 1, 1)$ can have only one occurrence in the chord. Therefore, the remaining order numbers of the hexachord's sections cannot contain another odd-number/subsequent-even-number pair. For instance, D_1 above possesses only two sections, 1 and 2, that have complementary instrumentations; and the remaining sections, 3, 5, 7, and 9, are all odd parity. Further, if a hexachord D_i is all-interval in this particular GIS, then the hexachord that is comprised of the instrumental complements of D_i is also all-interval. Figure 7.8 demonstrates this relationship for two all-interval hexachords, D_1 and D_2, in the piece.

7.5.   Conclusions

The study of all-directed-interval chords adds to the theory of all-interval chords in several useful ways. First, the incidence structure of planar difference sets relates to musical invariants, which are of considerable theoretical interest [1]. As a block design, the set of translations of an all-directed-interval chords possesses a consistent pattern of invariants, making possible certain musical-growth processes. Second, all-directed-interval chords are more intervallically efficient than all-interval chords that are not all-directed-interval. The latter include more than one occurrence of directed intervals that are involutions, such as the two occurrences of -6 (mod 12) in the traditional all-interval tetrachords of pitch-class set theory. Such efficiency is useful to composers and music analysts in providing a complete set of intervals in as small a space as possible. Finally, the theory of all-directed-interval chords presented here enables us to explore the musical aspects of these structures in relation to existing mathematical results, thereby furthering the interdisciplinary connections between mathematics and music.

Bibliography

[1] M. Babbitt, Twelve-tone invariants as compositional determinants, *Musical Quart.* **46** (1960) 246–259.

[2] L. D. Baumert and D. M. Gordon, On the existence of cyclic difference sets with small parameters, in *High Primes and Misdemeanours*, eds. A. van der Poorten and A. Stein (American Mathematical Society, 2004), pp. 61–68.

[3] A. Forte, *The Structure of Atonal Music* (Yale University Press, 1973).

[4] C. Gamer and R. Wilson, Microtones and projective planes, in *Music and Mathematics*, eds. J. Fauvel, R. Flood and R. Wilson (Oxford University Press, 2003), pp. 149–162.

[5] M. Hall Jr, *Combinatorial Theory* (Blaisdell Publ. Co., 1967).

[6] I. Isaacs, *Character Theory of Finite Groups* (AMS Chelsea Publishing, 1976).

[7] D. Jungnickel, A. Pott and K. Smith, Difference sets, in *Handbook of Combinatorial Designs*, 2nd edn. eds. C. Colburn and J. Dinitz (Chapman & Hall/CRC, 2007), pp. 419–435.

[8] D. Lewin, The intervallic content of a collection of notes, intervallic relations between a collection of notes and its complement: an application to Schoenberg's hexachordal pieces, *J. Music Theory* **4** (1960) 98–101.

[9] D. Lewin, Generalized interval systems for Babbitt's lists, and for Schoenberg's string trio, *Music Theory Spectrum* **17** (1995) 81–118.

[10] D. Lewin, *Generalized Musical Intervals and Transformations* (Oxford University Press, 2007).

[11] G. Perle, *Twelve-Tone Tonality*, 2nd edn. (University of California Press, 1996).

[12] J. Rahn, *Basic Atonal Theory* (Longman, Inc., 1980).

[13] J. Singer, A theorem in finite projective geometry and some applications to number theory, *Trans. Amer. Math. Soc.* **43** (1938) 377–385.

Chapter 8

Harmonious Opposition

Richard Plotkin

*University at Buffalo, The State University of New York,
New York, USA*

8.1. Introduction

This chapter works to resolve a deceivingly simple question: "What is the opposite of a parsimonious transformation?" Something un-parsimonious is, by definition, not parsimonious—but is it in opposition? And even if we say that the opposite of a parsimonious transformation is a transformation of maximal difference, by what criteria is that difference to be judged?

8.1.1. *A historical framework for parsimony*

Let us begin by historically situating the term *parsimonious transformation*. *Parsimonious voice-leading* was introduced into contemporary music theory by Richard Cohn [4], with Cohn citing a historical backing for the term in the work of Schoenberg (the "law of the shortest way") and Hotinský (using "parsimony"). According to Cohn, the defining feature of parsimonious voice-leading (in a triad) is double common-tone retention. Additionally, Cohn observes that only sc(037) maintains parsimonious voice-leading[a] under the

---

[a] *Voice-leading* is a bit of a misnomer. The objects being examined are set classes under octave equivalence, and the voice-leading aspect of parsimony assumes reordering of pitch-classes for the most efficient voice leading at any moment.

PLR-family operations. With emphasis on PLR, Cohn's discussion centers on parsimonious transformations where triadic set class is preserved, and where two common-tones are retained and the third tone changes by ic1 or ic2.

Adrian Childs [1] follows Cohn's work with an approach to parsimonious seventh-chord transformations on sc(0258). As in [4], set class is preserved in a parsimonious transformation, though in Childs' approach, a total of two pcs change in every parsimonious transformation. Jack Douthett and Peter Steinbach [7] observe that, in the case of both Cohn and Childs, two common tones remain fixed, and the other tones move by no more than ic2. Following from this, Douthett and Steinbach propose a relation $P_{m,n}$, where m represents the quantity of pitch-classes changing by ic1, and n represents the quantity of pitch-classes changing by ic2. They can then define the parsimonious transformations from Cohn's study as $P_{1,0}$ and $P_{0,1}$, and the parsimonious transformations from Childs' study as $P_{2,0}$.

Douthett and Steinbach consider symmetry in the chromatic universe (U_{12}) as another important aspect of parsimony, implying that such symmetry contributes to set class preservation:

> Parsimonious structures (e.g., *Tonnetze*) generally have high degrees of symmetry and are independent of tonal centers... With this new emphasis on symmetry, it is not surprising to discover that pitch-class collections from [the hexatonic, octatonic, and enneatonic set classes] — sets abundant in their degrees of symmetry — are associated with parsimony...

Although Cohn only stipulated double-common-tone retention as the defining feature of parsimony, the historical position on parsimonious transformations seems to be that parsimony is not only defined by a maximum number of retained tones (provided as a value of two for both triads and seventh chords), but is also *strongly* guided by set-class preservation and range-limited motion of ic1 and ic2.

8.1.2. *Scales, not set classes*

A problem with the historical approach to parsimonious transformations is that it leaves out some transformations that we musically intuit as parsimonious. Consider a G+ harmony.[b] If you expect G+ to be followed by a parsimoniously related triad, you might list the following possibilities: *P* (yielding G−), *L* (yielding B−), and *R* (yielding E−). Absent from your list would be B°, for although this harmony preserves two common tones, it does not preserve set class. However, if you were to consider this same G+ harmony in the context of the key of C-major, it would seem odd to exclude B°, since B° and E− are the only two chords within the key that share two common tones with G+. To allow theoretical space for our diatonically-oriented intuition, we may expand our idea of parsimony such that, when a scalar context (such as C-major) is invoked, the historical expectation of set-class preservation is replaced by a new expectation where related chords will preserve a distributional pattern within a particular scale.

Returning to the initial question posed, this chapter posits that the opposite of a parsimonious transformation can be determined when that transformation is considered within a scalar context. Section 8.2 expands and clarifies the idea of transformations within scalar contexts, as defined by iterated maximally even distributions. Douthett and Steinbach reference high degrees of symmetry as an enabling aspect of parsimony, and iterated maximally even distributions allow us to have near-symmetry spread across multiple levels, instead of confining parsimony to near-even sets in U_{12}. Section 8.3 lays out a plan for finding the polar opposite of a parsimonious transformation, and provides analyses of two passages that make use of opposing harmonies and scales.

8.2. Parsimonious Transformations in Scalar Contexts

The algebraic determination of maximal evenness was originally presented by John Clough and Jack Douthett [2], and is expanded

---

[b]Throughout this chapter, a consistent orthography is used for identifying keys and triads. Keys are always designated as "{Letter}-major" or "{Letter}-minor", with *major* and *minor* fully spelled. Triads are always designated as "{Letter}+" for major, "{Letter}−" for minor, and "{Letter}°" for diminished.

by Douthett [6] using a geometric calculation called Filtered Point-Symmetry (FiPS). Plotkin [8] further explores FiPS as a means of examining transformations in *Tonnetz*-like networks generated through FiPS configurations, and Plotkin and Douthett [9] show the isography of some of these configuration spaces to the parsimonious graphs of Douthett and Steinbach [7]. This section presents the relevant background information, followed by some considerations for transformations within a scalar context.

8.2.1. *Maximally even distributions*

A maximally even distribution is determined by

$$J_{c,d}^{m}(k) = \left\lfloor \frac{ck + m}{d} \right\rfloor, \quad 0 \le k < d, \tag{8.1}$$

where c is the cardinality of the set from which a maximally even distribution will be derived, and d is the desired cardinality of the maximally even subset. m is an offset that alters the specific elements of the determined subset, while preserving its distributional pattern. Examples of maximally even distributions in the chromatic universe include the pentatonic, whole tone, diatonic, octatonic, and enneatonic collections.

A maximally even subset can be derived from any superset, regardless of the superset's evenness. When the superset is maximally even, the resulting subset is said to be an *iterated maximally even distribution*. The equation for determining an iterated maximally even distribution is

$$J_{d_0,d_1,d_2,\ldots,d_n}^{m_1,m_2,\ldots,m_n}(k) = J_{d_0,d_1}^{m_1}\left(J_{d_1,d_2}^{m_2}\left(J_{d_2,d_3}^{m_3}\left(\ldots J_{d_{n-1},d_n}^{m_n}(k)\right)\right)\right) \tag{8.2}$$

where any pair d_{n-1}/d_n is in the same relationship as c and d above: d_n is the cardinality of the subset and d_{n-1} is the cardinality of the superset.

Geometrically, FiPS portrays a maximally even distribution as two concentric rings with a point-symmetric distribution of nodes around the ring, corresponding to cardinality c (or d_{n-1}) for the exterior ring and cardinality d (or d_n) for the interior ring. Beams connect nodes from the interior ring to the exterior ring, with each

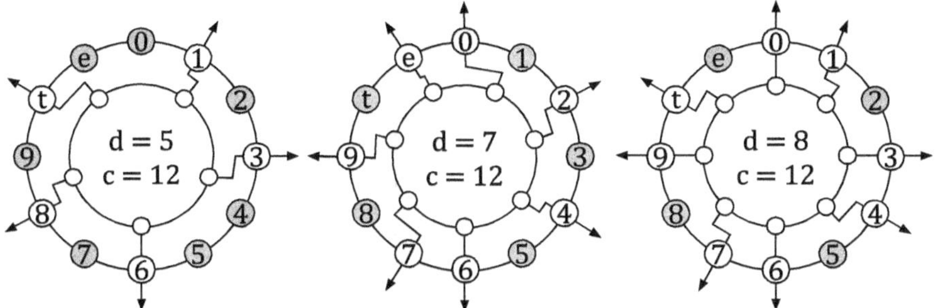

Fig. 8.1. Filtered point-symmetry showing the calculation of three maximally even distributions of varying cardinality within the chromatic universe of cardinality 12.

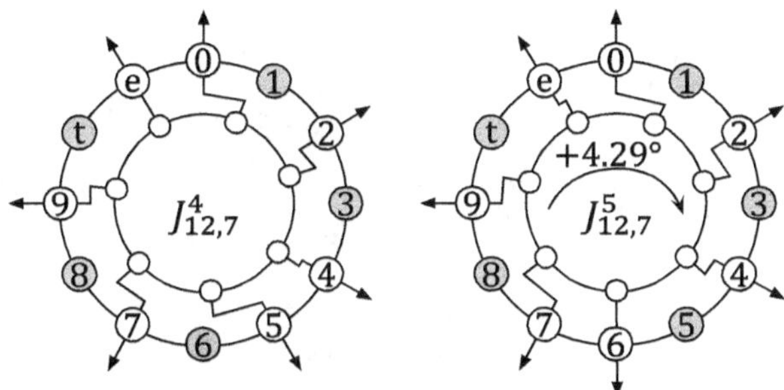

Fig. 8.2. A clockwise rotation of an interior ring, equivalent to an increment of m from 4 to 5 in the J function (where one m is equal, in degrees, to $360/c * d$).

beam extending normal to the interior ring, and traveling counter-clockwise until the beam arrives at a hole on the exterior ring. The counter-clockwise connection of the beams is the geometric equivalent to the floor function applied in Eq. (8.1). Figure 8.1 shows the FiPS calculation of pentatonic, diatonic, and octatonic collections. The rotational offset of a ring corresponds to the value of m, as shown in Fig. 8.2. Figure 8.3 shows an iterated maximally even distribution of $3 \rightarrow 7 \rightarrow 12$, which will always give a major, minor, or diminished triad in a specific diatonic key.

8.2.2. *Contextualized triads*

Major/minor triads appear as maximally even distributions in the pentatonic, diatonic, and octatonic collections (which makes each

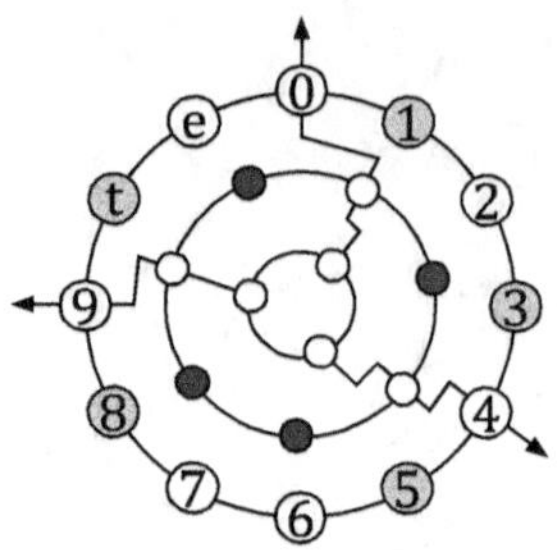

Fig. 8.3. A maximally even distribution of $3 \to 7 \to 12$.

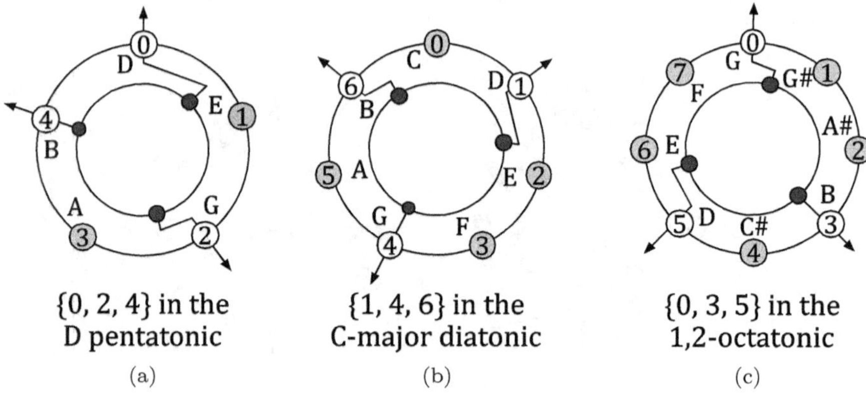

{0, 2, 4} in the
D pentatonic

(a)

{1, 4, 6} in the
C-major diatonic

(b)

{0, 3, 5} in the
1,2-octatonic

(c)

Fig. 8.4. Three maximally even expressions of a G+ harmony: as a subset of
the D-pentatonic, as a subset of the C-major diatonic, and as a subset of the
1,2-octatonic.

triad an *iterated* maximally even distribution of $3 \to n \to 12$).
Figure 8.4 shows a G+ triad in each collection. Despite the identical
pitch-class content, the scale-degree distribution of the pitch-classes
is distinct in each case.

In Fig. 8.4(c), there are six chords (shown in Table 8.1) sharing
two common tones with the G+ harmony in which the altered
pc moves as little as possible, by one increment clockwise (cw)
or counter-clockwise (ccw). Of the six chords in Table 8.1, only
G− and E− remain maximally even. In fact, any incremental move
from one maximally even chord to another within an octatonic
collection will be via P or L, and will consistently satisfy all of the
historical conditions for voice-leading parsimony.

Table 8.1. The harmonies in an octatonic collection that preserve two common tones with a given maximally even (MaxE) distribution (the G+ triad, $\{0, 3, 5\}$ mod 8).

Altered pc	Direction	Chord mod$_8$	Chord U$_{12}$	MaxE
G	cw	$\{\mathbf{1}, 3, 5\}$	$\{G\#, B, D\}$	n
G	ccw	$\{3, 5, \mathbf{7}\}$	$\{B, D, F\}$	n
B	cw	$\{0, \mathbf{4}, 5\}$	$\{G, C\#, D\}$	Y
B	ccw	$\{0, \mathbf{2}, 5\}$	$\{G, Bb, D\}$	n
D	cw	$\{0, 4, \mathbf{6}\}$	$\{G, B, C\#\}$	n
D	ccw	$\{\mathbf{2}, 4, 6\}$	$\{G, B, E\}$	Y

Table 8.2. The harmonies in a diatonic collection that preserve two common tones with a given maximally even (MaxE) distribution (the G+ triad, $\{1, 4, 6\}$ mod 7).

Altered pc	Direction	Chord mod$_7$	Chord U$_{12}$	MaxE
G	cw	$\{1, 5, 6\}$	$\{D, A, B\}$	n
G	ccw	$\{1, 3, 6\}$	$\{D, F, B\}$	Y
B	cw	$\{0, 1, 4\}$	$\{C, D, G\}$	n
B	ccw	$\{1, 4, 5\}$	$\{D, G, A\}$	n
D	cw	$\{0, 4, 6\}$	$\{C, G, B\}$	n
D	ccw	$\{2, 4, 6\}$	$\{E, G, B\}$	Y

Like Table 8.1, Tables 8.2 and 8.3 provide common-tone chord information for C-major (Fig. 8.4(b)) and D-pentatonic (Fig. 8.4(a)), respectively. For each collection, as it is with the octatonic, only two of the harmonies sharing common tones with G+ maintain a maximally even distribution.

The two evenness-preserving transformations within the octatonic collection preserved set class, but in both the diatonic and pentatonic, only $C+ \rightarrow E-$ preserves set class. The other evenness-preserving transformation in the pentatonic is $G+ \rightarrow Dsus4$, and in the diatonic it is the $G+ \rightarrow B°$ transformation problematized in Sec. 8.1.2. Expanding on the thoughts of 8.1.2, where it was proposed that parsimony is not guided by set-class preservation, but is instead guided by the expectation that related chords will preserve

Table 8.3. The harmonies in a pentatonic collection that preserve two common tones with a given maximally even (MaxE) distribution (the G+ triad, $\{0, 2, 4\}$ mod 5).

Altered pc	Direction	Chord mod_5	Chord U_{12}	MaxE
G	cw	$\{0, 3, 4\}$	$\{D, A, B\}$	n
G	ccw	$\{0, 1, 4\}$	$\{D, E, B\}$	n
B	cw	— *	—	—
B	ccw	$\{0, 2, 3\}$	$\{D, G, A\}$	Y
D	cw	$\{1, 2, 4\}$	$\{E, G, B\}$	Y
D	ccw	— *	—	—

*A change in set cardinality, which we are not considering for parsimony.

a distributional pattern, let us hypothesize that the preserved distributional pattern must be maximally even within a given scalar context. In the case of the octatonic, any maximally even trichord will always belong to sc(0,3,7), and any stepwise motion will be ic1 or ic2.[c]

If we consider as parsimonious the C-major diatonic transformation G+ $\rightarrow$ B° because evenness (within C-major) is preserved, we might notice that the historical guideline of stepwise motion by ic1 or ic2 still holds. This also seems to be the case for the D pentatonic transformation G+ $\rightarrow$ Dsus4... but the next evenness-preserving transformation in the D pentatonic would be from Dsus4 $\rightarrow$ Asus4, which involves the pitch-class G moving by ic3 to pitch-class E. Although this exceeds ic2, it seems a reasonable parsimonious move *due to the construction of the pentatonic scale*. Instead of the ic2 guideline in the historical approach to parsimony, we choose appropriate step size as the *generic interval of 1 in a given scalar context*.

The next section, and the theory this chapter pursues, are based on the assumed viability of this hypothesis regarding parsimony. At its core, the revised approach does not deviate from Cohn's initial

[c]In other words, the parsimonious transformation of 3-note maximally even sets in an octatonic scalar context will precisely correlate to the P and L transformations of neo-Riemannian theory. Furthermore, any change of octatonic collection (to 01-oct or 02-oct) will introduce the R transformation, as shown in the isography of a FiPS configuration space with the LPR chicken-wire torus [9].

stipulation that triadic parsimonious transformations preserve two common tones in a triad, nor does it deviate from the recognition that additional guidelines are necessary for finding musically meaningful transformations. Rather, this approach embraces new guidelines that include traditional neo-Riemannian theories while also incorporating scales.

8.3.   Scalar Polarity

8.3.1.   *Intra-collection cyclic groups*

Given a particular scalar context, a cyclic group of an order less than or equal to the cardinality of the elements of the scale can be arranged as a series of parsimonious transformations. Figures 8.5 and 8.6 show specific and generic chords in intra-collection pentatonic and diatonic cycles (intra-collection octatonic cycles are neo-Riemannian *PL* cycles). By "intra-collection", I mean chord cycles that do not leave a specific representation of a collection, e.g., "all the chords within C-major" or "all the chords within the D pentatonic". By generic chords, I mean chords represented by an ordered pair ⟨*scale degree, chrod quality*⟩, such that C+ in C-major is represented as ⟨$\hat{1}$, +⟩, but C+ in G-major would be represented as ⟨$\hat{4}$, +⟩. Things are somewhat cumbersome with the sus2 and sus4 chords in the pentatonic; this chapter consistently uses sus4 to represent sc(027).

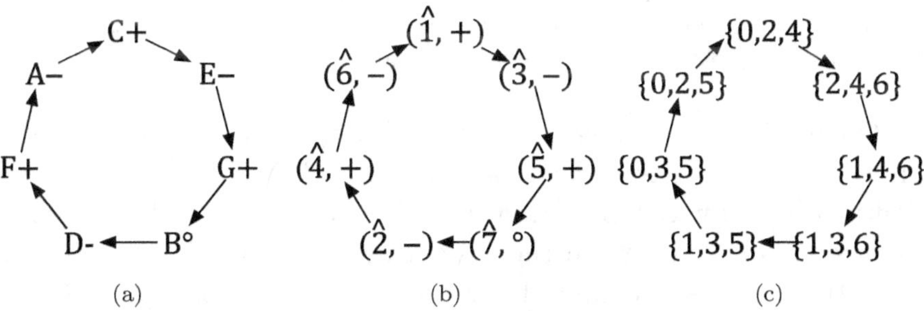

(a)   (b)   (c)

Fig. 8.5. Cyclic group Z_7 of parsimonious transformations within a specific diatonic key. (a) Parsimonious transformations of chords in the key of C-major. (b) Parsimonious transformations of chords in a generic diatonic key. (c) Parsimonious transformations of chords a generic diatonic key, as pcsets mod 7.

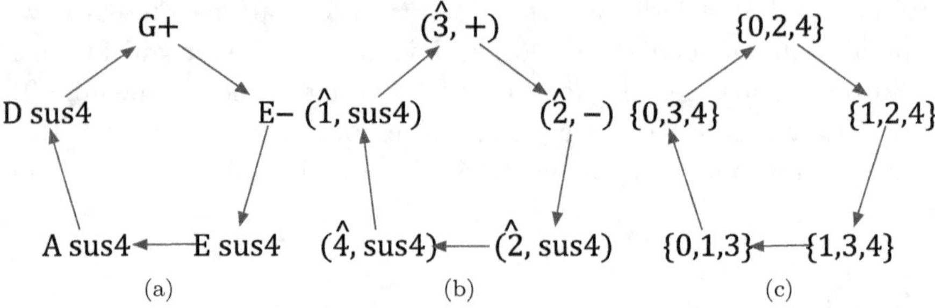

Fig. 8.6. Cyclic group $\mathbb{Z}_5$ of parsimonious transformations within a specific diatonic key. (a) Parsimonious transformations of chords in the D pentatonic. (b) Parsimonious transformations of chords in a generic pentatonic collection. (c) Parsimonious transformations of chords a generic pentatonic collection, as pcsets mod 5.

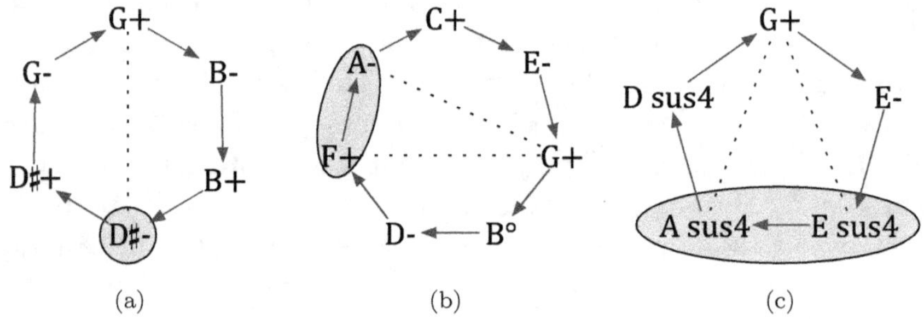

Fig. 8.7. Cyclic groups showing maximal displacement of same-cardinality trichords in a specific scalar context. (a) Parsimonious transformations of chords in the $\{1, 2\}$-octatonic, with the maximal displacement of G+ indicated as D#− (also known as a *hexatonic pole*). (b) Parsimonious transformations of chords in the C-major diatonic, with the maximal displacement of G+ indicated as either A− or F+. (c) Parsimonious transformations of chords in the D pentatonic, with the maximal displacement of G+ indicated as either Asus4 or Esus4.

Once these cyclic diagrams are made, it is trivial to identify the triads at the opposite ends of each diagram. Figure 8.7 makes these identifications for the G+ triad in each of the three scalar contexts under examination. While the visualization is straightforward, the theoretical leap is substantial: Fig. 8.7(a) shows a pairing of G#− and D#−. This pairing, called a *hexatonic pole* in neo-Riemannian literature, is first described in [3] and used historically and analytically in [5].

The existence of the hexatonic pole in Fig. 8.7(a) is something both the same as, and different from, the neo-Riemannian formulation, because in this figure that polarity arises from a harmonic distinction happening *because of* the specific nature of the $\{1, 2\}$-octatonic scalar context. That neither the G+ nor D#− triads occur in any of the other two octatonic collections is peculiar to the octatonic distributions. In the case of Fig. 8.7(b), which shows chords in the key of C-major, both F+ and A− are in equal polar opposition to G+.[d] If the key were altered to G-major, then G+ would be opposed by F#° and A−; and if the key were altered to D-major, then G+ would be opposed by F#− and A+. By establishing a scalar context for parsimonious transformations, we not only have a broader avenue for exploring parsimony, but also a means of defining the polar opposite of a parsimonious transformation.

8.3.2. *Intra-scale opposition in analysis*

The passage in Fig. 8.8, from Debussy's *Pour le piano*, "Sarabande," is a particularly dense selection of six tetrachords[e] in repeated succession. One way of partitioning the passage for analysis is by successively treating subsets from the series of chords as sets in polar opposition within a melodic minor context. Figure 8.8 identifies the B, A, and C# melodic minors for consecutive chords in the series.

Even though melodic minors are not maximally even distributions of seven in the chromatic universe, the tetrachords in use in this passage are maximally even within each melodic minor collection.

---

[d]It would be reasonable to make a conjecture that the polar opposite of G+ in C-major is the union of F+ and A− as a IV^7 chord, situated between the two triads. Such a move would reflect Hauptmann's construction of seventh chords. However, altering the cardinality of the originating set also seems to expand the discussion of polarity into something less stable — at least something that needs more exploration on its own. In the pentatonic, a union of Asus4 and Esus4 would yield a set that is *more* similar to G+, so it might make more sense to situate the symmetric difference of the sets in between. This would make at least some correspondence between a C: V $\rightarrow$ C: IV^7 polarity and a pentatonic unequal-cardinality polarity, in the sense that the polar opposite is the scalar complement.
[e]I hear the G# that initiates each measure in the soprano as a non-chord tone leading to the F#.

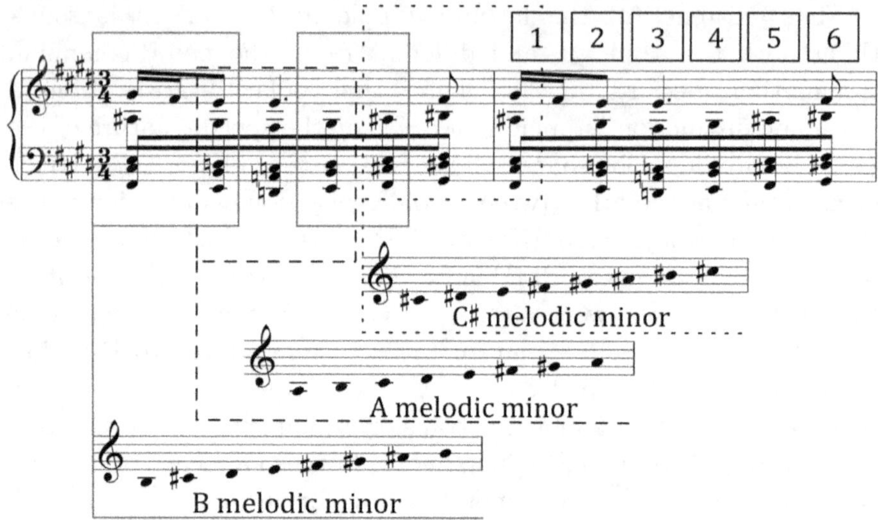

Fig. 8.8. Debussy *Pour le piano*, "Sarabande," with three melodic minor scales shown as supersets of chord sequences in the excerpt. The second measure, identical to the first, numbers the chords for reference in Fig. 8.10.

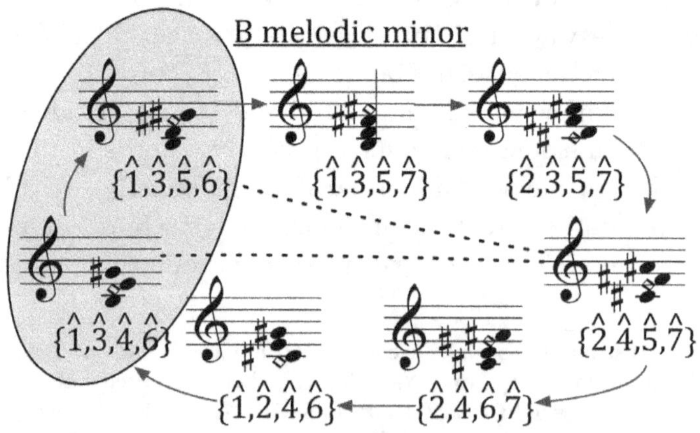

Fig. 8.9. A parsimonious tetrachord cycle, by way of maximally even distributions of scale degrees, in B melodic minor.

Figure 8.9 shows a parsimonious cycle of maximally even tetrachords within B melodic minor; the two chords in polar opposition to the Bm7 chord are attached by a dashed line and encased a shaded ellipse.

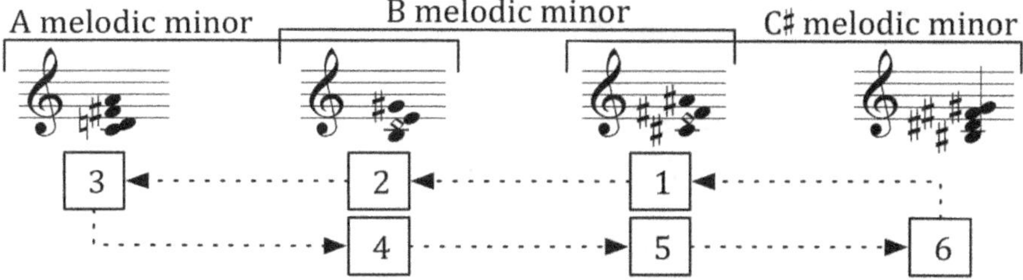

Fig. 8.10. A series of tetrachords in intra-scale polar opposition through three overlapping melodic minor scales. The numbers correspond to those provided in Fig. 8.8.

Figure 8.10 extracts the tetrachords from the excerpt, and continues to use dashed arrows to show tetrachords in polar opposition within the labeled scalar context. We see that this excerpt is a series of intra-scale oppositions occurring within three maximally similar scales. Any particular melodic minor scale shares harmonies with only two others scales — specifically, those related by whole step share 5 common tones and 2 harmonies. This passage, by using all available scales closely related to the B melodic minor, represents an exceptional positioning of continuity against opposition.[f]

8.3.3.   *Inter-scale opposition in analysis*

Harmonious opposition need not be contained *within* scales. Figure 8.11 gives the circle of fifths for major keys, with three keys in opposition to C-major. The astute reader will notice that G♭/F#-major is in exact polar alignment with C-major, and wonder why there are two additional keys in opposition. Since parsimony is

---

[f]The common tone relationship by whole step results in a 6-member cycle of closely-related melodic minor scales, and a separate T_1-related 6-member cycle. Curiously, all scales in the T_1-related cycle share four common tones with *any* of the scales in the other cycle. So, although within a 6-member cycle there is the familiar pattern of decreasing similarity from the starting element as one moves farther from it, any melodic minor scale related by an odd number of chromatic steps will have as many common tones as those items T_4/T_8 from the starting scale, and *more* common tones than the scale T_6 from the starting scale.

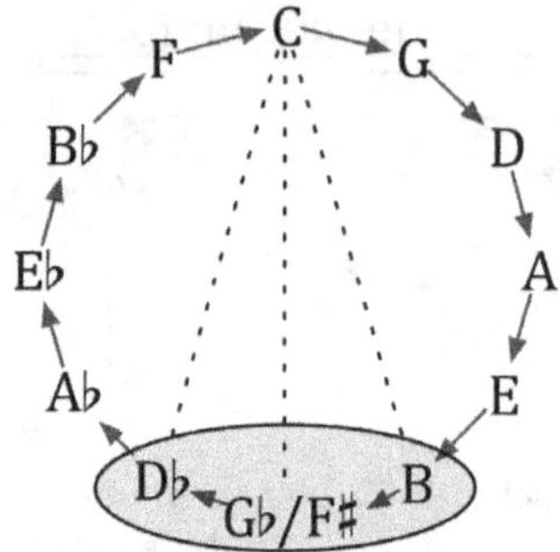

Fig. 8.11.   A parsimonious cycle of diatonic scales. The polar opposite of C-major is shown by dotted lines attached to the keys chromatically and tritone-related to C; this is called the *tritone halo*.

a measure of similarity, the extent of *dis*similarity is limited by the cardinality of the set being measured. In the case of the diatonic collection, the maximum dissimilarity of 2 common tones is achieved after five moves in either direction on the circle of fifths. Although the dissimilar notes are different, both chromatically related keys are as dissimilar as the tritone-related key. Because of this, I refer to the shaded area as the *tritone halo*, representing the full region of maximal dissimilarity.

Chopin's Prelude Op. 28, No. 17 in A♭-major has a very compelling, passionate, and heavily chromatic passage from mm. 18–36. Figure 8.12 analyzes the passage through a number of key areas, some of which are certainly audible, but many of which are suggested by dominants that do not bring with them a strong sense of key. The passage opens on the tonic of A♭-major, quickly tonicizes A-major — a key in the tritone halo of A♭-major, as indicated by the dashed arrow — and then moves back via the same maximal displacement to A♭. The next trip to A-major passes through E-major, where a climactic IV→V7 transformation is an *intra*-key polar opposition that, in its resolution, completes the large-scale inter-key opposing move from A♭-major to A-major (via the local parsimonious transformation of E-major to A-major). After the arrival in A-major at m. 24, every successive chord is an unresolved dominant in a key that sits in the tritone halo of the preceding key. The parsimonious transformation of E-major to A-major at m. 24 is

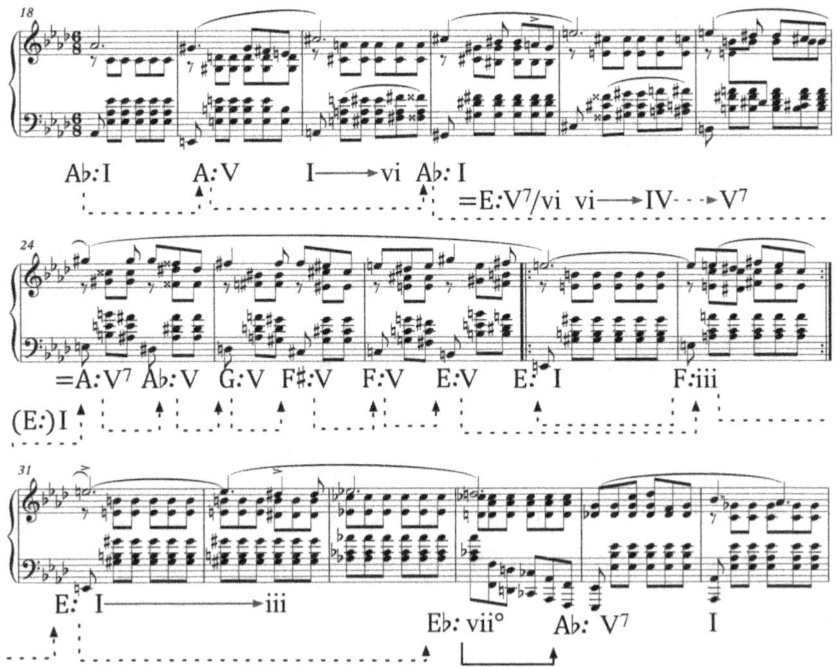

Fig. 8.12. Chopin's Prelude Op. 28, No. 17 in A♭-major, mm. 18–36, with tritone halo opposition indicated with dashed arrows, and parsimonious transformations indicated with solid arrows, at both intra- and inter-key levels.

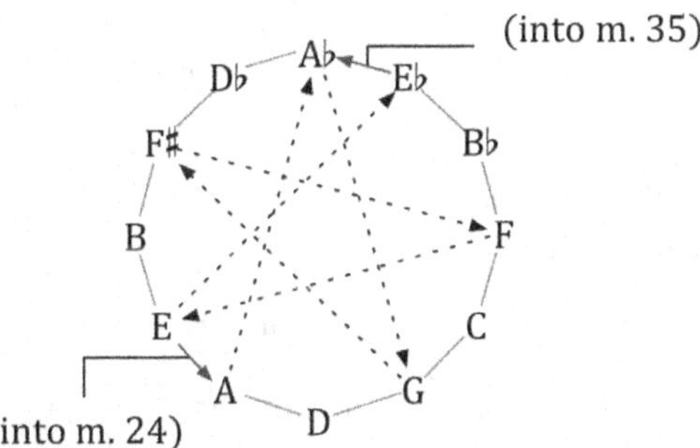

Fig. 8.13. Tritone halo oppositions between key areas as they occur in mm. 24–35 of the Chopin excerpt in Fig. 8.12.

mirrored in a final move from E♭-major to A♭-major at m. 35. The dashed lines in Fig. 8.12 are redrawn on the parsimonious circle of fifths in Fig. 8.13 to better illustrate this striking series of tritone halo oppositions.

8.4.   Conclusions

At the start of this chapter, the question was put forward, "What is the opposite of a parsimonious transformation?" We have established that, within a specific scalar context, the opposite of a parsimonious transformation can be identified with a clear, reusable methodology. Such a perspective is only possible when parsimony includes scale-based criteria. Two short analyses showed how one might use the idea of opposition to explore some densely chromatic passages. In the case of the Debussy excerpt, observing the play between the maximally similar melodic minors and the maximally different harmonies increases our ability to explain the drama in the passage. In the Chopin excerpt, especially with respect to the sequence of unresolved dominants, the idea of harmonious opposition provides language and analytical tools for understanding the patterned motion of the passage away from, and back to, its original key.

Bibliography

[1] A. Childs, Moving beyond neo-Riemannian triads: Exploring a transformational model for seventh chords, *J. Music Theory* **42** (1998) 181–193.

[2] J. Clough and J. Douthett, Maximally even sets, *J. Music Theory* **35** (1991) 93–173.

[3] R. Cohn, Maximally smooth cycles, hexatonic systems, and the analysis of late-romantic triadic progressions, *Music Anal.* **15** (1996) 9–40.

[4] R. Cohn, Neo-Riemannian operations, parsimonious trichords, and their *Tonnetz* representations, *J. Music Theory* **41** (1997) 1–66.

[5] R. Cohn, Uncanny resemblances: Tonal signification in the Freudian age, *J. Amer. Musicological Soc.* **57** (2004) 285–323.

[6] J. Douthett, Filtered point-symmetry and dynamical voice-leading, in *Music Theory and Mathematics: Chords, Collections, and Transformations*, eds. J. Douthett, M. M. Hyde and C. J. Smith (University of Rochester Press, Rochester, NY, 2008), pp. 72–106.

[7] J. Douthett and P. Steinbach, Parsimonious graphs: A study in parsimony, contextual transformations, and modes of limited transposition, *J. Music Theory* **42** (1998) 241–263.

[8] R. Plotkin, Transforming transformational analysis: Applications of filtered point-symmetry, dissertation, University of Chicago (2010).

[9] R. Plotkin and J. Douthett, Scalar context in musical models, *J. Math. Music* **7** (2013) 103–125.

Section II

Chapter 9

Orbifold Path Models for Voice Leading: Dealing with Doubling

James R. Hughes

Mathematical Sciences Department, Elizabethtown College,
Elizabethtown, PA 17022, USA
hughesjr@etown.edu

9.1. Introduction

Voice leading is the means by which one harmonic situation changes to another through specific part assignments. In the Bach passage below, the harmonic changes are already interesting as chord progressions, with secondary dominants, diminished chords, and half-diminished seventh chords occurring in quick succession. But the changes become even more complex and interesting when one takes into account their realization through individual sung lines. In particular, there are many instances of more than one voice singing the same pitch class, commonly called doubling (even when there are more than two voices on the same class). There are also instances of voice crossing (e.g., alto and tenor in the third measure of the passage), voice crossing modulo octave (alto and tenor in the sixth measure), and voice crossing through a doubling (bass and tenor in the first measure, and elsewhere).

Voice leading has been modeled mathematically in many different ways. In this chapter, we further develop the idea presented by the author at MCM2015 [3] of using the orbifold fundamental group (or groupoid) of a chord space to model voice leadings. This idea is based on the geometric modeling schema of Callender, Quinn, and Tymoczko [1], in which chord spaces are modeled as quotients of $\mathbf{R}^n$. We review how voice leadings are modeled (e.g., in Tymoczko's book [5]) as certain kinds of paths, called "generalized line segments", in these chord spaces. We then describe how ambiguities can arise in these models in situations where pitch class doubling occurs, and how these ambiguities can be resolved by using the orbifold fundamental group or groupoid of the chord space. Finally, we exhibit some examples of different doubling situations drawn from Bach vocal passages.

9.2.   Chord Spaces and Orbifolds

We begin by reviewing the definition of n-voice chord space C^n as given in [1]. We model pitch space by the real numbers $\mathbf{R}$ (e.g., by associating a pitch frequency f with the real number $\log_2(f/440)$). We then model ordered n-voice pitch space by $\mathbf{R}^n$, with each factor of $\mathbf{R}$ corresponding to a voice or part that sounds one pitch at a time. Next, motivated by the musical idea that a harmonic situation — i.e., a chord — consists of an unordered set of pitch classes, we introduce two group actions on $\mathbf{R}^n$. First, to pass from pitches to pitch classes, we impose octave equivalence on each voice, which mathematically corresponds to an action of $\mathbf{Z}^n$ on $\mathbf{R}^n$, where the action in each factor is translation (musically: transposition) by an integral number of octaves. Second, to pass from ordered sets to unordered sets, we make use of the action of the group Σ_n of permutations of n items on $\mathbf{R}^n$ by permuting coordinates. Combining the two actions gives an action of $\Gamma = \mathbf{Z}^n \rtimes \Sigma_n$ on $\mathbf{R}^n$. Finally, we define C^n to be the quotient space $\mathbf{R}^n/\Gamma$.

A useful characterization of C^n, described in detail by Tymoczko [5], is as the product of an $(n-1)$-simplex with a closed interval I, with ends attached after cyclic permutation of the n simplicial faces (cells) of dimension $n-2$. Hence C^2 is a Möbius strip, and C^3 is a prism with ends attached after a 120-degree twist. We observe the following musical interpretations of this characterization of C^n: First, translation along the interval I corresponds to uniform transposition (i.e., of all voices simultaneously). Second, each simplicial cross section contains all chord types. Of particular interest to us is that the boundary of a simplicial cross section corresponds to chords with doublings. Specifically, a dimension 0 cell corresponds to a "chord" with only one pitch class (i.e., all n voices on the same pitch class), and the interior of a dimension $k > 0$ cell of a simplicial boundary consists of chords with $k+1$ distinct pitch classes. Crucial for our purposes is the fact that for $0 < k < n - 2$, the pattern of multiplicities varies from cell to cell. For example, for $n = 4$ and $k = 1$, a chord can have "three of a kind" or "two pair". More precisely, in moving from one 0-cell (vertex) to another along a 1-cell (edge), either three voices descend uniformly by 3 semitones while one ascends by 9 semitones (three of a kind), or three voices ascend uniformly by 3 semitones while one descends by 9 semitones (also three of a kind), or two voices ascend by 6 semitones while two voices descend by 6 semitones (two pair).

As a quotient space of $\mathbf{R}^n$ (or of the n-torus $\mathbf{T}^n$), C^n is a developable orbifold. Recall that, roughly speaking, an *orbifold* is like a manifold, except quotient spaces $\mathbf{R}^n/G$ (where each acting group G is either finite or acts with finite isotropy) replace $\mathbf{R}^n$ as the local models. An orbifold is *developable* if it is a global quotient. It is important to note that an orbifold carries more information than its underlying topological space. The quotient maps are part of the data defining an orbifold, so each point of the orbifold retains some memory of its heritage as an orbit. It is this aspect of orbifolds that can be exploited to resolve voice leading ambiguities.

9.3.   Voice Leadings as Paths in Chord Spaces

In n-voice pitch space (modeled by $\mathbf{R}^n$), to specify a voice leading, it suffices to give the initial point and terminal point, since each part assignment is completely determined by the starting and ending pitches in the corresponding coordinate. But in n-voice chord space C^n, with its built-in ambiguities of register and ordering, there are multiple voice leadings between a given initial point and terminal point, and strong musical motivations for distinguishing between them. As Tymoczko emphatically states, "The specific path matters!" [5]. Callender, Quinn, and Tymoczko [1] define a "path" between points in a chord space C^n to be the image, under the canonical projection $p : \mathbf{R}^n \to C^n$, of a line segment in $\mathbf{R}^n$. It is important to note that two line segments in $\mathbf{R}^n$ that are related by the uniform operation of Γ will have the same image in C^n under p, but it is possible for two line segments in $\mathbf{R}^n$ that have equal initial points but distinct terminal points in $\mathbf{R}^n$ related by the operation of Γ to have distinct images in C^n with the same endpoints. Thus, to enumerate voice leadings between points x and y in C^n, we choose a particular point (i.e., a base point) $\tilde{x}_0 \in p^{-1}(x) \subset \mathbf{R}^n$; each line segment in $\mathbf{R}^n$ from $\tilde{x}_0$ to a member of $p^{-1}(y)$ then corresponds to a voice leading from x to y. An image under p of such a line segment is called a "generalized line segment" (GLS) [5]. The GLS model for voice leadings has some geometrical advantages with respect to length and angles, but there are some disadvantages. In particular, the concatenation of two GLSs is not necessarily a GLS, and, as discussed below, the corresponding enumeration of voice leadings is incomplete if one of the endpoints lies on the boundary of C^n (i.e., is a chord with doubling).

The correspondence between generalized line segments from x to y in C^n and members of $p^{-1}(y)$ in $\mathbf{R}^n$ is strongly reminiscent of the path lifting theorem and homotopy lifting theorem in the study of covering spaces, by which preimages of the terminal point of paths from x to y in a base space correspond with homotopy classes of those paths. This observation suggests the use of homotopy classes of paths in place of generalized line segments. With voice leadings from x to y in C^n defined to be homotopy classes of paths from x to y, if $p : \mathbf{R}^n \to C^n$ were a covering map, then the set of voice leadings from a particular chord x_0 to itself could be identified with the fundamental group $\pi_1(C^n, x_0)$ of C^n, since $\mathbf{R}^n$

is simply connected. An immediate advantage would be that we would not only obtain an enumeration of such voice leadings, but a concatenation operation and corresponding group structure on them. For voice leadings between arbitrary chords (i.e., not necessarily from a chord to itself), we could still utilize homotopy classes of paths and concatenation, with the algebraic structure of the fundamental groupoid $\pi(C^n)$.

A crucial observation at this juncture — in fact, the main point of the chapter — is that the canonical map $p : \mathbf{R}^n \to C^n$ is not, of course, a covering map. There are singular points of C^n, namely those whose preimages in $\mathbf{R}^n$ have non-trivial isotropy. These singular points of C^n are the chords with doublings, whose preimages in $\mathbf{R}^n$ are left fixed by the action of a non-trivial subgroup of Γ. If $x \in C^n$ is such a point, then the correspondence between homotopy classes of paths (or generalized line segments) from x to y and points of the fiber $p^{-1}(y)$ fails to be one-to-one. More specifically, suppose x and y are points of C^n and x is singular but y is not (i.e., the chord x has doubling and y does not), x_0 is the chosen base point in $p^{-1}(x)$, $\sigma \in \Gamma$ is in the isotropy group of x_0, and $y_0 \in p^{-1}(y)$. Then a path (or the line segment) from x_0 to y_0 and a path (or the line segment) from x_0 to $\sigma \cdot y_0$ will have the same image in C_n under p, so it is possible for multiple distinct points of the fiber $p^{-1}(y)$ (enumerated by the isotropy group of x_0 if the isotropy group of y_0 is trivial) to correspond to the same path homotopy class (or GLS) in C^n. As we shall see in the examples below, there are strong musical reasons for maintaining distinctions between based paths to such points of $p^{-1}(y)$ as voice leadings, but neither the fundamental groupoid of C^n nor generalized line segments can do so.

9.4.   Orbifold Fundamental Group and Groupoid

One of the main motivating ideas for the orbifold fundamental group is to generalize covering space theory to orbifolds, by appropriately defining orbifold covering maps and orbifold covering transformations. This is done by Thurston [4], who then *defines* the orbifold fundamental group of a connected orbifold Q, denoted $\pi_1^{\mathrm{orb}}(Q)$, to be the group of orbifold covering transformations of the universal orbifold cover $p : \tilde{Q} \to Q$. In the case of C^n, the universal orbifold cover is $\mathbf{R}^n$ and the group of orbifold covering transformations is the acting group Γ, so $\pi_1^{\mathrm{orb}}(C^n) = \Gamma$.

The seeming simplicity of the above characterization of the orbifold fundamental group of C^n as Γ $(= \mathbf{Z}^n \rtimes \Sigma_n)$ masks some of its subtlety. If the base point $x \in C^n$ has trivial isotropy, then elements of $\pi_1^{\mathrm{orb}}(C^n)$ are in one-to-one correspondence with path homotopy classes (or line segments) from a selected, fixed base point x_0 in the fiber $p^{-1}(x) \subset \mathbf{R}^n$ to another point (possibly x_0 itself) in the fiber. In this case, the situation is the same as that of an ordinary covering space, and generalized line segments are in one-to-one correspondence with elements of the orbifold fundamental group. However, if the base point $x \in C^n$ is a point with non-trivial isotropy (i.e., a chord with doubling), then the correspondence between elements of $\pi_1^{\mathrm{orb}}(C^n)$ and path homotopy classes (or line segments) from x_0 to a point in the fiber (possibly x_0 itself) is not one-to-one. Rather, for each such path homotopy class (or line segment), there will be $|\Gamma_x|$ elements of $\pi_1^{\mathrm{orb}}(C^n)$ (where Γ_x is the isotropy group of x). This multiplicity, which is the key difference between the orbifold fundamental group and the ordinary fundamental group, leads to some counter-intuitive situations. For a specific (extreme) example, suppose we take as the base point $x \in C^n$ a chord with all pitch classes equal to one another (i.e., a 0-cell of the boundary of a cross-sectional simplex). Choose a base point x_0 in the fiber consisting of n identical pitches, and consider the path homotopy class of the constant path at x_0. Corresponding to this path homotopy class are $|\Sigma_n|$ elements of $\pi_1^{\mathrm{orb}}(C^n)$. Allowing for $n!$ different voice leadings in a situation where all n voices start and end on the same pitch may seem nonsensical, but doing so is what enables unambiguous composition of voice leadings through chords with doublings.

The situation is similar with the orbifold fundamental groupoid $\pi^{\mathrm{orb}}(C^n)$, but it is more natural to describe it in terms of *orbifold paths*, defined and discussed in detail in [2]. Given endpoints x and y in C_n, elements of $\pi^{\mathrm{orb}}(C^n)$ with initial point x and terminal point y are homotopy classes of orbifold paths connecting two fixed lifts x_0 and y_0 in $\mathbf{R}^n$ of the points x and y, respectively. The homotopy classes of orbifold paths are pairs (p, g) where p is an ordinary path homotopy class of a path from x_0 to y_0, and $g \in \Gamma$. In our case, since $\mathbf{R}^n$ is simply connected, these pairs are in one-to-one correspondence with elements of Γ. If both x_0 and y_0 have trivial isotropy, the pairs are also in one-to-one correspondence with path homotopy classes of paths (or line segments) from x_0 to a point in the orbit of y_0, and therefore also in one-to-one correspondence with generalized line segments. If one or both of x_0 and y_0 have non-trivial isotropy, then there may be multiple members of $\pi^{\mathrm{orb}}(C^n)$ corresponding to a single generalized line segment.

9.5.   Examples

In this section we present some occurrences in actual music of voice leadings involving chords with various kinds of doubling. The examples are sufficiently commonplace to illustrate that voice leadings involving doublings are not rare or pathological, and therefore should be incorporated into mathematical modeling of voice leading.

For our first example, we refer back to the passage shown above in Fig. 9.1. The first chord in the boxed excerpt has a doubling; both the tenor and soprano are assigned to the pitch class F# (in fact, the same pitch F#4). The soprano then ascends five semitones to B4 while the tenor descends four semitones to D4. If the soprano were to descend to D4 and the tenor ascend to B4, the result (ignoring the other voices for now) would be a path in 2-voice pitch space $(\mathbf{R}^2)$ that differs from the path corresponding to the passage as written. The situation in $\mathbf{R}^2$ is illustrated in Fig. 9.2 below. Note that both paths map to the same path in the chord space C^2, so in the generalized line segment model they would be considered the same voice leading. Similarly, both paths correspond to the same element of the ordinary fundamental groupoid of C^2. However, they

Fig. 9.1. J.S. Bach: Erfreut euch, ihr Hertzen from Kantate zum 2. Osterfesttag, BWV 66, m. 235–239.

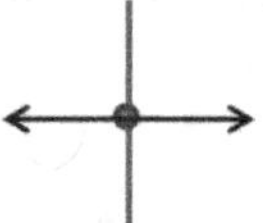

Fig. 9.2. Two paths emanating from a point with Σ_2 isotropy.

correspond to distinct elements of the orbifold fundamental groupoid. Specifically, if we choose (F#4, F#4) and (B4, D4) as fixed lifts in $\mathbf{R}^2$, then the two paths represent elements of $\pi^{\mathrm{orb}}(C^2)$ that differ by the element of Γ that takes (B4, D4) to (D4, B4). Note that the two elements are enumerated by the isotropy group of (F#, F#), which is Σ_2.

A similar situation occurs with the alto and tenor at the end of the boxed excerpt. Both voices are on the pitch class G#; the alto then moves down to F# and the tenor moves up to A. If we are not able to distinguish this from the other possibility (tenor moving down to F# and alto moving up to A), then we will be unable to detect whether a voice crossing occurs in the composition of voice leadings before and after the doubling. Such voice crossings through doublings are not unusual (in fact there are several other examples just in Fig. 9.1), so it is reasonable to require a voice leading model to detect them.

Fig. 9.3. BWV066, measure 245.

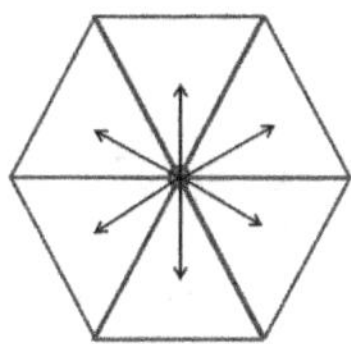

Fig. 9.4. Six paths emanating from a point with Σ_3 isotropy.

Our next example, the boxed excerpt in Fig. 9.3, initially has the soprano, alto, and tenor all on pitch class F#, and subsequently moving to D, F#, and B respectively. In this case there are six path homotopy classes (or line segments) in $\mathbf{R}^3$ that map to the same path homotopy class (or generalized line segment) in C^3; these are enumerated by the isotropy group of (F#, F#, F#), which is Σ_3. The situation in $\mathbf{R}^3$ is illustrated in Fig. 9.4.

Our final example, shown in Fig. 9.5, initially has the soprano and tenor on the pitch class A, and the alto and bass on the pitch class F#. In this case there are four path homotopy classes in $\mathbf{R}^4$ that map to the same path homotopy class (or generalized line segment) in C^4; these are enumerated by the isotropy group of (A, F#, A, F#), which is $\Sigma_2 \times \Sigma_2$.

Fig. 9.5. BWV066, measure 252.

Bibliography

[1] C. Callender, I. Quinn and D. Tymoczko, Generalized voice-leading spaces, *Science* **320** (2008) 346–348.

[2] G. Dragomir, Closed geodesics on orbifolds, Ph.D. thesis, McMaster University (2011).

[3] J. Hughes, Using fundamental groups and groupoids of chord spaces to model voice leading, in *Mathematics and Computation in Music: 5th International Conference, MCM2015, Proceedings*, T. Collins, D. Meredith and A. Volk, eds., Lecture Notes in Artificial Intelligence Vol. 9110 (Springer, 2015).

[4] W. Thurston, *The Geometry and Topology of 3-Manifolds* (Princeton University Press, Princeton, NJ, 1997).

[5] D. Tymoczko, *A Geometry of Music: Harmony and Counterpoint in the Extended Common Practice* (Oxford University Press, Oxford, 2011).

Chapter 10

Reflections on the Geometry of Chords

Thomas A. Ivey

Department of Mathematics, College of Charleston,
66 George St., Charleston, SC 29424, USA
iveyt@cofc.edu

10.1. Introduction

This chapter is about my responses to and reflections on a pair of (by now) well-known papers by Tymoczko [4] and Callendar, Quinn and Tymoczko [2]. These papers were first brought to my attention by a colleague while I was preparing to teach a course on math and music at the College of Charleston in 2009. Since my field is differential geometry, I was especially intrigued by the title *The Geometry of Musical Chords* of the first paper, and have spent some time thinking about how Tymoczko's sense of geometry fits into standard mathematical theories of geometry and topology. In particular, the notions of *metric geometry* in chord spaces proposed by Tymoczko seem to be less relevant (both musically and mathematically) than the *Kleinian geometry* of these spaces, and specifically their geometry as orbifolds.

I am grateful to the organizers Mariana Montiel and Robert Peck of the Special Session on Mathematics and Music at the 1117th meeting of the American Mathematical Society (held at the University of Georgia in March 2016) for the invitation to give the talk on which this chapter is based. In turn, that talk was based on a summer research project carried out in 2015 with undergraduate Omar Valencia. Omar and I read these papers together, and also digested much of the accompanying online material, which is where many of the mathematical details are fleshed out. Omar also created visualizations for several chord spaces using Mathematica, and his background as a musician and performer also helped us understand the musical relevance of the features of these spaces. Omar's participation in this project was supported by a stipend from the School of Science and Mathematics at the College of Charleston.

10.2.   Equivalence and Group Actions

The chord spaces, and voice-leading spaces defined in the second paper [2] are founded on the basic mathematical notion of an *equivalence relation*. In particular, these spaces are quotient spaces in which the points are equivalence classes; moreover, most of these equivalence classes are the orbits of the actions of various groups of transformations. These transformations act on n-tuples $(x_1, \ldots, x_n)$ of chromatic pitches, each of which is represented by an integer x. (For example, 0 represents C4 (middle C on the piano), the integer 7 represents the G seven semitones above middle C, the integer -6 represents the F$\sharp$ six semitones below middle C, and so on.)

Callendar *et al.* [2] identify five musically relevant notions of equivalence, four of which are generated by group actions on the set $\mathbb{Z}^n$ of (ordered) n-tuples of integers:

O octave transposition $x \mapsto x \pm 12$, acting on individual pitches within an n-tuple;

P permutations in the symmetric group S_n, shuffling n-tuples of pitches;

T semitone transpositions $x \mapsto x \pm 1$, acting uniformly on all pitches in an n-tuple;

I inversion $x \mapsto a - x$, acting uniformly;

C adding or removing a doubled pitch (or pitch class) in a chord.

As these authors point out, musicians also apply the first four transformations to "horizontal" sequences of pitches — i.e., melodies or motifs. However, for the rest of this talk, we will think of these n-tuples as "vertical" collections of pitches, sounding simultaneously — i.e., chords. For example, we will refer to points in $\mathbb{Z}^n$ as *voiced chords*, where instrument #1 plays pitch x_1, instrument #2 plays pitch x_2, and so on. While the first four equivalence relations are generated by group actions on $\mathbb{Z}^n$ for fixed n, C-equivalence connects n-tuples for different values of n. As a relation on a set, C-equivalence is defined on the space of voiced chords with arbitrary numbers of pitches, i.e.,

$$\mathcal{Z} = \bigcup_{n=1}^{\infty} \mathbb{Z}^n.$$

10.3.   Quotient Spaces and Orbifolds

One can consider quotients of $\mathbb{Z}^n$ by different combinations of these equivalence relations, depending on what one is interested in: just as taking the quotient of the set $\mathbb{Z}$ of pitches by O-equivalence yields the set $\mathbb{O} \cong \mathbb{Z}_{12}$ of pitch classes, similarly taking the quotient of $\mathbb{Z}^n$ by O-equivalence yields the set $\mathbb{O}_n$ of ordered n-tuples of pitch classes (e.g., tone rows). If we further quotient by P-equivalence, we obtain the set of unordered n-tuples of pitch classes. Because repeats are allowed, these unordered n-tuples of pitch classes are multisets, and for the sake of convenience we will refer to them simply as n-note chords, denoted by $\mathbb{OP}_n$. If we further quotient by T-equivalence, we obtain the set $\mathbb{OPT}_n$ of n-note chord types. For example, when $n = 3$ all major triads are represented by a single point in this set, and all minor triads by another point; if we further quotient by I-equivalence then all major and minor triads are identified. If we quotient the set $\mathcal{Z}$ by O-, P- and C-equivalence, we obtain the set of all chords, containing any number of distinct pitch classes, ignoring doublings.

Remark. As the title of [2] indicates, these equivalence relations can also model the space in which voice-leading takes place. If we define the k-fold Cartesian product

$$\mathcal{W}_k^n = \mathop{\text{\Large $\times$}}_{j=1}^{k} \mathbb{Z}^n,$$

then we obtain the set of length k sequences of n-note chords. If we apply O-equivalence and P-equivalence *uniformly* across sequences of voiced chords, then we obtain the space of voice-leadings. Note that we do not apply O-equivalence to individual pitches within chords, since moving an individual note up or down an octave in a voice part can change an easy step into an awkward leap for the singer.

For the rest of this chapter, we will focus on the sets $\mathbb{OP}_n$ and $\mathbb{OPT}_n$, of n-note chords and chord types respectively. How do we make these sets into spaces, endowing them with geometry and topology? Once we apply octave equivalence, the quotient of $\mathbb{Z}^n$ is $\mathbb{O}_n$, a finite set with no distinguished topology. Instead, we have to extend the above transformations to the ambient space $\mathbb{R}^n$. Then the quotient $\mathbb{R}^n/O$ is an n-dimensional torus $\mathcal{T}^n$, which we can represent as an n-dimensional cube with the opposite sides identified, and inside this the set $\mathbb{O}_n$ sits as a cubical lattice. Similarly, $\mathbb{OP}_n$ sits inside an n-dimensional space $\mathcal{U}^n = \mathbb{R}^n/OP$, and $\mathbb{OPT}_n$ sits inside an $(n-1)$-dimensional space $\mathcal{V}^{n-1} = \mathbb{R}^n/OPT$. These spaces $\mathcal{U}^n$ and $\mathcal{V}^{n-1}$ are not manifolds but *orbifolds*, and we will describe their geometry in detail below for low values of n.

An n-dimensional orbifold (see [1]) is a space that is locally identifiable with a "model" which is a quotient of Euclidean n-space by a finite group of isometries. The simplest examples are "good" orbifolds, which are globally quotients of Euclidean space by a discrete group action. Two-dimensional orbifolds can have three kinds of singular points: reflection points (where the local model is the quotient of the plane by a reflection), cone points (where the model is the quotient by a finite group of rotations about a point), and reflection-corners (where the model is the quotient by a dihedral group, generated by reflections in intersecting lines); in each case the singular point corresponds under the quotient map to a point with a non-trivial stabilizer under the group action. As we will see, higher-dimensional orbifolds can have more exotic singular points.

As indicated above, more interesting chord spaces arise when we quotient by P-transformations (i.e., permutations of the n-tuples). In the online supplementary material to [4], Tymoczko defines a fundamental domain $\mathcal{D}$ for the OP-action by

$$0 \leq \sum_{i=1}^{n} x_i < 12, \qquad x_1 \leq x_2 \leq \cdots \leq x_n \leq x_1 + 12. \tag{10.1}$$

For example, when $n = 2$ the OP-action is generated by 12-unit horizontal and vertical translations in the $x_1 x_2$ plane (the O-action), together with reflection across the line $x_1 = x_2$. If we take the quotient of $\mathbb{R}^2$ by just the O-action, a fundamental domain is a 12×12 square with opposite edges identified by translations, resulting in a smooth two-dimensional torus $\mathcal{T}^2$. Because the OP-action also includes reflections, its fundamental domain $\mathcal{D}$ is a diamond-shaped region centered on the vertical axis in upper halfplane. The set $\mathbb{OP}_2$ of dyads appears as a lattice within this diamond, with dyads consisting of a single pitch class double represented by points along the bottom-right and top-left edge (see Fig. 10.1). The OP group includes reflections along the lines through these edges, and thus points along the top-right edge are identified with points along the bottom-left edge, but with a twist. Thus (as Rachel Hall observed [3]) the quotient of the torus $\mathcal{T}^2$ by the P-action is homeomorphic to a Möbius strip. However, here we see the difference between an orbifold and its underlying topological space. While points along the top-left and bottom-right edges form the boundary of the Möbius strip, in the orbifold $\mathcal{U}^2$ these are reflection points, and the orbifold has no boundary.

In what sense does the set of dyads inherit a *geometry* from the orbifold in which it lies? In the supplementary material to [4], Tymoczko makes it clear that there is no preferred notion of distance between chords. He does suggest several possibilities, some of which are based on taking a norm of the vector of differences between pitches in two n-tuples (e.g., the Euclidean norm, the L^∞ or "max" norm, the L^0 or "taxicab" norm); the distance between two chords

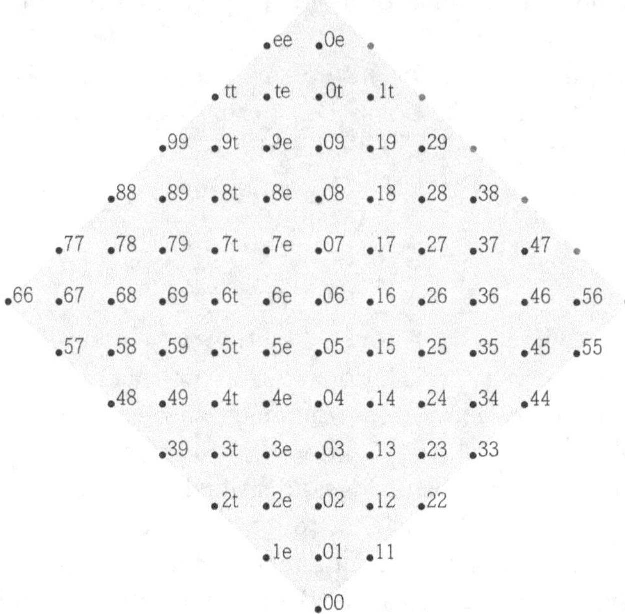

Fig. 10.1. The set of dyads inside the orbifold $\mathcal{U}^2$. Pitch classes are notated modulo 12, where 't' stands for 10 and 'e' stands for 11.

is then taken to be the smallest distance between all possible pairs of lifts into the space of voiced chords. Nonetheless, in the basic sense of the word "geometry", i.e., measuring distances, there seems to be no preferred way of doing this within chord spaces.

Meanwhile, the orbifold has a natural topology, and distinguished singular points; it thus seems natural that discrete sets of chords inside the orbifold should have distinguished points if they coincide with singular points of the orbifold. However, again there is no useful topology inherited by this subset. What seems to play the role of topology for a "chord space" is its structure as a *graph*, where the edges of the graph connect chords that differ by modifying exactly one pitch class by a semitone. (This graph naturally arises as the image of the cubical lattice in $\mathbb{R}^n$ under the map to the quotient space.) Moreover, singular points of the orbifold (when they belong to the chord space) are distinguished within the graph as vertices of unusually low degree.

10.4. Triad Spaces

To visualize this graph, we now review the construction of the orbifold $\mathcal{U}^3$ in which the set $\mathbb{OP}_3$ of triads sits. (Later in this section we will also construct $\mathcal{V}^2$ and interpret its orbifold fundamental group.) As in (10.1), a fundamental domain for the OP-action on $\mathbb{R}^3$ is given by

$$0 \leq x_1 + x_2 + x_3 < 12, \qquad x_1 \leq x_2 \leq x_3 < x_1 + 12. \qquad (10.2)$$

These inequalities define $\mathcal{D}$ as a prism with edges parallel to the line $x_1 = x_2 = x_3$, with cross sections that are equilateral triangles. By octave translation, the "bottom" face of $\mathcal{D}$ (along which $x_1 + x_2 + x_3 = 0$) is identified with the "top" face with a 120-degree twist. In other words, if we take a corner of the bottom face and transpose it up 4 semitones, that corner on the top face is O-equivalent to the next corner on the bottom (see Fig. 10.2):

$$(0, 0, 0) + 4 \text{ semitones} = (4, 4, 4) \overset{O}{\sim} (-8, 4, 4)$$

$$(-8, 4, 4) + 4 \text{ semitones} = (-4, 8, 8) \overset{O}{\sim} (-4, -4, 8)$$

$$(-4, -4, 8) + 4 \text{ semitones} = (0, 0, 12) \overset{O}{\sim} (0, 0, 0).$$

(Here and below, $\overset{O}{\sim}$ denotes O-equivalence.)

Thus, the quotient of $\mathbb{R}^3$ by the OP-action is a compact three-dimensional orbifold $\mathcal{U}^3$; the three rectangular sides of the prism form a single surface within $\mathcal{U}$, which is homeomorphic to a Möbius strip. This surface is not the boundary of the orbifold; rather, it consists of reflection points; for example, paths in the orbifold do not end when they encounter this surface, but rather are reflected back into the non-singular part of the orbifold. Within $\mathcal{U}^3$ the points of $\mathbb{OP}_3$ form a cubic lattice (see Fig. 10.3(a)). It may be useful to visualize these points as grouped into layers in $\mathcal{D}$, according to the value of the sum $x_1 + x_2 + x_3$; thus, the bottom face is layer zero, while the top face is layer 12 (which is not part of $\mathcal{D}$ but forms its boundary). Non-singular points of the lattice have 6 neighbors, but lattice points lying on the reflection surface (which with represent triads with one

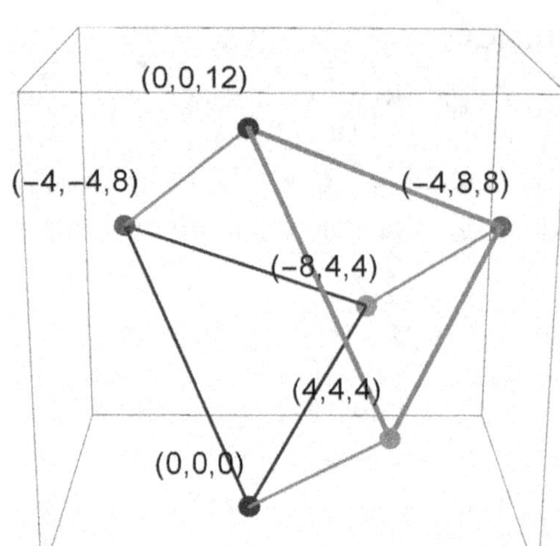

Fig. 10.2. The fundamental domain defined by (10.2). The bottom face (dark triangle) is identified with the top face (light triangle) with a 120-degree twist, matching vertices with the same color.

pitch-class doubled) have only 4 neighbors. The edge of the Möbius strip consists of reflection corner points, and lattice points on this edge (which represent triads consisting of just one pitch-class tripled) have only 2 neighbors (see Fig. 10.3(b)).

Remark. At the outset of the research project with Omar, we speculated that distinct styles of harmony would correspond to graphs inside chords spaces that would have some distinguishing geometrical features. If that is the case, then these features are either quite subtle, or just amount to the obvious feature that the chords used in "common practice" harmony avoid semitone distances, and so appear as points near the center of the triad space. In fact, the major and minor triads, along with augmented triads, form the vertices of a stack of cubes whose diagonals lie along the centerline of the prism (see Fig. 10.3(c)). If we plot just the diatonic triads used in a particular major key, they form a sort of spiral of points winding around this centerline (see Fig. 10.4(a)). But this spiral does not really match the root progressions used in common practice.

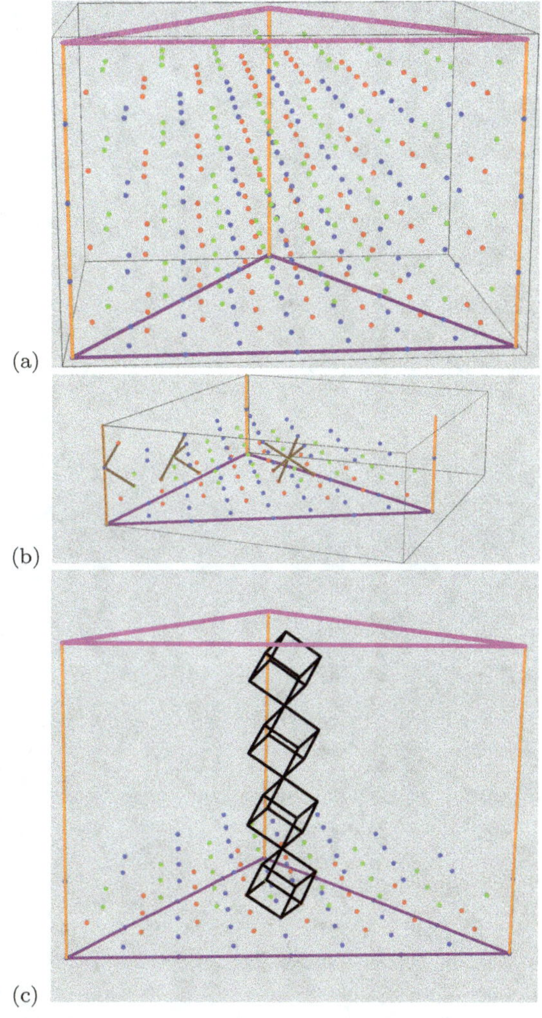

Fig. 10.3. The set $\mathbb{OP}_3$ of triads within the orbifold $\mathcal{U}^3$, modeled by a cubical lattice inside a triangular prism. (The top and bottom faces of the prism are to be identified with a 120-degree twist, as in the previous figure, but here the viewpoint is rotated so that the faces appear horizontal.) At top, the points of $\mathbb{OP}_3$ are grouped into horizontal layers $0, 1, 2, \ldots$ (according to the sum of pitch classes mod 12) and colored cyclically blue, red, and green. The middle figure shows that interior lattice points have 6 neighbors, points on the reflection planes have 4 neighbors, and reflection corner points have two. The bottom figure shows common-practice triads sitting in the center of the prism as a stack of cubes, whose vertices represent augmented triads (top and bottom corners, in blue layers), minor triads (in red layers) and major triads (in green layers).

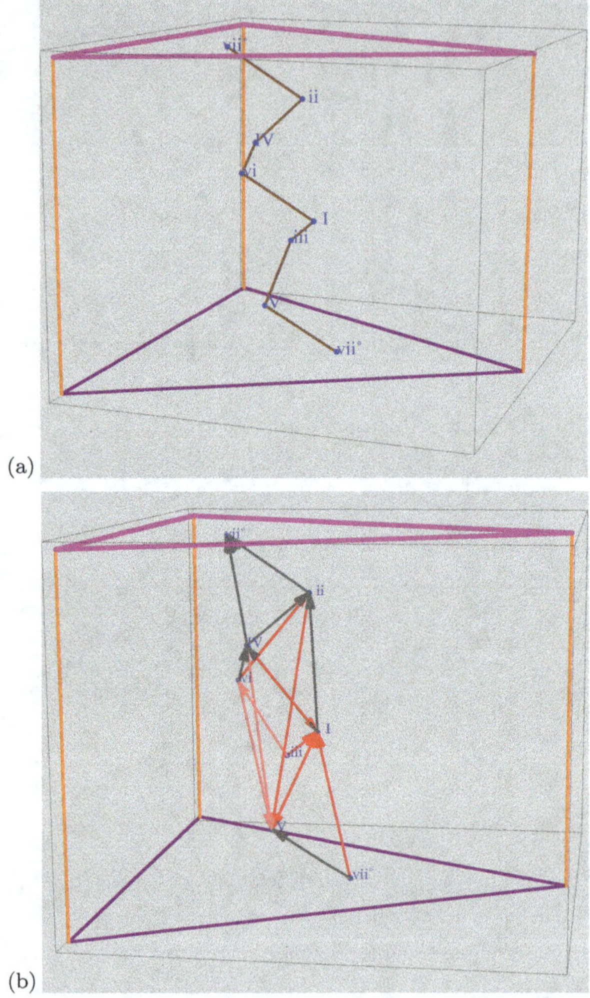

Fig. 10.4.

For example, we may incorporate these points into a directed graph that reflects the typical root progressions of a particular style (see Fig. 10.4(b)). In this figure, the red arrows indicate root movements that happen more than 50% of the time, the pink arrows indicate movements that happen 20–50% of the time, and the gray arrows movements that happen 10–20% of the time. (This is based

on data from Mozart's major-key piano sonatas, see [5, Fig. 7.1.6].)
Again, there seems to be nothing discernibly special about this shape;
but that may not be surprising since in the Euclidean geometry of
$\mathbb{Z}^3$ that underlies this picture, two chords are closest when they differ
by a single semitone in one voice, transitions which are rare in music
of the classical era, while the transitions that are most common
correspond to wider jumps within triad space. (However, these might
not look so wide if we replaced our notion of adjacency based on
semitone modifications with, say, adjacency based on modifications
by a fifth.)

Returning to our construction of orbifolds associated to the triad
spaces, recall that T-equivalence means uniform transposition by
semitones; thus, any pair of chords are T-equivalent if they partition
the octave into the same ordered set of intervals. When we pass
from the quotient space by OP-equivalence (i.e., chord space) to
the quotient under OPT-equivalence, we obtain the space of *chord
types*. In the case of triads, applying T-equivalence to $\mathcal{U}^3$ collapses the
prism onto its bottom face ("layer 0"); but because the bottom face
is OP-equivalent to the top face ("layer 12") rotated by 120 degrees,
the bottom face is OPT-equivalent to itself rotated 120 degrees.
Thus, the space $\mathcal{V}^2 = \mathbb{R}^3/OPT$ is a two-dimensional orbifold whose
fundamental domain is the quotient of an equilateral triangle by a
120-degree rotation. But to understand the singular points of this
orbifold it is helpful to apply the equivalence relations in a different
order. First, applying T-equivalence to $\mathbb{R}^3$ gives $\mathbb{R}^2$, since we can
take the "layer 0" plane $x_1 + x_2 + x_3 = 0$ as a fundamental domain
for this action. The group of OP transformations is generated by
the mappings $\sigma : (x_1, x_2, x_3) \mapsto (x_2, x_1, x_3)$ and $\rho : (x_1, x_2, x_3) \mapsto$
$(x_3 - 12, x_1, x_2)$. If we let Δ denote the equilateral base in layer 0 of
the prism $\mathcal{D}$, then these transformations correspond respectively to
reflection in one of the sides of Δ and a 120-degree rotation about
its center.

We will choose as a fundamental domain for OPT-action a
kite-shaped quadrilateral whose vertices are one vertex of Δ, the
midpoints of the adjacent sides, and the center of Δ. The short sides

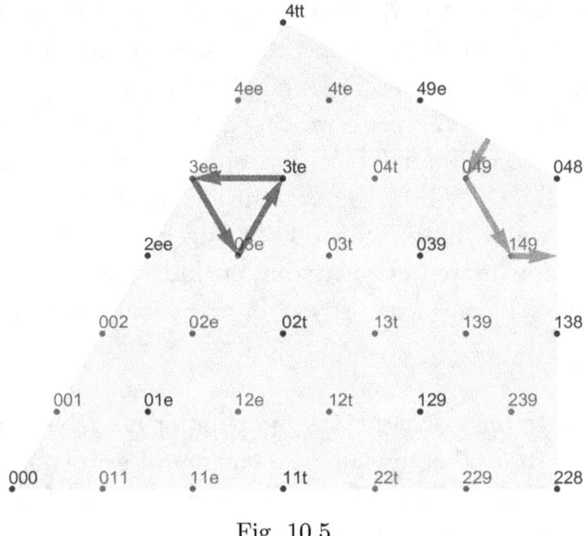

Fig. 10.5.

of the kite (adjacent to the center) are identified by the rotation ρ, so that the center is a 120-degree cone point of $\mathcal{V}^2$. Because of the reflection σ and its conjugates, the points along the sides of Δ become reflection points in $\mathcal{V}^2$, and they meet at a reflection corner. Within the orbifold, the set of triad types sits as an equilateral triangular grid, including the cone point (representing augmented triads) and the reflection corner (representing unison triads); triads with one pitch class doubled sit along the reflection edges (see Fig. 10.5).

While its underlying topological space is a closed disc, the orbifold $\mathcal{V}^2$ has a finer invariant, the orbifold fundamental group [1], which tracks its geometry. Like the usual fundamental group, it is generated by loops in the orbifold, with the usual definition for composition, but a loop is equivalent to the identity if its lift to the *orbifold universal cover* is homotopically trivial. In this case, $\mathcal{V}^2$ is a "good" orbifold: it is the quotient of its universal cover $\mathbb{R}^2$ by a discrete group G of Euclidean isometries generated by the rotations about the center of Δ and reflections in one of its sides. The orbifold fundamental group G is a semidirect product of $\mathbb{Z}^2$ with the permutation group S_3. To see this, note that the union of Δ with its reflection in one side is a rhombus $\mathcal{R}$, and within G there is a normal subgroup H generated by translations that identify opposite sides of $\mathcal{R}$. Thus, the quotient map from $\mathbb{R}^2$ to $\mathcal{V}^2$ factors through a covering $\mathbb{R}^2 \to \mathbb{R}^2/H$, and the intermediate space $\mathbb{R}^2/H$ is a smooth two-dimensional torus $\mathcal{T}^2$ with fundamental group $H \cong \mathbb{Z}^2$; then it is an easy exercise to also check that $G/H \cong S_3$. Moreover, we can identify the torus as the quotient of $\mathbb{R}^3$ by OT-equivalence, and then the quotient map $\mathcal{T}^2 \to \mathcal{V}^2$ is the natural map taking OT-equivalence classes to OPT-equivalence classes.

We will focus on the generators of the "finite part" G/H of the orbifold fundamental group of $\mathcal{V}^2$, and show that these have interesting musical realizations. It turns out that this part of the group is generated by a loop ℓ_1 around the cone point and a loop ℓ_2 which passes through a reflection point (see Fig. 10.5), and these turn out to have orders 3 and 2 respectively. To see why, we will lift each of these loops into $\mathbb{R}^3$ (i.e., making a sequence of voiced chords), and then apply OT-equivalence to project into the torus. (One can also find lifts that generate the infinite part $\mathbb{Z}^2 \subset G$, but these just consist of sequences of chords where one voices rises chromatically through an octave.)

For example, suppose we lift ℓ_1 into $\mathbb{R}^3$ — i.e., choose a starting chord that realizes the chord type of our starting point in $\mathbb{OPT}_3$, and then do semitone moves in one voice at a time, to produce lifts of the successive points on the path. The result is a modulation that ends with a chord that differs (modulo O-equivalence) by 4-semitone transposition from where we started, but with a different voice taking the root of the triad (see musical example below). When we quotient by OT-equivalence, the result is a path in $\mathcal{T}^2$ that starts and ends at the same triad type, but with the voices cyclically permuted; thus, we have a curve which returns to the same point in the cover $\mathcal{T}^2$, but only after 3 iterations of the lift. (In other words, if we went around the base curve in $\mathcal{V}^2$ three times, its lift into $\mathcal{T}^2$ would close up only after the third time around.) Similarly, the lift of ℓ_2 into $\mathbb{R}^3$ is a sequence of three chords, starting and ending at minor/major seventh chord (with the fifth omitted), with the root moving up two semitones.

However, the voices that take the root and the 7th of the chord swap roles, so that after taking the quotient by OT-equivalence, the result is an open curve in $\mathcal{T}^2$, but which closes up after another iteration of the lift.

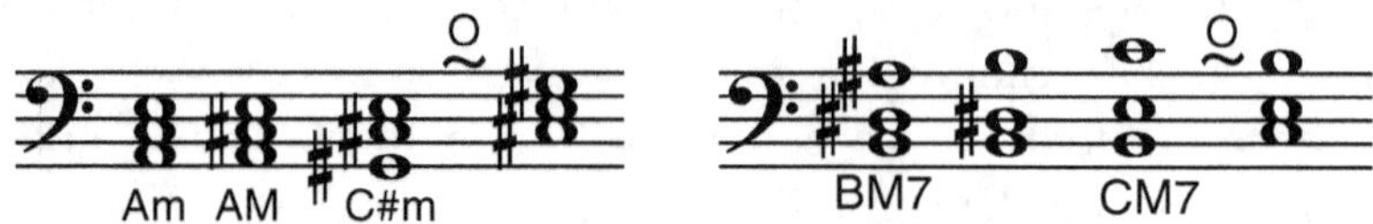

10.5. Tetrad Spaces

The set $\mathbb{OP}_4$ of tetrads (i.e., four-note chords, modulo permutation and octave equivalence) sits inside the fundamental domain $\mathcal{D} \subset \mathbb{R}^4$ defined by the inequalities (10.1) for $n = 4$. Since the inequalities $x_1 \leq x_2 \leq x_3 \leq x_4 \leq x_1 + 12$ define an intersection of four half-spaces, a cross section of $\mathcal{D}$ (defined by intersecting with a hyperplane $x_1 + x_2 + x_3 + x_4 = c$) is a solid tetrahedron; for this reason, we will think of the domain as a prism with a tetrahedral base. Again, we divide the chord set into layers 0 through 11, depending on the value of c. For example, the vertices of the tetrahedron forming layer zero are

$$a = (0,0,0,0), \quad b = (-9,3,3,3), \quad c = (-6,-6,6,6),$$
$$d = (-3,-3,-3,9).$$

(Notice that this tetrahedron is not equilateral, since edges ac and bd are longer than the other four.) Operation T transposes a chord up one semitone, so this moves us from layer 0 to layer 4, layer 1 to layer 5, and so on. Transposing by three semitones each of the chords in layer 0 gives a chord in layer 12, which is O-equivalent to a chord in layer zero. Under this identification, $T^3a \overset{O}{\sim} b$, $T^3b \overset{O}{\sim} c$, $T^3c \overset{O}{\sim} d$, and $T^3d \overset{O}{\sim} a$. Thus, the quotient space $\mathcal{U}^4 = \mathbb{R}^4/OP$ can be visualized as the product of a solid tetrahedron with a closed interval, where the ends are identified by a mapping that cyclically permutes the vertices of the tetrahedron. Note that $\mathcal{U}^4$ is again an orbifold, in which points where two coordinates coincide (e.g., $x_1 = x_2$) are reflection points, and points where three coordinates coincide are reflection corners.

It is, of course, easier to visualize a three-dimensional object, so we will pass to considering the set $\mathbb{OPT}_4$ of tetrad types, which lies inside the quotient of $\mathcal{U}^4$ by the T-action. The resulting orbifold $\mathcal{V}^3$ is the quotient of a solid tetrahedron by a rigid motion that cyclically permutes its vertices. We can visualize this by projecting points (x_1, x_2, x_3, x_4) from $\mathcal{D}$ into $\mathbb{R}^3$, using for example

$$\pi : (x_1, x_2, x_3, x_4)$$
$$\mapsto \tfrac{1}{2}\left(x_2 + x_4 - x_1 - x_3, x_1 + x_4 - x_2 - x_3, x_3 + x_4 - x_1 - x_2\right).$$

This mapping collapses each layer onto the same solid tetrahedron in $\mathbb{R}^3$, with vertices $A = \pi(a)$, $B = \pi(b)$, $C = \pi(c)$ and $D = \pi(d)$. On this tetrahedron the cyclic permutation is generated by composing a reflection in the plane parallel to line segments AC and BD with a 90-degree rotation about the axis connecting the midpoints of these segments (see Fig. 10.6). We will let this composition be denoted by ρ.

We take as a fundamental domain for this action the solid convex polyhedron $\mathcal{P}$ with vertices C, M the midpoint of AC, E the midpoint of BM, F the midpoint of BC, G the midpoint of CD, and H the midpoint of DM. Inside of this polyhedron, $\mathbb{OPT}_4$ sits as a cubic body-centered lattice (see Fig. 10.7(a)), i.e., each point sits at the center of a cube with its 8 adjacent vertices sitting at the corners of the cube. If we think of this set as a graph, with two vertices connected by an edge if they differ by a semitone in one pitch class, then most vertices of the graph have degree 8, but the exceptions are points that coincide with singular points of the orbifold $\mathcal{V}^3$. (As we will see, $\mathcal{V}^3$ is obtained from $\mathcal{P}$ by gluing the face EMQ onto FMQ, and $EFRQ$ onto $RGHQ$, where Q and R are the midpoints of EH and FG respectively.)

As an orbifold, $\mathcal{V}^3$ inherits some singular points from $\mathcal{U}^4$, namely the reflection points along its trapezoidal faces $CFEM$ and $CGHM$ and its triangular face CFG. Points along these faces represent tetrads with one pitch-class doubled (hence, vertices of degree 6),

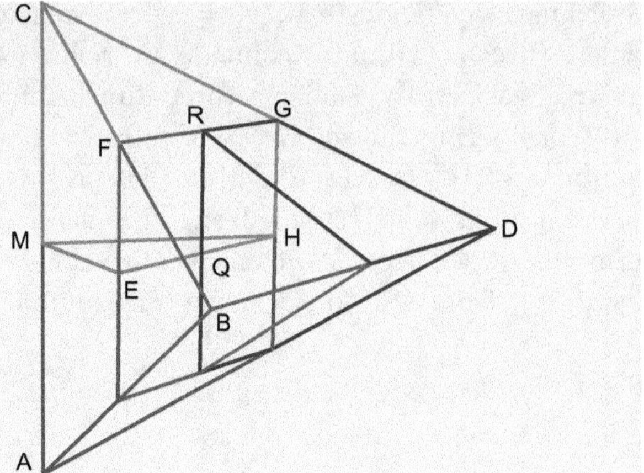

Fig. 10.6.

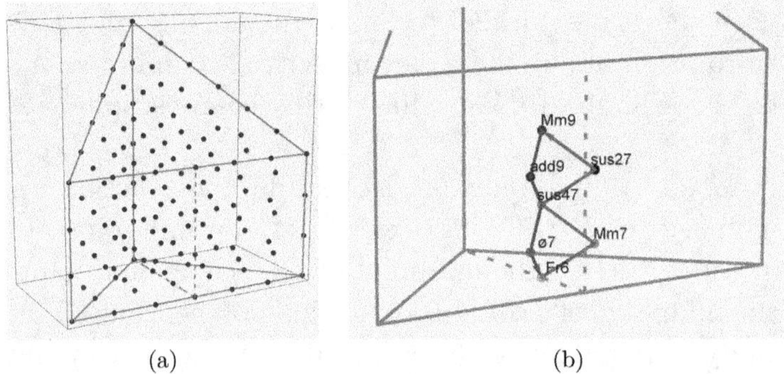

(a)   (b)

Fig. 10.7.

while points along the edges CF and CG represent tetrads with one pitch-class tripled (hence, of degree 4) and points along CM represent tetrads with two pitch classes doubled (also of degree 4). But there are also singular points arising from taking the quotient of the tetrahedron $ABCD$ by the ρ-action, and these lie on the 'bottom' face EHM of $\mathcal{P}$.

For example, ρ^2 identifies $\mathcal{P}$ with its image under a 180-degree rotation about the axis running through M and Q. This axis runs through the "bottom" face of $\mathcal{P}$, and points on the face that are on one side of the axis are identified, by the rotation, with points on the other side. Thus, points along the line MQ are cone points of the orbifold, subtending a solid angle of 2π (i.e., the area of the quotient of the unit sphere by a 180-degree rotation). However, at every other point of the bottom face, away from the line segments EM and HM, the orbifold is locally Euclidean. Tetrad types that lie along MQ are those which, when realized as chords, are symmetric under a 6-semitone transposition: the French 6th, which partitions the octave into $4, 2, 4, 2$ semitone intervals, and a less easily identified chord which partitions the octave into $5, 1, 5, 1$ semitone intervals. Within the graph, these vertices have degree 4; for example, the four neighbors of the French 6th are the half-diminished 7th (partition $3, 3, 4, 2$), its inversion the major-minor 7th $(4, 3, 3, 2)$, a kind of major chord with a suspended 4th and flattened 9th $(5, 1, 4, 2)$ and its inversion $(4, 1, 5, 2)$.

The fully diminished 7th chord $(3, 3, 3, 3)$ is represented by point Q. This lies the center of the tetrahedron $ABCD$, and is fixed by ρ. If we take a small spherical neighborhood of Q inside the tetrahedron, the sphere can be cut into four quarters, each subtending a solid angle of π, and each identified with the next by ρ. Thus, Q is a kind of cone point (with degree two in the graph) but with the difference that the orbifold is not orientable in a neighborhood of Q. This is best visualized in terms of a gluing pattern on the non-reflection faces of $\mathcal{P}$. While one half EMQ of the "bottom" face is glued to the other half HMQ by ρ^2, the square $EFRQ$ forming one half of the "front" face is glued by ρ to the other half $RGHQ$, with vertices corresponding in the order listed. (The resulting orbifold would be a strange environment in which to live: for example, if you exit $\mathcal{P}$ through the 'window' $EFRQ$ standing upright, you immediately re-enter $\mathcal{P}$ through the adjacent 'window' $RGHQ$, but rotated by 90 degrees; furthermore, if you are right-handed before you exit, when you re-enter you are left-handed due to the reflection.)

Again, the orbifold universal cover of $\mathcal{V}^3$ is $\mathbb{R}^3 = \mathbb{R}^4/T$, and the quotient map $\mathbb{R}^3 \to \mathcal{V}^3$ factors through the natural mapping $\mathbb{R}^3 \to \mathbb{R}^4/OT$, whose image is a 3-dimensional torus $\mathcal{T}^3$. Thus, the orbifold fundamental group G of $\mathcal{V}^3$ contains $\pi_1(\mathcal{T}^3) = \mathbb{Z}^3$ as a normal subgroup (which we will again denote by H), and it is easy to check that the quotient group is isomorphic to S_4. In what follows, we will again realize generators of the finite part G/H of the orbifold fundamental group by lifting loops in $\mathcal{V}^3$ into the space of voiced chords, and mod out by OT-equivalence.

For example, a loop ℓ_1 in $\mathcal{V}^3$ that bounces off a reflection point has a lift into $\mathcal{T}^3$ that interchanges the bottom and top voices of the chord.

Thus, only after lifting two iterations of the loop in $\mathcal{V}^3$ do we get a loop in $\mathcal{T}^3$ that closes up. Next, consider a loop ℓ_2 in $\mathcal{V}^3$, starting at a major/minor 9th chord, that exits trough one half of the bottom face EHM and re-enters through the other half before closing up (see Fig. 10.7(b)). This loop has a lift to $\mathcal{T}^3$ in which the lower two voices change roles (as do the upper two voices) by the time we return to a major/minor 9th chord.

Thus, the lift closes up only after one more iteration, and we see that ℓ_2 (like ℓ_1) is an element of order two in the fundamental group. Finally, consider a loop ℓ_3 that connects the point on edge EQ representing a minor 7th chord, passes through a point representing a half-diminished 7th, and then closes up by arriving at the point on RQ that also represents the minor 7th (due to the identification of EQ with RQ by ρ). When we lift this closed loop to a sequence of voiced chords in $\mathcal{T}^3$, the voices are cyclically permuted.

Thus, the lift must be repeated four times in order to close up, and ℓ_3 is an element of order 4 in the fundamental group. It is now easy to verify that the three loops we have described together generate $G/H \cong S_4$.

10.6.   Conclusions

To sum up, we have discussed how chord and chord-type spaces are discrete subsets inside orbifolds. It is perhaps an exaggeration to say that these sets have a geometry, but they can be given the structure of a graph, where adjacency indicates a single-semitone difference between two chords. The orbifolds $\mathcal{U}^n$ and $\mathcal{V}^{n-1}$ in which these sets are embedded do have a complicated geometry, mitigated by the fact that their orbifold universal covers are easily understood. In the case of $\mathcal{V}^{n-1}$, the generators of the orbifold fundamental group G generators lift to the space of voiced chords to give stepwise modulation sequences that effect an overall transposition, but within these sequences the voice-leading realizes generators of the symmetric group S_n, the "finite part" of G. This relationship between mathematically relevant information and musical relevant phenomena seems novel and intriguing.

Bibliography

[1] M. Boileau, S. Maillot and J. Porti, *Three-Dimensional Orbifolds and their Geometric Structures*, Panorames et Synthèses # 15 (Societé mathématique de France, 2004).

[2] C. Callendar, I. Quinn and D. Tymoczko, Generalized voice-leading spaces, *Science* **320** (2008) 346–348.

[3] R. Hall, Geometrical music theory, *Science* **320** (2008) 328–329.

[4] D. Tymoczko, The geometry of musical chords, *Science* **313** (2006) 72–74.

[5] D. Tymoczko, *A Geometry of Music* (Oxford University Press, 2011).

Chapter 11

Theoretical Physics and Category Theory as Tools for Analysis of Musical Performance and Composition

Maria Mannone

School of Music, University of Minnesota,
2106 4th St., Minneapolis,
MN 55455, USA
manno012@umn.edu

11.1. Introduction

Musical gesture is a growing field of research. There are theoretical studies about gestures in topology [33], homotopy theory [7, 10], and category theory [18]. There are also technological applications, where gestures can modify sound synthesis [6], or new musical instruments and new music are conceived, as well as gesture synthesis development in parallel with sound synthesis for music performance simulation [8, 9]. Gestural approaches are also applied to learning and performing contemporary music [2]. Gestures are also involved in musical perception [29]. Performance gestures, also in the case of not-directly touching the instrument as in theremin performance playing, are a fundamental resource for shaping sounds and expressivity.

However, the influx of gesture into music dates back far before from electricity, computer, and advanced technology. We can think, for example, of conducting gestures, another topic of recent computational study [28], and also to the great dependence of timbre from the performers' chosen gestures. A subtitle of this paper may be 'Gestural Similarity in Music applied to the Orchestra.' Here, the general idea is primarily the creation of a simplified mathematical model to describe musical performance in orchestras, with gestural interaction and communication between performers and conductor — ideally moving from the composer to the conductor to the listener. The second idea is the comparison of musical gestures with 'gestures' of other artistic fields, such as visual arts. The first topic involves new developments of the mathematical theory of musical gestures. The second topic requires the fusion of this theory with an artistic technique to transform tridimensional images into music [14]. This can be done by mapping a selection of points of the image into three axes: time, loudness, and pitch. Because a drawing can be seen as the result of a drawing gesture — and a sculpture as the result of a sequence of sculpting gestures, and so on — we can study visual arts in the frame of gesture theory. Category theory is the environment where mathematical gesture theory is defined; thus our aim is to build the basis of a categorical model of connections between music and image.

Successful experiments of translation from music to visuals and vice versa present crossmodal correspondences [30] between auditory and visual stimuli. This implies that we may try to build a bridge between category theory, interdisciplinary studies between math, music, and art, and elements of psychology. Analogies between gestures on different musical instruments, or in different fields, can be studied with the formal tools of homotopy and of comparison between musical spectra. We will give later a provisional definition of Gestural Similarity [15]. This reminds us of iconicity [5, 24], icons, and gestural analogies between sound and movement [35]. As an application of the theoretical research, we end the paper with some reference to a piece for chamber orchestra, soprano, and piano built upon these ideas [16].

11.2. State of Art About Mathematical Theory of Musical Gestures

After musical gestures were generically defined as a mapping from a skeleton (a directed digraph) to a body (a system of continuous curves in a topological space) [20], the theory progressed. After the first studies for piano gestures [17][a], recent developments involve conducting gesture [21] and singing in terms of (inner) vocal gestures [21]. The general formalism used in [20], depending on the choice of the topological space, can be adapted to a variety of musical instruments. Other developments involve the transformation of symbolic gestures hidden in the scores, into physical gestures of the real performance [17]. This describes the process of performance starting from a score. An inverse process, such as from improvisation to composition, should have inverted arrows.

Former studies on gestures involved theoretical physics with world-sheets [17, 22]. Quantum field theory can be well adapted to describe the transition from symbolic to physical gestures. Moreover, category theory has promising connections with physics [3] and with music [18], and, in general, diagrammatic thinking [1] fits very well with science and art. Musical acoustics is the final result of a musical gesture. A gesture in itself can be seen as a curve in space and time, and it can be investigated with functional analysis.

11.3.　The Categorical Model

Let us start with a classical reference in musical scores. When a composer asks a flutist and a singer to play without vibrato (or with vibrato), he or she is asking for a similar sound effect from different performers. Vibrato sounds in different musical instruments have something in common, if we analyze their spectra. But they are also different in the characteristic shapes, that give the different timbre and allow us to recognize the instrument. We can imagine a square diagram: vibrato flute transformed into vibrato voice, and non-vibrato flute into vibrato voice. And another pair of arrows transforming vibrato in non-vibrato. For each sound, there is a generating gesture. So our diagram becomes tridimensional, as shown in Fig. 11.1. Is this always commutative? It is not. For example,

---

[a]There was a missing decay term in the Poisson equation. However, the numerical solution approximates very well the surface in that article.

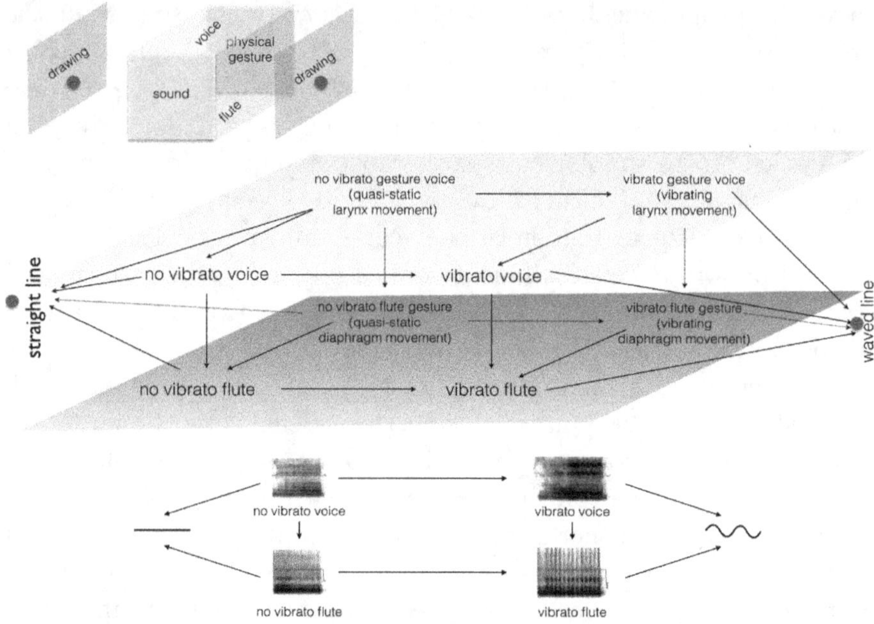

Fig. 11.1. Categorical scheme including the level of sound, of gestures to produce them, their respective analogies, and their approximative graphical descriptions.

a sound synthetically produced, and impossible to be played by any human performer or on any musical instrument is part of a non-commutative diagram. Now, we will add visual dimension. If we ask non-musicians to draw something to indicate non-vibrato and vibrato, they will be likely to draw a straight curve and a waved one, respectively. This is why Fig. 11.1 shows the two lines at the extremes of the cubic diagram. We will now shortly summarize some ideas also described in a recent study [15]. We are using here the mathematical formalism for gestures given in [20], indicating the skeleton with Greek capital letters, and the body with capital Latin letters, with a superimposed arrow to represent the set of curves in the space. Hypergestures are defined as gestures of gestures, and they allow an easy description of nested structures. Let us start with the forte of a pianist. A pianist can play with an unspecified dynamic, making the gesture g (Fig. 11.2, right side). But then he or she can modify the gesture into a forte one, g_F. We can describe this as

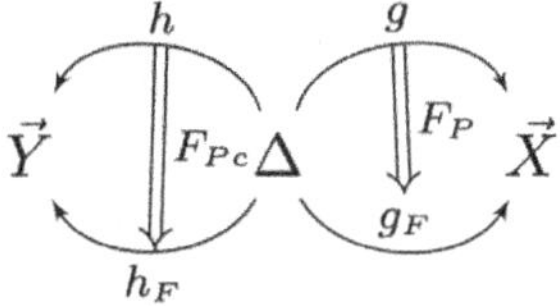

Fig. 11.2. 2-Category of similar gestures: *forte* for the pianist, and *forte* for the percussionist.

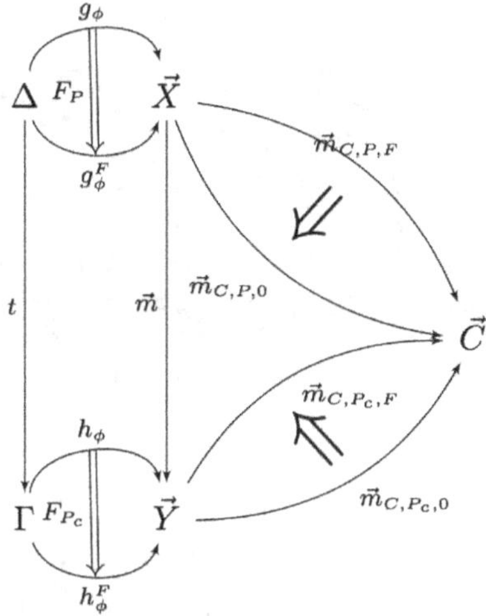

Fig. 11.3. Diagram including conducting gestures.

the effect of a F_P operator within the formalism of 2-categories. If we describe a similar situation with a percussionist, we can use the same formalism and a similar operator F_{Pc}, see Fig. 11.2, left side, with the two gestures sharing the same skeleton. We have to carefully choose the operators depending on the different spaces. For example, if we apply the same operator for percussion on a flute, we do not get a forte on a flute, but we only break the flute. In this case, the forte for the flute can be given by an augmented diaphragmatic pressure. In this frame, we can include conducting gestures, as shown in Fig. 11.3. We can easily define vertical and horizontal composition (see Fig. 11.4). In our system, horizontal composition modifies the topological space, moving a gesture into another (or the skeleta, if also the structure of the gesture changes);

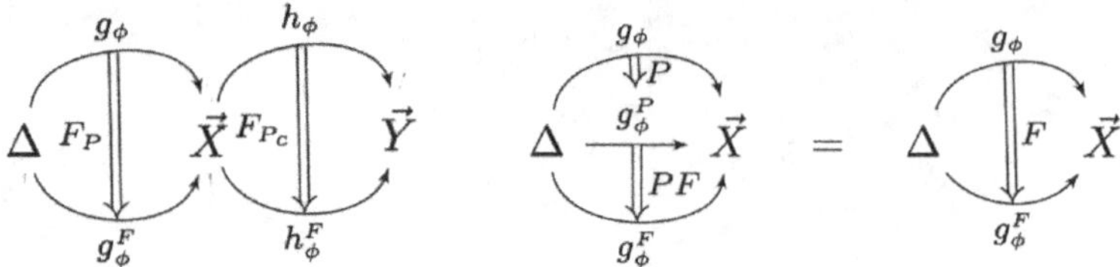

Fig. 11.4. Horizontal composition (on the left): change from one instrument to another; vertical composition (on the right): combination of dynamical levels in a *crescendo* in the given example.

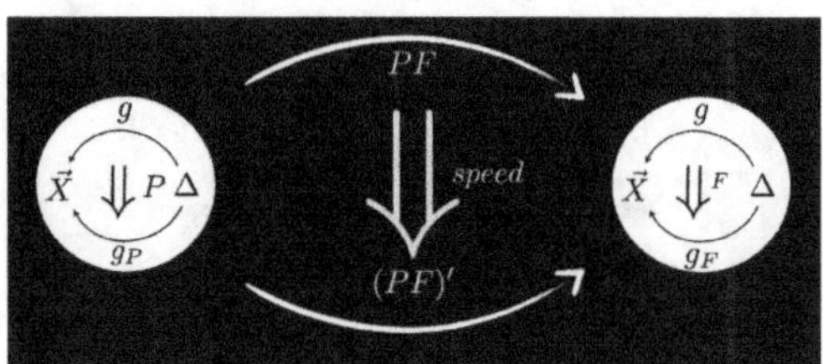

Fig. 11.5. A natural transformation modifying a slow *crescendo* functor (PF) into a fast one (PF)'.

vertical composition deals with the change of dynamic. Is this all? We can include in our model also the change of dynamic. For example, a faster accelerando can be 'transformed' into a slower accelerando via a time operator.[b] Mathematically, it is a natural transformation, see Fig. 11.5. A time operator for a violin can be transformed into a time operator for a cello, and so on, progressively moving toward more complex and higher hierarchies. It is clear that a simplified but precise description of music requires nested categories and higher functors. This is coherent with the concept of hyper and hyper-hypergestures [20]. Attempting to draw the lines of a general model, the concept of n-category helps as the natural environment for such an approach. Theoretically, we can refer to infinite categories [13]. Now, we are ready to give a very first definition of gestural similarity (Fig. 11.6). We say that two gestures are similar if they share the same generator, and their acoustical results are similar. More

---

[b]We are not referring here of tempo operator of note's performance theory [19], but it is evident that the theory at the level of notes can and should be extended at the level of gestures.

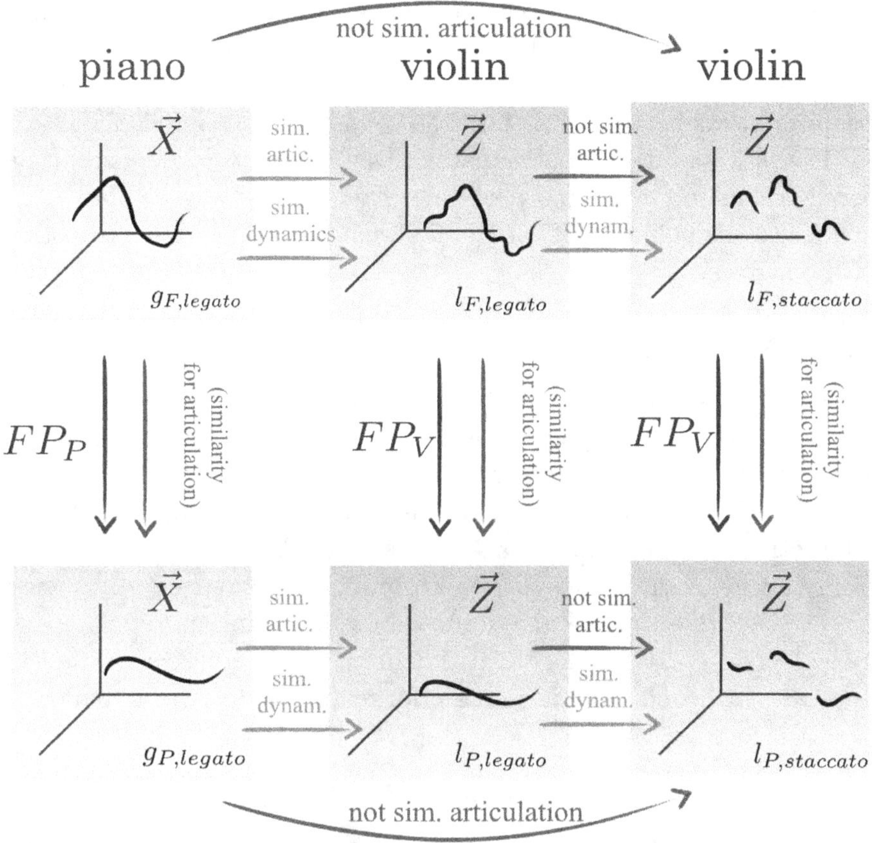

Fig. 11.6. Gestural similarity for loudness and articulation.

precisely, two gestures are similar if they can be transformed the one into the other in a homotopic way, and if the resulting sound spectra show similarities. We can, for example, have two gestures that are similar in articulation — i.e., both staccato — but different in loudness. See Fig. 11.6 for a graphical explanation.[c] Now, we are ready for a conjecture. We can describe conductor, orchestra, and listener in the frame of the same model, seeing conducting

---

[c]Following the observation by Dmitri Tymoczko, homotopy must be adapted to the musical instrument considered. For example, in the case of the trombone, the staccato is given by the lips, as similarly happens in flute playing with tonguing. Inner gestures are also contained in the theory, such as the gestures for singing.

gestures as the colimit of orchestral gestures, and listener's 'gestures' as their limit. Why? Conducting gesture 'contains' all the gestures of the musicians, i.e., their most important elements. The listener projects his or her thoughts into the sound. When there is feedback, for example, a conductor changes his or her gesture depending on orchestral sound, or a solo performer changes the way to play as an answer to the behavior of the public, we can just invert the direction of the arrows.

11.4.   Music and Image

Jackson Pollock throwing the colors into the canvas, and Cecil Taylor throwing notes on the piano keyboard have something in common. We describe their gestures in the same commutative diagram of Fig. 11.7. It is time to apply the categorical structure to the interactions between music and visual arts. Examples are endless: from the program music to not-so-conscious references, to movies. One scene for all: the knife scene in Psycho joins movements and violent string bowing. Gestural similarity may be used as a tool to analyze synchronization in movies, for example.[d] Gestural analogies

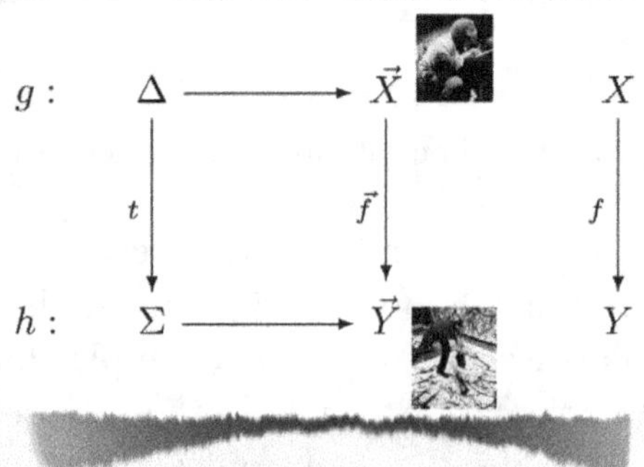

Fig. 11.7. Similarity of gestures in piano playing and painting. The two chosen examples are Cecil Taylor and Jackson Pollock.

---

[d]In movies, also 'anti-similarity' may be defined, to increase the tension: it is not expected to hear a soft music for a violent scene, and vice versa. But the perceptive criteria are still important, and the increase of tension may also be due to unexpected things.

Fig. 11.8. Inner recursion in the fugue: subject inside the subject, as an hyper-gesture. Mm. 63–67, flute, oboe, clarinet, bassoon, horns in F, trumpet in B flat, castanets, snare drum, triangle. For the complete score of this, and the following musical examples: see [17], https://conservancy.umn.edu/handle/11299/188931.

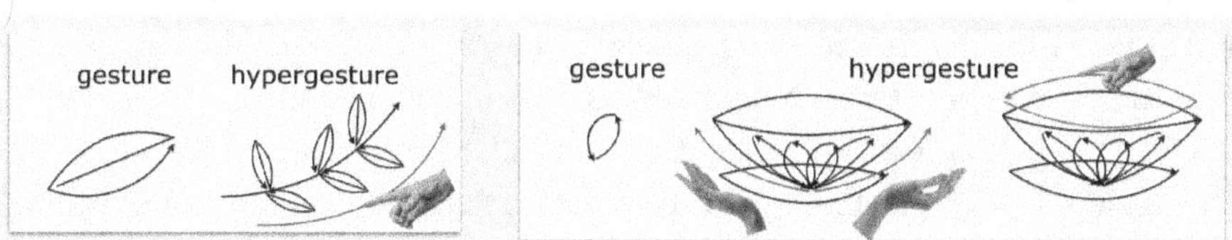

Fig. 11.9.   Tridimensional shape of plants and nymphaea flowers seen as the result of drawing gestures.

are important for artistic expressivity. In fact, non-musically trained people can also describe music via visual references or gestural images. They may say: 'This sounds soft.' And maybe people think of a caressing gesture, in correspondence to a soft piano touch and, consequently, a soft timbre. As discussed in the introduction, it is possible to map two- and three-dimensional images into sound [14, 34]. Some experiments of this second technique involved the

```
N[Table[Sin[n/4 Pi], {n, 0, 4}]]
{0., 0.707107, 1., 0.707107, 0.}

Sound[{SoundNote[0], SoundNote[{7}], SoundNote[{10}], SoundNote[{7}], SoundNote[{0}]}]
```

```
N[Table[-Sin[n/4 Pi], {n, 0, 4}]]
{0., -0.707107, -1., -0.707107, 0.}

Sound[{SoundNote[0], SoundNote[{-7}], SoundNote[{-10}], SoundNote[{-7}], SoundNote[{0}]}]
```

```
Sound[{SoundNote[0], SoundNote[{7, -7}], SoundNote[{10, -10}], SoundNote[{7, -7}],
    SoundNote[{0}]}]
```

Fig. 11.10. Simple Mathematica instructions to get the notes for the leaves.

Sagrada Família in Barcelona, and a 'Concetto Spaziale' by Lucio Fontana [14, 23]. In the second case, the irregular holes in the canvas are musically rendered via staccato and acciaccature.[e] There is a variety of possible mappings, but not all mappings are effective for the listener. Why? The more effective mappings verify crossmodal correspondences between sound and image. This can be described in the Gestalt approach. Several examples are also known under the classical name of synesthesia.[f] Let us discuss a known experiment involving sound and image. In 1924 Uznadze, and Köhler in 1929, [32] proposed an experiment comprised of the presentation of two graphics, one with smooth lines, the other with angles. The person taking the test is asked to connect the two images with one of

---

[e]Italian plural for 'acciaccaturas.'

[f]The word 'synesthesia' is currently used to indicate a neurological condition.

Fig. 11.11. From the Fugue: subject shaped as the bay branch (see the circles). Mm. 71–81, first violins.

the two words 'takete' and 'maluma', respectively. These words do not have real meanings, but they are very different in terms of articulation. Most of the participants associate the first word with the angled image, and the second word with the smooth one. A recent study [25], between linguistics, acoustics and psychology shows an analogy between the angles and the explosive pronunciation of the dental 't' and of guttural 'k', as opposed to nasal 'm' and liquid 'l', giving a plausible reason for this 'obvious' association. Vocal gestures, both in speaking and in singing, can be framed in the context of category theory and mathematical theory of gestures [21]. Intermediate sound/visuals can be described via fuzzy logic, and are also the object of research in psychophysics [15, 31]. Several studies in the field of psychophysics reveal that the link between sound, space localization and visual shape is very deep [26, 27, 30]. There are developments of these topics in psychophysics, cognitive science and neuroscience [4, 11, 12].

11.5.   Examples

We will present here an original creative application of the described ideas: the composition of a piece for small orchestra, piano, and soprano voice [16]. The piece is titled 'Genesis of Music from Gestures,' and it is constituted by two movements, Fantasia and

Fig. 11.12. Nymphaea. Drawing by Maria Mannone.

Fugue.[g] The main concept used in the Fantasia is gestural similarity: the sounds build up little by little starting with the voice, and proceeding via analogies, such as tremolo on strings and on piano and timpani, high staccato with voice and percussion, and so on. The main concepts used in the Fugue are the contraposition between gestures and hypergestures, and the musical rendition of visual shapes. From the Fugue, let us see an example of hypergesture as inner recursion: theme into the theme, see Fig. 11.8. Let us now describe the visuals involved in the piece. The considered visual shapes are an olive branch and a bay branch, and also nymphaea's petals. They are seen as the result of drawing gestures and hypergestures, see Fig. 11.9.

---

[g]The recording can be found here: first part, https://soundcloud.com/maria-mannone/mathnone-genesis-of-music-from-gestures-first-part-fantasia; second part, https://soundcloud.com/maria-mannone/mannone-genesis-of-music-from-gestures-second-part-fugue.

Fig. 11.13. Nymphaea. Made by Maria Mannone.

The notes of the bay branch have been found with software Mathematica, see Fig. 11.10. Higher transpositions give the idea of an ascending branch.

Figure 11.11 shows the (rhythmically variated) subject of the Fugue shaped inside the bay branch notes. Red circles indicate notes of the upper side of a bay leaf.

Musical rendition of visual shapes in terms of drawing gestures opens the way to deeper discussions about discrete and continuous gestures. In fact, a gesture is continuous, but a sequence of its photo shots is discrete. This reminds us of French chronophotography and to futuristic Italian painting (Giacomo Balla for example). When we describe a bay branch as a gesture, we can ideally make the jump to continuous gestures: the leaves will be the 'discrete shots' of a continuous shape. Continuity is rendered in 'Genesis of Music from Gestures' through a glissando on strings, connecting the melodies corresponding to two leaves. This idea allows a connection between continuity and discrete systems such as L-system theory, developed by the botanist Aristid Lindenmayer to study plants' growth and development.

Fig. 11.14. Scheme of the musical rendition of the nymphaea.

We end the chapter with a last musical example, the musical rendition of a nymphaea shape. We start from the image of Fig. 11.12, we make a simplified tridimensional version (Fig. 11.13), and we write a single melodic sequence for big petals, and a single melodic sequence for small petals. We choose a different orchestration for each petal. To give the idea of rotation, or better, or 'walking through' the petals, we use a Doppler-like effect: we change pitch and loudness of the melody in each petal once we approach or we move from a petal, giving in this way the feeling of movement. See Fig. 11.14 for a graphical explanation.

11.6.   Conclusions and Developments

We started from mathematical theory of musical gestures, and we extended it to cover orchestral playing and orchestral conducting.

A key concept used are 2-categories and infinite-categories. Our approach is meant to simplify the understanding of music for scientists, as well to simplify the understanding of science for artists. Further developments may involve measures and degrees of homotopic similarity [2]. Applications of these ideas inside music can improve the development of musical creativity and musical analytical skills, as well as possible improvements of pedagogical skills. How many times has a music teacher tried to explain a specific way to move the hands by referring to some other field? Other extra-musical applications may be suitable for psychology, psychophysics, neuroscience, cognitive sciences. All of these interdisciplinary studies and creative applications have mathematics as language of unification and exchange between different fields and activities.

Bibliography

[1] C. Alunni, *Penser par le diagramme—de Gilles Deleuze à Gilles Châtelet,* chap. Diagrammes et Catégories comme prolégomènes à la question: Qu'est-ce que s'orienter diagrammatiquement dans la pensée? Presse Universitaire de Vincennes (2004).

[2] P. Antoniadis and F. Bevilacqua. Processing of symbolic music notation via multimodal performance data: Brian Ferneyhough's Lemma-Icon-Epigram for solo piano, phase 1. *Proceedings of TENOR* 2016, http://tenor2016.tenor-conference.org/papers/18_Antoniadis_tenor2016.pdf (2016).

[3] J. Baez and A. Lauda, A prehistory of n-categorical physics, in *Beyond the Hilbert Space Formalism: Category Theory,* ed. H. Halvorson (Cambridge University Press, 2011), pp. 13–128.

[4] F. Barbieri, A. Buonocore, R. Dalla Volta and M. Gentilucci, How symbolic gestures and words interact with each other, *Brain Lang* **110**(1) (2009) 1–11.

[5] P. Beim Graben and R. Potthast, Inverse problems in dynamic cognitive modeling, *Chaos* **19**(1) (2009) 015103.

[6] A. Bergsland and R. Wechsler, Composing interactive dance pieces for the MotionComposer, a device for persons with disabilities, *Proc. Int. Conf. New Interfaces for Musical Expression,* online (2015).

[7] F. Bevilacqua, N. Schell, N. Rasamimanana, B. Zamborlin and F. Guedy, in Online Gesture Analysis and Control of Audio Processing, *Musical Robots and Interactive Multimodal Systems* (Springer, 2011), pp. 127–142.

[8] A. Bouënard, S. Gibet and M. Wanderley, Hybrid inverse motion control for virtual characters interacting with sound synthesis — application to percussion motion, *Visual Comput. J.* **28**(4) (2012) 357–370.

[9] A. Bouënard, M. Wanderley, S. Gibet and F. Marandola, Virtual gesture control and synthesis of music performances: Qualitative evaluation of synthesized Timpani exercises, *Comput. Music J.* **35**(3) (2011) 57–72.

[10] J. Françoise, B. Caramiaux and F. Bevilacqua, A hierarchical approach for the design of gesture-to-sound mappings, *Proc. 9th Sound and Music Conf. (SMC)* (2012).

[11] M. Gentilucci and P. Bernardis, Imitation during phoneme production, *Neuropsychologia* **45**(3) (2007) 608–615.

[12] M. Gentilucci, R. Dalla Volta and C. Gianelli, When the hand speak, *J. Physiol.* **102** (2008) 21–30.

[13] J. Lurie, What is an ∞-category? *Notices AMS* **55**(8) (2008) 949–950.

[14] M. Mannone, *Dalla Musica all'Immagine, dall'Immagine alla Musica — Relazioni matematiche fra composizione musicale e arte figurative* (*From Music to Images, from Images to Music — Mathematical relations between musical composition and figurative art*) (Edizioni Compostampa, Palermo, 2011).

[15] M. Mannone, Introduction to gestural similarity in music. An application of category theory to the orchestra, *J. Math. Music.* (2018).

[16] M. Mannone, Musical gestures between scores and acoustics: A creative application to orchestra, PhD dissertation, University of Minnesota (2017).

[17] M. Mannone and G. Mazzola, Hypergestures in complex time: Creative performance between symbolic and physical reality, in *Proc. MCM 2015 Conf.* (Springer, Heidelberg, 2015), pp. 137–148.

[18] G. Mazzola, *The Topos of Music* (Birkhäuser Basel, 2002).

[19] G. Mazzola, *Musical Performance*, Computational Music Science (Springer, Heidelberg, 2011).

[20] G. Mazzola and M. Andreatta, Diagrams, gestures and formulae in music, *J. Math. Music.* **1**(1) (2007) 23–46.

[21] G. Mazzola, R. Guitart, J. Ho, A. Lubet, M. Mannone, M. Rahaim and F. Thalmann, *The Topos of Music III: Gestures* (Springer, Heidelberg, 2017).

[22] G. Mazzola and M. Mannone, Global functorial hypergestures over general skeleta for musical performance, *J. Math. Music.* **10**(3) (2016) 227–243.

[23] G. Mazzola, M. Mannone and Y. Pang, *Cool Math for Hot Music*, Computational Music Science (Springer, Heidelberg, 2016).

[24] W. Mayerthaler, System-independent morphological naturalness, in *Leitmotifs in Natural Morphology*, eds. W. U. Dressler, W. Mayerthaler, O. Panagl and W. U. Wurzel Studies in Language Companion Series, Vol. 10. (Benjamins, 1987), pp. 25–58.

[25] L. Nobile, Words in the mirror: Analysing the sensorimotor interface between phonetics and semantics in Italian, in *Iconicity in language and literature 10: Semblance and Signification*, eds. F. O. Michelucci, Pascal and C. Ljunberg (John Benjamins, Amsterdam/Philadelphia, 2013), pp. 101–131.

[26] J. Ohala, The voice of dominance, *J. Acoust. Soc. Amer.* **72**(S66) (1982). https://asa.scitation.org/doi/abs/10.1121/1.2020007.

[27] S. Roffler and R. Butler, Factors that influence the localization of sound in the vertical plane, *J. Acoust. Soc. Amer.* **43** (1968) 1255–1259.

[28] A. Sarasua, Context-aware gesture recognition in classical music conducting, *Proceedings of the 21st ACM International conference on Multimedia,* pp. 1059–1062 (2013).

[29] M. Schutz and S. Lipscomb, Hearing gestures, seeing music: Vision influences perceived tone duration, *Perception* **36** (2007) 888–897.

[30] C. Spence, Crossmodal correspondences: A tutorial review, *Attention, Perception Psychophy.* **73** (2011) 971–995.

[31] P. Thompson and Z. Estes, Sound symbolic naming of novel objects is a graded function, *Quart. J. Exper. Psycho.* **64**(12) (2011) 2392–2404.

[32] D. Uznadze, Ein experimenteller beitrag zum problem der psychologischen grundlagen der namengebung (an experimental contribution to the problem of the psychological foundations of naming), *Psychol. Forschung* **5**(1–2) (1924) 24–43.

[33] F. Visi, R. Schramm and E. Miranda, Gesture in performance with traditional musical instruments and electronics: Use of embodied music cognition and multimodal motion capture to design gestural mapping strategies, *Moco '14: Proc. 2014 Int. Workshop on Movement and Computing.* Paris (2014).

[34] I. Xenakis, *Formalized Music* (Pendragon rev. edn., New York, 2001).

[35] L. Zbikowski, *Conceptualizing Music.* AMS Studies in Music (Oxford University Press, 2002).

Chapter 12

Intuitive Musical Homotopy

Aditya Sivakumar and Dmitri Tymoczko

Princeton University, 310 Woolworth Center,
Princeton, NJ 08544, USA
dmitri@princeton.edu

12.1. Introduction

Intuitively, a *voice leading* is a way of moving from one chord to another. Formally, it can be represented as a vector or path in the configuration space of possible chords [2, 16]. The collection of all bijective voice leadings from a chord to itself is the *orbifold fundamental group*, a generalization of the fundamental group to singular quotient spaces [8]. This can be extended to the *voice-leading group*, modeling the bijective voice leadings linking transpositionally related chords within some scale (or, in the limit, continuous unquantized chord space); a further extension allows us to consider non-bijective voice leadings in which one or more notes are doubled. In each case, voice leading is closely analogous to homotopy, an exploration of loops and paths in an abstract configuration space. Contemporary mathematics provides tools for describing these loops, allowing us to understand the contrapuntal routes from a chord to itself, or from one chord to another.

For centuries keyboard players have employed physical heuristics to accomplish a similar purpose. David Heinichen, in *Der General-Bass in der Composition*, recommended that improvising keyboardists keep their right hands in "close position" when playing

Fig. 12.1. In close position, voices are ordered registrally and span less than an octave.

Fig. 12.2. Keyboard players find it difficult to distinguish situations in which different voices articulate registrally equivalent arrangements of notes. In this chapter, we will assume that this cannot be done, considering only situations in which voices are ordered in register.

Fig. 12.3. The three pairwise swaps that generate the subgroup of voice exchanges; these swap root and fifth, third and fifth, and root and third.

triads (Fig. 12.1, [1]). These close-position voice leadings comprise an important subgroup of the three-note voice-leading group, containing voice leadings with *no crossings in pitch-class space* [12, 13]. (Note that in this context, a keyboard player cannot physically differentiate between registrally equivalent arrangements of voices, such as those in Fig. 12.2; this limitation will remain in force throughout this paper.) This subgroup is the quotient of the voice-leading group by the *voice exchanges*, a normal subgroup generated by those voice leadings that swap two adjacent pitch classes along equal and opposite paths (Fig. 12.3). Thus musicians long ago identified simple, physically realizable gestures that generate a mathematically interesting subset of contrapuntal possibilities — namely those without voice crossings in pitch-class space.

In this chapter, we develop this connection between the practical and the theoretical, asking how musicians might understand and manipulate the large number of voice leadings theoretically available to them. We begin by analyzing the structure of the voice-leading group. We then generalize the notion of "close position" to five physical configurations — registral arrangements of upper voices, all musically familiar and common in many contrapuntal genres. We then ask "what portion of the theoretically possible voice leadings are reachable using these configurations?" Our main result is that these generalized configurations allow composers to reach a substantial proportion of the "nearby" possibilities, including all of those with less than four voice crossings in pitch-class space.

12.2.   The Voice Leading Group and its Extensions

A *path in pitch-class space* is an ordered pair whose first element is a pitch class (point on a circle) and whose second is a real number (element of the tangent space at that point) representing how the note moves [13, 16]. In this context there is an isomorphism between elements of the tangent space and homotopy classes of paths in the circle, so we can conceive of paths in pitch-class space as motions in the circle. A *voice leading* is a multiset of paths in pitch-class space, representing motion from one chord to another. Voice leadings can be transposed and inverted in the obvious way: transposition acts on a voice leading's pitch classes while leaving its real numbers unchanged; inversion inverts the pitch classes while multiplying the real numbers by –1 [13].

The collection of bijective voice leadings from a chord to itself form a group, the orbifold fundamental group for the configuration space of n-note chords, $\mathbb{T}^n/\mathcal{S}_n$. This can be written as $\mathcal{S}_n \ltimes \mathbb{Z}^n$, with $\mathcal{S}_n$ factor permuting the chord's notes and the $\mathbb{Z}^n$ factor displacing them by octaves [8]. We will find it useful to rewrite this as $\mathbb{Z} \ltimes (\mathcal{S}_n \ltimes \mathbb{Z}^{n-1})$, with $\mathbb{Z}$ the *crossing-free subgroup* and $(\mathcal{S}_n \ltimes \mathbb{Z}^{n-1})$ the subgroup of voice exchanges. As mentioned, the voice exchanges are generated by the n pairwise swaps that exchange adjacent notes in pitch-class space; together, they span an $(n-1)$-dimensional subspace of the configuration space of possible chords [2]. For a non-degenerate chord (i.e., chord whose pitch classes are all distinct), the voice-exchange subgroup is independent of the structure of the chord and the scale it is in; it depends only on the chord's cardinality. The voice exchanges contain all and only those bijective voice leadings $X \to X$ whose paths sum to 0, while the paths in the crossing-free subgroup sum to an integral number of octaves. From this it follows that $xyx^{-1} \in (\mathcal{S}_n \ltimes \mathbb{Z}^{n-1})$ for $x \in \mathbb{Z}$ and $y \in (\mathcal{S}_n \ltimes \mathbb{Z}^{n-1})$, and hence that $\mathcal{S}_n \ltimes \mathbb{Z}^{n-1}$ is normal. This mathematical result underwrites the common musical practice of ignoring or "factoring out" voice crossings and voice exchanges [5, 13].

The *voice-leading group* connects transpositionally equivalent chords, with voice leadings acting in the natural way: if $\mathcal{V}(a)$ is any voice leading from a to one of its transpositions, then $\mathcal{V}(\mathbf{T}_n(a))$ is defined as $\mathbf{T}_n(\mathcal{V}(a))$ [13], where $\mathbf{T}$ is the transposition operator. We can represent these voice leadings as abstract schemas such as "keep the root of the major triad fixed, move its third up by semitone, and its fifth up by two semitones", descriptions that will be unambiguous so long as the chord is not transpositionally symmetrical and contains no duplicate pitch classes. We can write the voice-leading group as $(\mathbb{Z} \times \mathcal{C}_{\gcd(c,n)}) \ltimes (\mathcal{S}_n \ltimes \mathbb{Z}^{n-1})$, with $\mathcal{C}$ the cyclic group and $\gcd(c, n)$ the greatest common divisor of the size of the chord (n) and the size of the enclosing scale (c).[a] This decomposition can be represented graphically as shown in Fig. 12.4, with a line winding n times around the circular dimension of an annulus, the c transpositions of the chord equally spaced along it. Chords sharing the same angular coordinate have pitch classes summing to the same value. The counterintuitive n-fold winding is a manifestation of the chord's n inversions, which are equivalent only assuming octave equivalence; in continuous space one can move orthogonally to the line of transposition between these n inversions, along a series of paths in pitch class space summing to 0.

Crossing-free voice leadings are represented by homotopy classes of paths in the annulus. To understand their structure it is useful

---

[a]For continuous space this becomes $\mathbb{R} \times \mathbb{C}_n$.

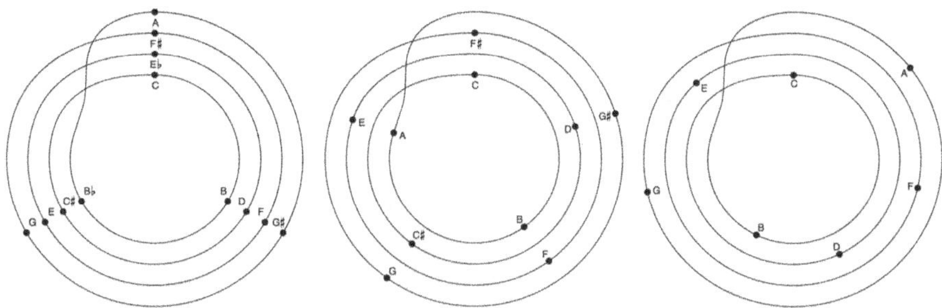

Fig. 12.4. The crossing-free voice leadings for four-note chords in twelve-note, ten-note, and seven-note space.

to adopt the conception of a scale as a metric, described in [13, Chapter 4]; this allows us to use scale-degree numbers even in continuous pitch-class space. Angular motion along the line corresponds to transposition along the scale, with clockwise motion $1/x$ of the way around the circle transposing downward by $1/nx$ steps, and radial motion between adjacent lines representing the "diagonal action" that transposes the chord up $1/n$ of an octave and down one chordal step, thus leaving the sum of its pitch classes unchanged. The resulting chord will not in general lie in the scale. However, by modeling scales as metrics we can conceive of these chords as points in continuous space, labeled with scale-degree numbers.

The subgroup $\mathcal{C}_{\gcd(c,n)}$ contains all the crossing-free voice leadings whose paths sum to zero; it is the only finite, crossing-free subgroup of the voice-leading group.[b] When n divides c there are n distinct copies of the chord at each radial position, related by transposition $1/n$, as in Fig. 12.4(a); here the subgroup $\mathcal{C}_{\gcd(c,n)}$ is the quotient of the crossing-free subgroup by transpositional voice leadings. When n and c share a common factor, there are $\gcd(c,n)$ copies of the chord sharing the same pitch-class sum, as shown in Fig. 12.4(b); here we quotient out just some of the scale's transpositions. And when

---

[b]It is clear that any finite subgroup must have paths summing to zero, since we can iterate any voice leading with sum x to obtain an infinite collection of voice leadings whose paths sum to ix, for $i \in \mathbb{Z}$.

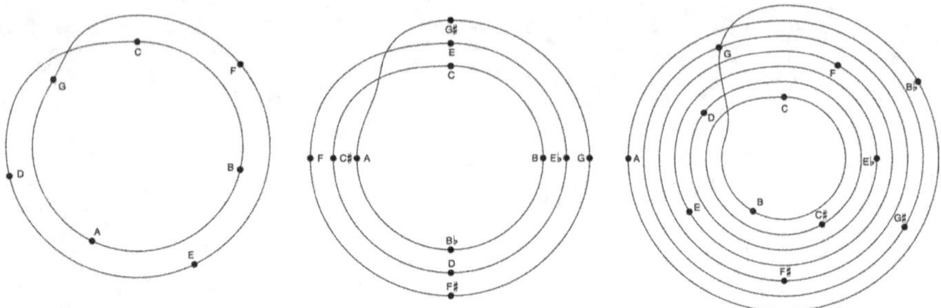

Fig. 12.5. The crossing-free voice leadings for two-note chords in seven-note space, three-note chords in twelve-note space, and seven-note chords in twelve-note space.

the sizes of chord and scale are relatively prime, the crossing-free subgroup is just $\mathbb{Z}$, whose generator is Hook's "signature transform" (Fig. 12.4(c), [6, 7, 15]). In this case, the voice-leading group is isomorphic to the orbifold fundamental group, even though it relates a larger collection of transpositionally related chords. Figure 12.5 shows a variety of graphs for chords and scales of various sizes, all clearly similar.

The annular structure of all of these graphs reflects the topology of the underlying orbifold $\mathbb{T}^n/\mathcal{S}_n$, whose non-singular interior is the twisted product of an $n-1$ simplex with a circle. Angular motion on our graphs corresponds to motion along the circular dimension of $\mathbb{T}^n/\mathcal{S}_n$. A voice leading that makes a complete circle in this dimension acts as a "transposition along the chord", moving voices by the same number of chordal steps — for example, from C up to E, from E up to G, and from G up to C in the major chord $\{$C, E, G$\}$. The radial dimension of the annular graph collapses the $n-1$ dimensions of the simplex; consequently, a sequence of purely radial motions $x_1 \rightarrow x_2 \rightarrow \cdots \rightarrow x_1$ will leave all voices where they began. Radial paths are therefore *contractible* whereas angular circles are not — as we can see both in the annular graphs and in the topology of the orbifolds they model.

A characteristic difference between the theory of voice leading and the earlier "transformational theory" is that the former

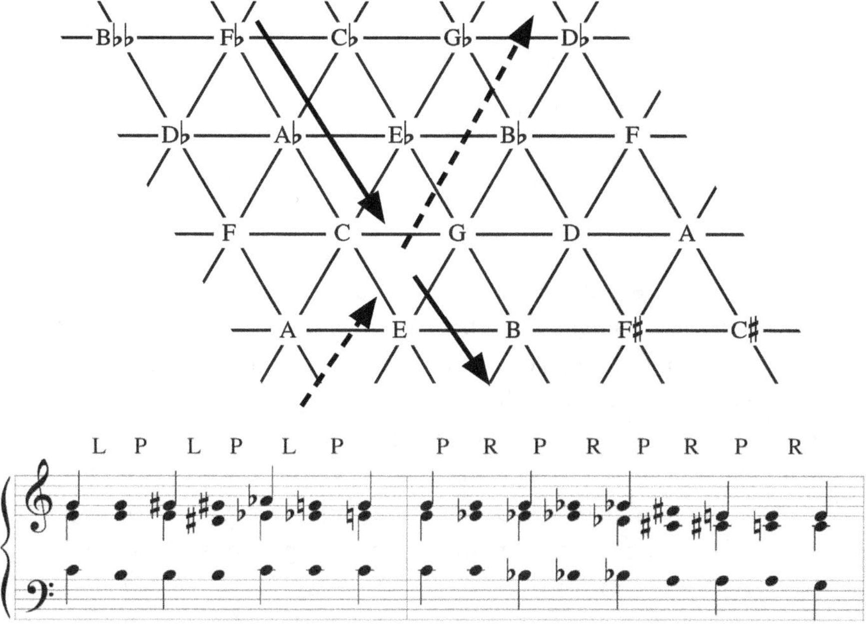

Fig. 12.6. The LP cycle (solid line) and PR cycle (dotted line) look similar on the *Tonnetz*, but are musically very different, the former returning all voices to their starting point and the latter acting as a one-step scalar transposition.

emphasizes the non-trivial effect of these non-contractible circles — representing non-trivial voice leadings from a chord to itself. Traditional "transformational" models sometimes incorporate these facts in a tacit or unrecognized way. For instance, an "LP cycle" on the familiar *Tonnetz* leaves voices unchanged, whereas a "PR cycle" acts as a one-step scalar transposition; this difference is evident when *Tonnetz*-motions are translated into music, even while remaining obscure in the geometry of the *Tonnetz* itself. (See Fig. 12.6 and [3, pp. 35–36].) The theory of voice leadings helps explain these puzzles: the LP cycle links C, E, and A♭ major triads, forming a contractible circle in the orbifold $\mathbb{T}^3/\mathcal{S}_3$, and represented solely by radial motion on Fig. 12.5(b), whereas the PR cycle links C, E♭, G♭, and A major triads, traversing a non-contractible circle in chord space — represented on Fig. 12.5(b) by a full clockwise turn. This distinction is obscured by the common belief that the

Fig. 12.7. A collection of zero-sum voice leadings that move the doubled note from root to third to fifth and back.

Tonnetz is a torus with two distinct non-contractible axes; cf. [14]. Many other transformational models are similar to the *Tonnetz* in implicitly encoding the structural relationships that emerge explicitly in the geometrical approach — details that, in the earlier theory, sometimes become evident only when we translate the model's abstract geometry into actual musical notation.

We can extend the voice-leading group to include non-bijective voice leadings, or voice leadings with doublings. For instance, consider the group of four-voice voice leadings that connect triads in the C-major scale, with one note of each chord doubled. Here there are three possible doublings corresponding to the three notes of the chord: for each particular doubling we have a subgroup of the four-note voice-leading group $\mathbb{Z} \ltimes (\mathcal{S}_4 \times \mathbb{Z}^3)$, equivalent to Fig. 12.4(a).[c] To these possibilities we need to add the voice leadings that move the doubling from one chordal element to another. A natural choice here are the zero-sum, crossing-free voice leadings that move the doubling from root to third to fifth and back (Fig. 12.7). This gives us the *four-voice triadic voice-leading group* $(\mathbb{Z} \times \mathcal{C}_3) \ltimes (\mathcal{S}_4 \times \mathbb{Z}^3)$. Extending this construction to arbitrary chords in arbitrary scales is an interesting problem which we leave for the future, as it is not relevant to what follows.

12.3. Five Registral Configurations

We are interested in the connection between simple physical heuristics, such as "keep the chords in close position", and the voice

---

[c]This is a subgroup because of the degenerate voice leadings that swap identical pitch classes.

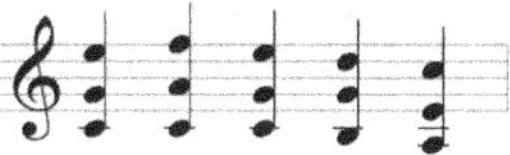

Fig. 12.8. An open-position version of the music in Fig. 12.1.

leading group. Specifically, we ask "what portion of the theoretically possible voice leadings are *performable* with a given set of physical gestures?" We are not, for the purposes of this chapter, interested in forbidden parallels or the avoidance of second-inversion triads: our only question is which voice leadings can be realized given a set of physical options. (The acceptability of those voice leadings, according to the tradition, is a separate question.) Furthermore, in keeping with Fig. 12.2, our notion of "performability" is keyboard-based, forbidding crossings between registrally indistinguishable chords.

For any nondegenerate chord in any scale, close-position voicings suffice to generate all the bijective crossing-free voice leadings. We can extend this result, in the triadic case, by adding the well-known "open" position, which separates the notes of triad so there are two triadic steps between registrally adjacent notes (Fig. 12.8). Moving from close to open position involves one pairwise voice exchange; furthermore, any pairwise voice exchange can be represented in this way by placing the relevant notes in the top and bottom voice. Thus, the second physical configuration expands our contrapuntal range: a composer who uses only close and open positions can access all and only those bijective, triadic voice leadings with at most one crossing.

We now describe a collection of possible configurations for the upper voices in four-voice triadic counterpoint, in which every sonority is a complete and exactly two voices sound the same pitch class. If we require that voices be separated by an octave or less, then there are just the five possibilities in Fig. 12.9: "doubled interval", where two voices sound the same note and the third is less than an octave away; "close position" where the voices sound three different pitches spanning less than an octave; "half-open", where the outer voices are octave apart and the middle voice is between them; "open position", where the voices sound a complete triad spanning more than an octave but with adjacent notes separated by less than an

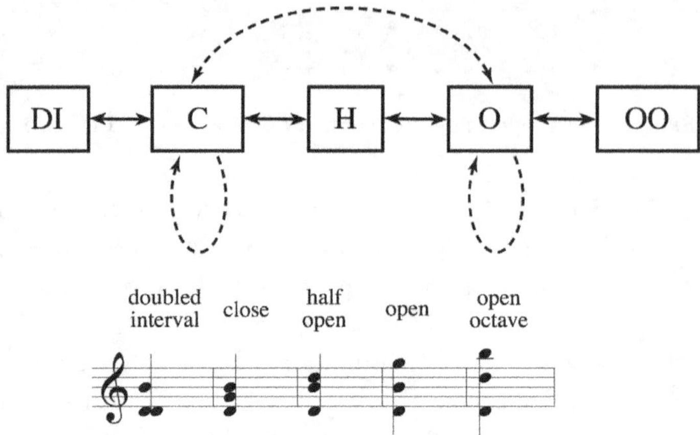

Fig. 12.9. Five upper-voice configurations.

Fig. 12.10. DI can be transformed into H, and H into OO, by moving the middle voice up an octave. The same is true of C and O.

octave; and "open octave", where two adjacent voices are an octave apart, the third less than an octave away from its nearest neighbor. These are arranged in order of increasing separation between voices, so that the "half-open" can genuinely be said to lie between close and open. Clearly, there are two categories here, the complete triadic voicings (C and O) and those with doublings (DI, H, OO). Figure 12.10 shows that these are related by octave displacements: we can turn DI into H, H into OO, and C into O, by transposing the middle voice up by octave. Since we require that all triads be complete, the pitch class of the fourth voice is determined by the content of the upper voices in DI, H, and OO position; in C or O position, the bass can sound any chord tone.

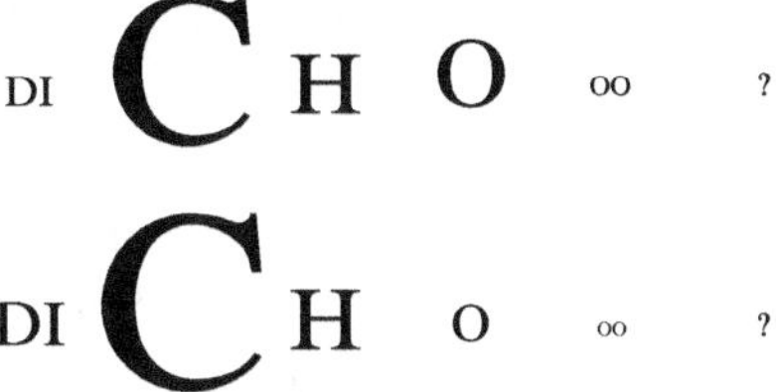

Fig. 12.11. Distribution of upper-voice configurations in Bach (top) and Palestrina (bottom). Character size is proportional to frequency of occurrence.

Table 12.1. The most common transitions in the chorales.

Voice leading	% of total
C→C	31.8
O→O	11.8
C→H	9.1
H→C	8.5
H→O	4.6
O→H	4.3
C→DI	3.6
DI→C	3.5
H→H	3.3
O→C	3.1
C→O	2.8
OO→O	1.7
O→OO	1.6

Figure 12.11 shows the distribution of upper-voice configurations in Bach's chorales and the four-voice passages of Palestrina's masses. Palestrina's upper voices are generally close together, concentrated toward the left side of the graph, while Bach features the C ↔ H ↔ O subsystem. (This may reflect the greater importance of voice-crossings in Palestrina's style, and their almost total absence from the chorales.) Table 12.1 shows the most popular triadic transitions in the chorales: almost a third of the voice leadings connect close-position triads, with another 10% connecting open position to open position; the seven next-most popular patterns connect adjacent positions on the map in Figure 12.9. Note the frequency of motion between C

and O, which is slightly more popular than motion to and from the open-octave position. These common motions are modeled by the lines on Figure 12.9.

Teachers can use these five positions to provide students with a simple and symmetrical set of contrapuntal guidelines: DI goes to C, C goes anywhere except OO, H goes either to C or O, O goes anywhere except DI, and OO goes to O.[d] Together, these rules cover about 85% of the four-voice triadic voice leadings in Bach's chorales and about 75% of those in Palestrina's masses, with Palestrina's greater use of pitch-space voice crossings (e.g., Fig. 12.2) accounting for much of the difference. The approach simplifies the long lists of voice leadings sometimes found in textbooks [9, 10], allowing students to produce idiomatic part-writing quickly and with a minimum of pain: the focus on upper-voice configurations helps them avoid parallels, encourages them to emphasize spacing, and gives them a *positive* set of options to think about. (They know, for example, that if they are in H position then they will be moving to either C or O on the next chord; furthermore they know that it is impossible for upper voices to form parallel octaves following these rules, and unlikely for them to form parallel fifths.[e]) It can also be useful to have students label the configurations in pre-existing music, so that they can see for themselves how they typically behave.

12.4. Performability of Voice Leadings

Suppose a keyboard player is generalizing traditional keyboard practices by using their right hand to articulate three voices in some subset of the DI, C, H, O, and OO, positions, with their left hand unconstrained. What voice leadings can be reached in this manner? That is, for a given set of upper-voice configurations, which voice leadings can and cannot be performed? In asking this question we

---

[d]Note that H $\rightarrow$ H voice leadings will have parallel octaves unless the doubled notes stay fixed.

[e]For example a DI $\leftrightarrow$ C transition can produce fifths only if moving between a close, root position triad and a fifth with one note doubled; H $\leftrightarrow$ O transitions can produce fifths only between adjacent voices.

will assume that the two hands can be separated so that their voices
do not cross in pitch space, or perhaps played on separate manuals of
a multi-manual instrument such as an organ; thus there is no question
of whether the bass crosses the upper voices. We also acknowledge
that the performability of the OO voicings is somewhat unrealistic,
requiring unusually large hands.

Our question fundamentally concerns the orbifold fundamental
group, the collection of voice leadings from a chord to itself. For
suppose we have a four-voice voice leading between two distinct triads
such that its upper three voices are free of crossings while also using
familiar positions. First, we can transpose the destination chord so
that it is equivalent to the starting chord (ignoring the doubling)
without changing the position of its upper voices, as in Fig. 12.12.
Second, we can transform the destination chord so that its doubling
is equivalent to that of the first chord by transposing all notes along
the triad, as in Fig. 12.13; this will again preserve the arrangement
of upper voices, sending DI to DI, C to C, H to H, O to O, and OO
to OO. It follows that we can restrict our attention to voice leadings
from a chord to itself with the very same doublings; in other words

Fig. 12.12. We can transpose the source and destination chord to C without
changing the upper-voice configurations.

Fig. 12.13. We can move the doubling to C without changing the upper-voice
configuration.

we consider the voice-leading group from the triadic multiset {C, C, E, G} to itself.

These manipulations indicate that we should not focus on the absolute distance moved by the voices, since that can be changed by adding an arbitrary transposition to the destination chord, but rather the number of *voice crossings in pitch-class space* contained by the voice leading. In pitch space, the number of voice crossings is equal to the number of crossed voices, but in pitch-class space two voices can cross multiple times: for instance, the voice (D, −14) crosses the voice (C, 1) twice. Geometrically, the number of voice crossings records the number of times the voice leading's path touches the singularities of the orbifold $\mathbb{T}^n/\mathcal{S}_n$. In the three-voice context, we have seen that close position by itself allows the expression of every voice leading with zero crossings, while close and open can express voice leadings with one or fewer crossings. More right-hand positions will allow us to perform voice leadings with more crossings.

Before considering four-voice triadic case, it is worth looking at the simpler situation of four-note chords with no doublings, such as seventh chords. Since there are no doublings, the upper three voices can be in just two positions: close (spanning less than an octave) and open (spanning more than an octave but with less than an octave between adjacent notes). Table 12.2 shows that all voice

Table 12.2. The number of performable and unperformable voice leadings, with various numbers of crossings for a four-note chord with no doublings.

Crossings	Total	Performable	Unperformable	Pct.
0	4	4	0	100%
1	16	16	0	100%
2	40	40	0	100%
3	80	80	0	100%
4	136	112	24	82%
5	208	128	80	62%
6	296	128	168	43%
7	400	128	272	32%
8	520	128	392	25%
9	656	128	528	20%
10	808	128	680	16%

leadings with fewer than four crossings are performable, with the number of performable voice leadings reaching a maximum of 128. The total number of voice leadings with n crossings, when divided by four, yields the series $1, 4, 10, 20, 34, 52, 74, 100, \ldots$, or $2n^2 + 2$ for $n > 0$. This is the number of exterior points on the tetrahedra representing the tetrahedral numbers $1, 4, 10, 20, 35, 56, 84, 120, 165, 220$; our sequence is equal to the nth tetrahedral number t_n for $n < 4$, and $t_n - t_{n-4}$ otherwise (Sloane's A005893; see [4, p. 126, 11]). The musical interpretation is as follows: crossing-free voice leadings whose paths sum to 0 are represented geometrically by voice leadings within a tetrahedral region that tiles three-dimensional space (the "cross section of four-note chord space" discussed in [13]), which contains four "modes" of each sonority, represented on Fig. 12.4(a) by chords sharing the same radial position. The voice leadings with n crossings are found within the tetrahedral regions that can be reached from some particular region by exactly n face-preserving reflections. These tetrahedral regions themselves form the exterior of a tetrahedron isomorphic to that which is used to represent the nth tetrahedral number.[f]

Turning now to four-voice voice leadings between triads: if we require that the left hand stay in just the C position (or equivalently, just O), then we can obtain three of the four crossing-free voice leadings from a doubled triad to itself. The outlier, shown previously as the second voice leading in Fig. 12.12, is what one of us has called a *non-factorizable* voice leading — that is, a voice leading between four-voice triads that does not contain three voices articulating two complete triads [13, Chapter 7]. If the right hand can use *both* C and O, then we can obtain all the one-crossing voice leadings, but we still cannot reach Fig. 12.12's non-factorizable voice leading. (That voice leading, in other words, is the only non-factorizable voice leading with at most one crossing.) If we allow, O, C, and H, then we can perform all the voice leadings with at most one crossing, but we cannot perform the two-crossing voice leadings in Fig. 12.14. And if

---

[f]Similarly, the three-voice voice leadings with n crossings correspond to the number exterior points on the triangle representing the nth triangular number.

Fig. 12.14. Three two-crossing voice leadings that are not performable using just C, H, and O.

Table 12.3. The percentage of voice leadings with various numbers of crossings performable with various upper-voice configurations.

Crossings	Count	C	C, O	C, H, O	All
0	4	75%	75%	100%	100%
1	9	22.2	100	100	100
2	20	10	45	85	100
3	34	5.9	23.5	58.8	100
4	54	3.7	14.8	33.3	79.6
5	74	2.7	10.8	24.3	67.6
6	103	1.9	7.8	19.4	48.5
7	133	1.5	6	13.5	39.1
8	166	1.2	4.8	10.8	30.1
9	206	1	3.9	9.7	24.3
10	250	0.8	3.2	7.2	20.8

we permit all of the five configurations, then we can obtain all the voice leadings with *three* or fewer voice crossings. Table 12.3 shows the total number of voice leadings for various numbers of crossings, and the percentage of these performable with various upper-voice configurations. Note that the doubling now obscures the connection to the tetrahedral numbers discussed earlier.

We conclude that relatively simple set of physical heuristics, all common in the musical literature, allow access to a substantial range of contrapuntal possibilities. More precisely, the five upper-voice configurations of Fig. 12.9 provide a conceptual tool sufficient for manipulating the 67 four-voice triadic voice-leading patterns with the fewest crossings, and about 75% of the 128 patterns with the next-fewest crossings. (Recall here that we are considering only voice leadings from a chord to itself with the very same doubling, as discussed in connection with Fig. 12.13: by allowing the position of the doubling to vary, or allowing chords to be transposed, we obtain many more possibilities.) Note furthermore that these voice leadings can be performed without voice-crossings in pitch space, such as those in Fig. 12.2; in this sense, crossings in pitch space are not necessary for exploring a wide-range of voice leadings. Our configurations are simple enough that they can be internalized by any musician, regardless of mathematical aptitude. These schemas thus provide a link between the embodied world of practical musicianship and the abstract space of topological possibility — showing us exactly what proportion of the homotopy group can be easily accessed by keyboardists.

Our approach treats the upper voices as a unit and the bass as an independent actor. An interesting question is whether this division is simply a matter of convenience or whether it grounded in identifiable features of musical practice. Historically, these ideas originate in figured-bass pedagogy, which described the C and H configurations as convenient hand positions allowing keyboardists to generate four-voice counterpoint in real time; in these styles, the $3 + 1$ division represents a genuine fact about the music, which treats of the bass as distinct from the chordal upper parts and assigns them to different hands. We have generalized the traditional categories of figured-bass pedagogy to encompass *all possible upper-voice configurations* in which the voices are registrally ordered and no more than an octave apart. This generalized strategy is potentially applicable to any music that limits the distance between voices, whether conceived in a $3 + 1$ fashion or not. It is therefore worth asking to what extent different musical repertoires support this partitioning into bass and upper voices, or whether it is merely an external framework that we have imposed on the music.

The answer is that the bass typically plays a exceptional role, though the degree of independence varies from repertoire to repertoire. In the four-voice chords found in Palestrina's masses, for example, the bass is most likely to sound chord roots (69% of all sonorities vs. 25–30% for the other voices), most likely to move by fifth (9% of all intervals vs. 4–5% for the others), most likely to have its note doubled by another voice (32% vs. about 23%) and is separated on average by the greatest distance from the nearest voice (by about half a diatonic step more than the others). These subtle asymmetries are larger in other genres, not just later music such as Bach's chorales, but also the more chordal genres of the sixteenth century — Frottole, for example, or Goudimel's 1560s harmonizations of the Geneva Psalter. In this sense, it seems that the $3 + 1$ division is not just a mathematical conceit, but a distinction supported by even the exemplars of Renaissance polyphony.[g] One might say that it points toward a subtle and implicit precursor to later chordal thinking.

Voice leading involves complicated mathematics, groups representing homotopy classes of paths in higher-dimensional, singular spaces. These spaces were explored by practical musicians long before mathematicians developed the concepts needed to describe them. Upper-voice configurations are an intuitive tool for manipulating these musical possibilities, learned implicitly and deployed not just by keyboard players but composers more generally. Modern music theory allows us to connect these two approaches, making explicit the musician's implicit knowledge, giving students new tools, and allowing us to appreciate the deep and inherently mathematical knowledge of the practical musician.

Bibliography

[1] G. Buelow, *Thorough-Bass Accompaniment According to Johann David Heinichen.* University of Nebraska Press, Lincoln, (1992).

[2] C. Callender, I, Quinn and D. Tymoczko, Generalized voice-leading spaces, *Science* **320** (2008) 346–348.

[3] R. Cohn, Neo-Riemannian operations, parsimonious trichords, and their 'Tonnetz' representations, *J. Music Theory* **41**(1) (1997) 1–66.

[4] E. Deza and M. Deza, *Figurate Numbers*, (World Scientific, Singapore, 2012).

---

[g]A further subtlety: in a genre that makes frequent use of voice crossings, there is a difference between *the lowest voice acting special* and *there being a single musical voice that acts like a special lowest part.* In Palestrina's music, for example, the tenor sometimes crosses below the bass and "acts like a bass voice" for a while.

[5] A. Forte and S. Gilbert, *Introduction to Schenkerian Analysis* (Norton, New York, 1982).

[6] J. Hook, Signature transformations, in *Music Theory and Mathematics: Chords, Collections, and Transformations*, eds. J. Douthett, M. M. Hyde and C. J. Smith. (University of Rochester Press, Rochester, 2008), pp. 137–160.

[7] J. Hook, Contemporary methods in mathematical music theory: A comparative case study, *J. Math. Music* **7**(2) (2013) 89–102.

[8] J. Hughes, Using fundamental groups and groupoids of chord spaces to model voice leading, in *Mathematics and Computation in Music, Proc. 5th Int. Conf.* eds. T. Collins, D. Meredith and A. Volk, (Springer, New York, 2015), pp. 267–278.

[9] S. Kostka and P. Dorothy, *Tonal Harmony*, 4th edn. (Alfred A. Knopf, New York, 2003).

[10] A. McHose, *The Contrapuntal Harmonic Technique of the 18th Century.* (Crofts, New York, 1947).

[11] OEIS Foundation, The on-line encyclopedia of integer sequences, (2017) http://oeis.org.

[12] D. Tymoczko, Scale theory, serial theory, and voice leading, *Music Anal.* **27**(1) (2008) 1–49.

[13] D. Tymoczko, *A Geometry of Music* (Oxford University Press, New York, 2011).

[14] D. Tymoczko, The generalized tonnetz, *J. Music Theory* **56**(1) (2012) 1–52.

[15] D. Tymoczko, Geometry and the quest for theoretical generality, *J. Math. and Music* **7**(2) (2013) 127–144.

[16] D. Tymoczko, In quest of musical vectors, in *Mathemusical Conversations: Mathematics and Computation in Performance and Composition*, E. Chew, G. Assayag and J. Smith, (Imperial College Press/World Scientific, Singapore, 2016), pp. 256–282.

Chapter 13

Geometric Generalizations of the *Tonnetz* and Their Relation to Fourier Phases Spaces

Jason Yust

Boston University School of Music,
855 Comm. Ave., Boston, MA 02215, USA
jyust@bu.edu

13.1.   Formulations of the *Tonnetz*

The *Tonnetz* originated in nineteenth-century German harmonic theory as a network of pitch classes connected by consonant intervals. Initially presented as a grid of perfect fifth and major third intervals, Arthur von Oettingen and Hugo Riemann used it to show acoustical relationships between tones, but in Ottokar Hotinský's reformulation, it became a triangular network of tones connected by perfect fifths, major thirds, and minor thirds. Riemann adopted this form of the network as a map of tonal relationships for analysis of chromatic harmony [8, 11]. The important feature of this triangular network is that each triangle contains all the pitch classes of a major or minor triad, and triads sharing two pitch classes are adjacent across a common edge.

In an influential article, Richard Cohn [7] suggests that the *Tonnetz* can be generalized by changing the intervals of the network, so that the triangles correspond to different possible trichord types. He also points out a feature unique to the triadic *Tonnetz*, not shared by other possible *Tonnetze* within the 12-tone system: for any two adjacent triads, the unique tones that distinguish them are always separated by a small distance, one or two semitones.

As a result of this long history, the *Tonnetz* has acquired, as Dmitri Tymoczko points out [18], at least three theoretical meanings: a map of acoustical relationships between tones, a map of common-tone relationships between triads, and a map of minimal voice-leading relationships between triads. Cohn's generalization makes little sense according to the acoustical understanding of the *Tonnetz*. It also prioritizes common-tone relationships over stepwise voice leading, which Tymoczko takes as the essential property of a *Tonnetz*. Nonetheless, Cohn's common-tone based formulation is of significant musical interest, and Tymoczko's voice-leading *Tonnetze* may in fact be understood as special cases of it. Considering the topic from a geometrical perspective, rather than a purely topological approach — as reflected in a number of recent mathematical papers on generalized *Tonnetze* [3, 4, 6] — helps to enrich their potential music-theoretic applications. In this sense the work below shares a kinship with Tymoczko's recent work, also fundamentally grounded in geometric considerations, but the geometries derived here have a different mathematical basis, the discrete Fourier transform (DFT).

For the *Tonnetz* as a network, the triangles are mathematically significant as its *maximal cliques* (maximal sets of vertices all of which are adjacent). The corresponding geometrical concept is the *simplex*. Thus, as Louis Bigo and others have observed [4], the *Tonnetz* may be understood as a simplicial complex, and generalized along these lines. From this perspective, Hotinský's "completion" of the network was necessary and inevitable, because Oettingen's original grid contains no cliques that could be identified with 2-simplexes (triangles). The *Tonnetz* is more than just a simplicial complex, however, because it also fills an entire space with 2-simplexes, making it a *triangulation* of that space. This will have important ramifications for three-dimensional *Tonnetze*.

13.2. Definition and Construction
of Two-Dimensional *Tonnetze*

We can define a *Tonnetz* geometry with some basic stipulations:

- There are a finite number of pitch classes corresponding to some $\mathbf{Z}_u$.
- Each pitch class is assigned a unique point in the space.
- *Transposability*: All translations of $\mathbf{Z}_u$ (musical transpositions) correspond to some rigid geometric transformation of the space that maps the pitch-classes appropriately.

The transposability condition is crucial; it demands that the cyclic structure of the pitch-class universe be reflected in the geometry as rigid transformations. It also implies that intervals of $\mathbf{Z}_u$ of the same size are represented by the same distances.

A straightforward construction that maps musical transpositions to translations in a cyclic space can immediately generate a family of spaces satisfying the above stipulations for any $\mathbf{Z}_u$. The procedure is as follows:

(1) Choose a number of dimensions.
(2) For each dimension, n, choose a mapping $1 \mapsto 2\pi x_n/u$ where x_n is an integer between 0 and $n - 1$, such that all of the x_n are mutually coprime to u.
(3) Place the pitch classes by iterations of interval $1 = (x_1, x_2, \ldots)$.

In step (3), and throughout, a constant multiplier of $2\pi/u$ is assumed. It is clear that this procedure will produce a geometric embedding of the pitch classes that satisfies the transposability condition, where translations of the space by $(kx_1, kx_2, \ldots)$ correspond to transposition by k. The coprime condition ensures that each pitch class has a unique position in the space.

Given a space satisfying the above conditions, we can define a generalized *Tonnetz* as the *triangulation* of the space whose vertices correspond precisely to the pitch classes, and where the dimension of the simplexes matches the dimension of the space. The following construction for the two-dimensional case will be generalized to 3 and n dimensions below.

(1) Define a two-dimensional space for the given $\mathbf{Z}_u$ as specified above.
(2) Choose a line segment that connects pitch-class 0 to any other pitch class without passing through another pitch class.
(3) Translate this line segment to all pitch classes.
(4) Choose another line segment from pitch-class 0 to any other that is not parallel to the first line segment and does not cross it or any of its translates. Translate this to all pitch classes. The result is a skewed square lattice.
(5) Choose a diagonal of one of the parallelograms and translate it to each pitch class.

A simple example is given in Fig. 13.1 for $\mathbf{Z}_6$, represented as a whole-tone universe. The pitch classes are set by choosing $1 \mapsto (1,2)$. We make a lattice by first connecting the major thirds with line segments $(2,-2)$ from each pitch class, then the major seconds with $(1,2)$ line segments. There are two possible diagonals to bisect this lattice, one of which (connecting the tritones) creates an (026)-*Tonnetz*. The other possibility would give an (024)-*Tonnetz* by creating a distinct set of major-second edges. This possibility of interval duplications is inherent to the formalism and will be further explored below.

Other *Tonnetz* formulations that have been proposed relate to this toroidal construction, although they may display other

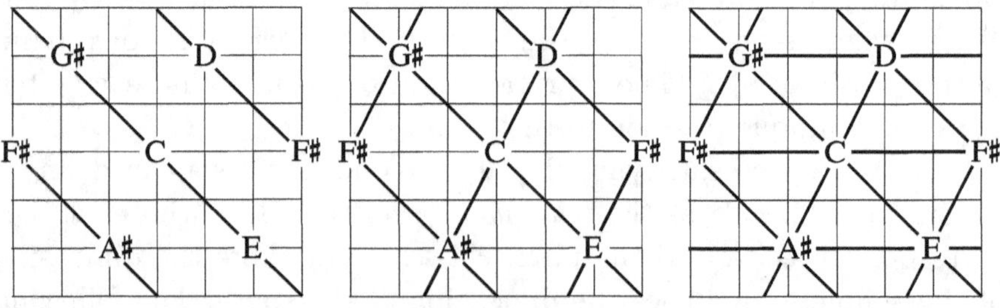

Fig. 13.1. Construction of a *Tonnetz* in a whole-tone universe.

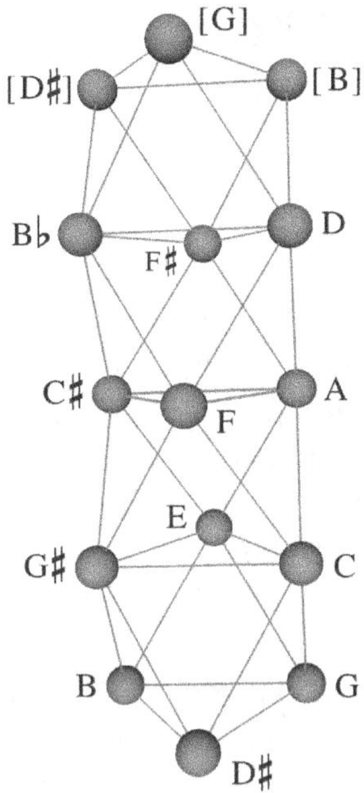

Fig. 13.2. Tymoczko's triadic *Tonnetz* in voice-leading space [18].

topological features. For instance, Tymoczko's triadic voice-leading *Tonnetz* (Fig. 13.2) is formulated in 3-note chord space, which is a bounded, non-orientable triangular prism whose ends are glued with a twist [17]. The geometry satisfies the transposability condition through translation along the central axis of the space and screw rotation around the axis. The dimensionality of the space can be reduced, however, by taking just the toroidal surface surrounding this axis where all the pitch classes lie, making it equivalent to one of the toroidal spaces derivable from the construction proposed above. The purpose of Tymoczko's extension of this space is to show the augmented triads as triangles in the added dimension cutting across the torus, essential to his minimal voice leading requirement.

Other kinds of spaces can be derived in certain cases by folding toroidal ones, resulting in non-orientable, bounded, and spherical topologies. Examples are described below.

13.3.   Geometric *Tonnetze* and Fourier Phase Spaces

Emmanuel Amiot [1] first noted a connection between Fourier phase spaces and the *Tonnetz*. Phase spaces are defined by applying the DFT to the characteristic function of a pitch-class set, then taking the phase values of certain coefficients as coordinates. Because phases are cyclic, the resulting spaces are toroidal. Amiot observed that using phases of the third and fifth coefficients yields an arrangement of major and minor triads mimicking the dual of the standard *Tonnetz*. I investigated this further [20] observing that the *Tonnetz* was a triangulation on the pitch classes, and that such a triangulation could be defined for any trichord in any two-dimensional phase space. Figure 13.3 shows the standard *Tonnetz* in a $\mathrm{Ph}_{3,5}$-space (where "$\mathrm{Ph}_{x,y}$" refers to a two-dimensional phase space on phases of coefficients x and y [21].) This space is optimal for this *Tonnetz*, but a well-formed *Tonnetz* for any trichord is possible in this space provided that $\mathbf{Z}_{12}$ is its minimal embedding universe. For example, we can draw an (013)-*Tonnetz*, although the $\mathrm{Ph}_{3,5}$-space is clearly not optimal for making compact (013) regions.

This naturally raises the question of whether the reasoning can be reversed, deriving the Fourier phase spaces from some elementary principles about the geometric embedding of a *Tonnetz*. The above

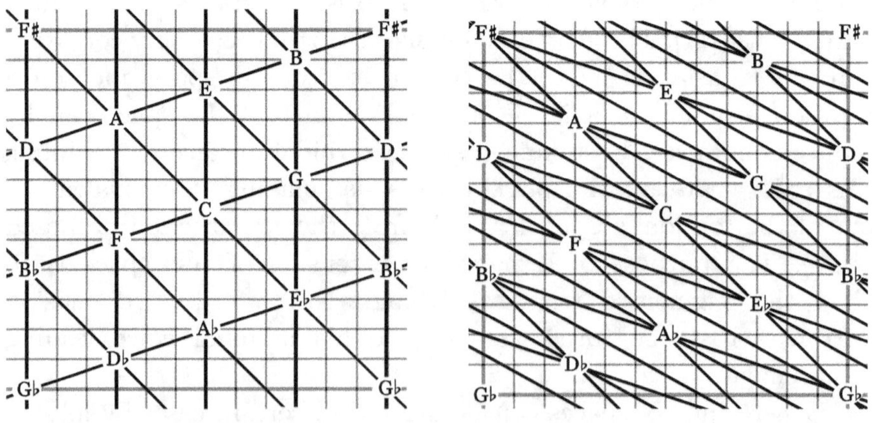

Fig. 13.3. The standard *Tonnetz* and (013)-*Tonnetz* in $\mathrm{Ph}_{3,5}$-space.

construction shows that this is indeed the case, provided we limit the transposability condition to translations, enforcing a simple toroidal topology. The one-dimensional cyclic spaces defined by the condition $1 \in \mathbf{Z}_u \mapsto 2\pi x_n/u$ are precisely the possible one-dimensional phase spaces for the DFT of a pitch-class set in universe u, and an n-dimensional phase space is a direct product of n one-dimensional spaces.

The relation to phase spaces may be a key to successful music-analytic application of non-triadic *Tonnetze*. Thus far, such applications have been limited; examples include an (013)-*Tonnetz* analysis of a Bach fugue subject by David Lewin [12, 15], Van den Toorn and McGinness's octatonic-diatonic (025) *Tonnetz* [19], Stephen Brown's ic1/ic5 *Tonnetz* [5], and Siciliano's [16] Cohn-cycle analyses which could be readily plotted in (026) and (014) *Tonnetze*. These analyses typically require a high degree of saturation of a musical passage with a single trichord type, and must leave by the wayside aspects of the harmony not accounted for by that trichord. Yet recent research has indicated a wide range of analytical applications of Fourier phase spaces. [2, 21–23] The inherent limitation of the *Tonnetz* as a network to single trichord-type may be loosened up by the embedding in phase spaces, which situate it in a conceptually richer music-theoretic context. In addition, Amiot's [2] application of the DFT to beat-class sets introduces the possibility of rhythmic *Tonnetze*.

13.4.   Intervallic Duplications and Foldings

One important feature of toroidal spaces is that there are an infinite number of conceivable line segments connecting any two given points. This means that a musical interval may be identified with a vector in a phase space but not necessarily a specific realization of that vector as a path. I have previously [20] proposed an expanded concept of musical interval that allows for such distinctions, and a corresponding concept of *intervallic axes* in phase spaces. The possibility of multiple realizations of a given interval has important consequences for generalized *Tonnetze*.

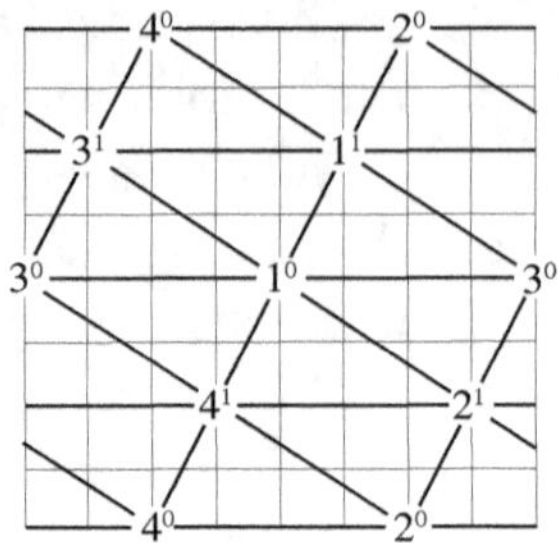

Fig. 13.4. A rhythmic *Tonnetz.*

Consider the (026)-*Tonnetz* of Fig. 13.1. Each tritone occurs twice as an interval in the network and the distinction is significant. The tritone F♯–C on the left, for instance, is filled in differently (by G♯ or A♯) than the one oriented as C–F♯ on the right (which is filled in by D or E). Figure 13.4 provides a similar example in the rhythmic domain, where numbers refer to beats of a 4/4 measure and superscripts to eighth-note offsets. The chosen phase space represents temporal proximity on the x-axis and the possible syncopations of the half-note pulse on the y-axis. The *Tonnetz* is made up of ♩.♪♩ and ♪♩.♩ rhythms shifted to different places in the measure. A given half-note pulse is represented in two forms differing in which half of the measure is filled.

The necessity of duplicated intervals in toroidal *Tonnetze* explains why Cantazaro [6] finds non-toroidal topologies in his computations, because his procedure does not recognize the possibility of duplicated intervals. In many situations, the distinction between forms of the same interval may be musically significant, and hence it is preferable to retain the toroidal space. Yet, if the distinctions are not significant, it is possible to fold the space over a given intervallic axis, creating a bounded space. For example, the triadic $\mathbf{Z}_7$-*Tonnetz* in Fig. 13.5 has two copies of each third, and, because the triad is symmetrical in this universe, there are also two copies of each triad. Therefore, it is possible to shear and fold the space over the fifths axis such that the two forms of third map onto each other and the equivalent triads. The result is the triadic Möbius

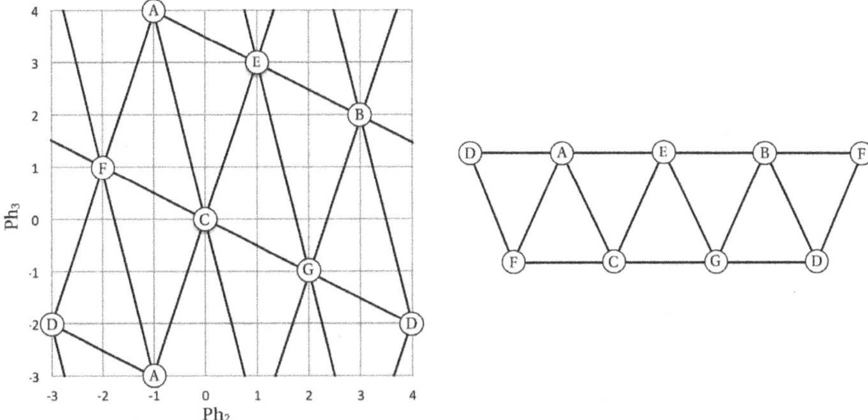

Fig. 13.5. A triadic *Tonnetz* in $\mathbf{Z}_7$ and a folding of it into a triadic Möbius strip.

strip discussed by Muzzulini [14] and Mazzola [13]. The same kind of folding can be performed in any instance where a symmetry exists in the *Tonnetz*, and the result is a bounded space that will be orientable in an even-cardinality universe and non-orientable in an odd-cardinality universe. Muzzulini's folding simplifies the triadic *Tonnetz*, but the toroidal version might also be of theoretical value. One might, for instance, treat the upwards oriented third as major and the downward oriented one as minor, so that a distinction between major (upwards pointing) and minor (downwards pointing) triads can be reflected in the *Tonnetz* despite the fact that chromatic variants of a given generic note-name are not distinguished in the space.

The $\mathbf{Z}_4$ *Tonnetz* presents a unique case in which a spherical topology is possible through folding. The full *Tonnetz* is shown in Fig. 13.6 as a *Tonnetz* on (036) subsets of a diminished seventh chord. Due to the symmetry of the trichord, a folding is possible over the tritone axis. Because the resulting boundaries are tritone axes, they also contain duplications that can be identified. This folding reduces the edge count by two without changing the number of vertices or triangles. Hence, the Euler characteristic of the space changes from 0 (the value for all toroidal *Tonnetze*, which in the

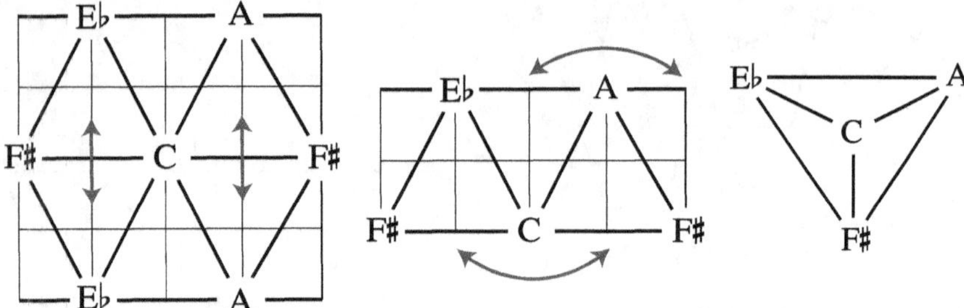

Fig. 13.6. Deriving a spherical *Tonnetz* through folding of the $\mathbf{Z}_4$ *Tonnetz*.

two-dimensional case have u vertices, $3u$ edges, and $2u$ triangles) to 2. The spherical *Tonnetz* satisfies the transposability conditions through rotations of the sphere, a cyclic subgroup of the tetrahedral symmetry group.

13.5. Optimization

The observation that any *Tonnetz* can be drawn in any phase space for the given universe motivates the search for a method of optimizing the space to the *Tonnetz*. The obvious criterion for optimization is to seek the most compact spaces, avoiding stringy regions like those in the (013)-*Tonnetz* of Fig. 13.3. However, this could potentially be operationalized in different ways. One method would be to minimize the perimeter of the regions, which amounts to minimizing the number of times each axis cycles the space. However, this measure turns out to be somewhat coarse. The connection to the DFT offers another more sensitive method, which is to take the magnitude of the DFT components for the given trichord (in the two-dimensional case). These indicate how compact the trichord is in each dimension, so maximizing them optimizes the fit.

The existence of duplicated intervals in the set class, however, complicates the optimization process, because the space must allow for two different line segments representing the same interval, such that neither is excessively long. This is made possible by choosing

Table 13.1. Optimum *Tonnetz* spaces for each trichord of $\mathbf{Z}_{12}$.

Trichord	Best space	DFT mag's	Wrapping	Other space	DFT mag's	Wrapping
(012)	$\mathrm{Ph}_{1,6}$	2.73, 0.73	16			
(013)	$\mathrm{Ph}_{1,4}$	2.39, 1.73	14	$\mathrm{Ph}_{1,5}$	2.39, 1.51	16
(014)	$\mathrm{Ph}_{1,3}$	1.93, 2.24	14	$\mathrm{Ph}_{3,4}$	2.24, 1.73	14
(015)	$\mathrm{Ph}_{2,3}$	2, 2.24	14			
(016)	$\mathrm{Ph}_{1,2}$	1, 2.65	16			
	$\mathrm{Ph}_{2,3}$	2.65, 1	16			
	$\mathrm{Ph}_{2,5}$	2.65, 1	16			
(025)	$\mathrm{Ph}_{4,5}$	1.73, 2.39	14	$\mathrm{Ph}_{1,5}$	1.51, 2.39	16
(027)	$\mathrm{Ph}_{5,6}$	2.73, 0.73	16			
(037)	$\mathrm{Ph}_{3,5}$	2.24, 1.93	14	$\mathrm{Ph}_{3,4}$	2.24, 1.73	14

one dimension that *minimizes* the DFT value of this interval. A complete optimization strategy then is first to ensure that there is a minimizing dimension for each duplicated interval, then to choose the remaining dimensions so as to maximize the DFT components of the set class.

Table 13.1 shows the results for each trichord type of $\mathbf{Z}_{12}$. The wrapping number gives the total number of times all of the axes cycle the space in either dimension. In two cases, (014) and (037), the DFT magnitudes distinguish between two possibilities with the same wrapping number. The association of trichords with optimum spaces (excluding degenerate spaces $\mathrm{Ph}_{1,1}$ and $\mathrm{Ph}_{5,5}$) is as close to one-to-one as possible given that there are more possible spaces than trichords. $\mathrm{Ph}_{1,2}$ and $\mathrm{Ph}_{2,5}$ appear as alternate spaces for the same trichord, (016), and $\mathrm{Ph}_{1,5}$ and $\mathrm{Ph}_{3,4}$ appear only as second-best spaces for two possible trichords.

Figure 13.7 shows an (012) *Tonnetz* in $\mathrm{Ph}_{1,6}$ space. The (012)s are not very compact in the Ph_6 dimension, but this is optimal because it allows for two distinct ic1 intervals splitting the Ph_6 cycle in half. The upward-pointing and downward-pointing (012)s are enharmonically the same, but are distinguished by spelling, taking advantage of the redundancy.

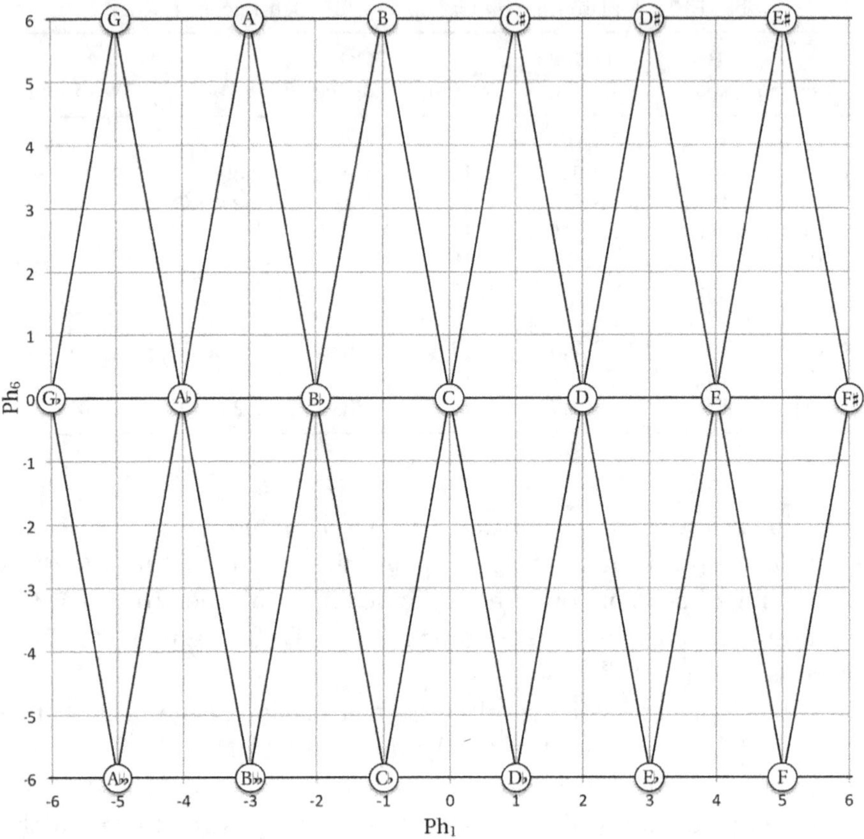

Fig. 13.7. An enharmonic (012) *Tonnetz*.

13.6.   Three-Dimensional *Tonnetze*

The idea of a *Tonnetz* as a triangulation gives new perspective on
a persistent question of how to generalize the *Tonnetz* to tetrachords. Edward Gollin [10] proposed a three-dimensional network of
tetrahedra as a seventh-chord *Tonnetz*. However, Gollin (and others)
have also assumed that a *Tonnetz* must be specific to a single set
class, which means that the (0258) tetrachords of Gollin's *Tonnetz*
sometimes share a complete triangular face with another tetrachord,
but most of the time link to one another in, at best, a shared edge,
leaving a lot of empty space between tetrahedra. Generalizing the

idea of a complete triangulation of a toroidal space instead motivates discarding the requirement of a single tetrachord type.

The construction of a three-dimensional *Tonnetz* follows roughly the same procedure as the construction of the two-dimensional *Tonnetz*:

(1) Choose a three-dimensional phase space for $\mathbf{Z}_u$.
(2) Choose any line segment from pitch class 0 to some pitch class as an edge that does not pass through any other pitch class. Translate it to each pitch class in the space.
(3) Choose another line segment from 0 to some pitch class that is not parallel to and does not cross any of the previous line segments, and translate it to each pitch class.
(4) Repeat the previous step. The result will be a skewed regular cubic lattice partitioning the space into u regions (parallelepipeds).
(5) Choose a parallelepiped incident upon pitch-class 0. For each of its three faces, add a diagonal from 0, and translate these to each other pitch class in the space.
(6) Add a diagonal for the entire parallelepiped from 0, and translate this similarly.

The entire process results in a total of seven distinct sets of intervallic axes defining the three-dimensional *Tonnetz*. This means that for $u = 12$, at least one interval type must be duplicated, since there are only six interval classes. Figure 13.8 illustrates the process schematically through a particular choice of intervals. The cubic lattice of step (4) defines three planes and their intersections with three intervallic axes. These three intervallic axes (which correspond to interval classes) plus the choice of orientation for each (which correspond to intervals proper) determine the rest of the process, and hence the entire *Tonnetz*. Choosing the same interval classes with different orientations would amount to orienting the parallelepiped in Fig. 13.8 from a different corner, and would result in different *Tonnetze*. However, reversing orientations of *all* of the intervals (orienting from the F♯) results in the same *Tonnetz*.

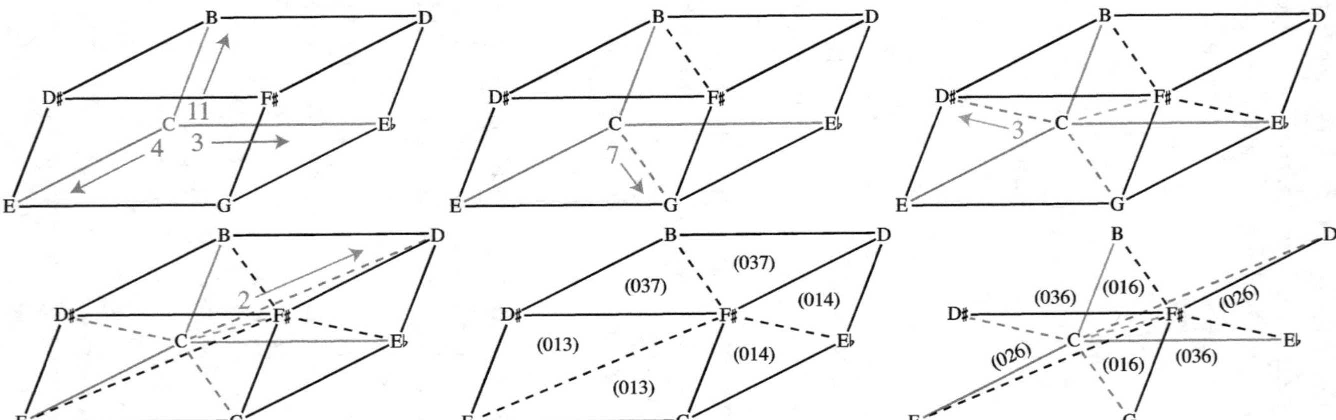

Fig. 13.8. Diagram (not to scale) of the construction of a three-dimensional *Tonnetz*.

Figure 13.8 shows steps 5–6 of the process in three stages. The first interval (7) defines a new plane that cuts the parallelepipeds into two triangular prisms. It also completes a triangulation of the plane made by intervals 3 and 4 into a *Tonnetz*, which happens to be a triadic *Tonnetz* in this case. The next stage cuts (0147) tetrachords (BCD♯F♯ and CE♭F♯G) out of the corner of each triangular prism and also triangulates the plane defined by intervals 4 and 11. The new axis is a distinct axis associated with interval class 3 — here understood as an augmented second, since it makes an (014) *Tonnetz*. Note that the D♯ in the upper left corner must be the same point in the space as the E♭ in the lower right corner, connected to C twice from different directions. The two planes that have been added at this stage intersect in the middle of the parallelepiped, in the tritone axis going through C–F♯. This tritone axis, corresponding to the interval added in step 6 of the process, also triangulates the two new planes. Finally, in the third stage of step 5 one more plane is added that cuts the each of the remaining pyramid-shaped regions each into two tetrahedra and also triangulates the plane made by intervals 3 and 11. This plane also intersects the others created in step 5 in the central tritone. The new tetrahedra correspond to (0137) — CEF♯G, BCDF♯ — and (0236) — CD♯EF♯, CDE♭F♯ — tetrachords.

The entire process thus defines three tetrachord types (each appearing in two inversions) to make a complete simplicial partition. It also creates, in total, six planes, each of which houses its own two-dimensional *Tonnetz*. The choice of intervals in Fig. 13.8 results in (037), (013), (014), (016), (026), and (036) *Tonnetze*. The number of planes could also be determined combinatorially by noting that each of the three tetrahedron types has four distinct faces, and each face is shared by two tetrahedra. Note also that the construction satisfies the requirement of a Euler characteristic of zero: u vertices, $7u$ edges, $12u$ triangles, and $6u$ tetrahedra: $\chi = u - 7u + 12u - 6u = 0$. The odd number of edges is perhaps at first surprising, until we note that the process creates two types of edges. The intervallic axes added in steps 4 and 6 belong to all tetrachord types, whereas those added in step 5 belong only to two out of the three (or four out of the six, if we distinguish the rotationally related tetrahedra corresponding to inversionally related tetrachords). Thus, four of the interval classes—minor third, major third, semitone, and tritone in the example—occur at the intersection of six tetrachords (all types) where three—augmented second, perfect fifth, and major second— occur at the intersection of four (two types). This also checks out combinatorially ($6 \times 6 = 4 \times 6 + 3 \times 4$).

The example given in Fig. 13.8 duplicates a single interval class, ic3. The duplication may be theoretically justified by a spelling distinction — the ic3s that occur in (037)s and (013)s are understood as minor thirds, whereas those occurring in (014)s are augmented seconds. The (036)s have both types, making them asymmetrical. The (0137) tetrachords may therefore be understood as key-defining diatonic subsets, the (0147)s as triad + leading-tone subsets of harmonic major/minor scales, and the (0236)s as scale segments containing an augmented second. One can thus imagine potential analytical uses and theoretical interest for this *Tonnetz* dealing with harmonic major and minor scales.

For a given space and universe, the number of possible *Tonnetze* is technically infinite, since the initial intervallic axes may be freely chosen and in a toroidal space any two points may be connected by an infinite number of lines. These may be reduced to a finite number by equating *Tonnetze* that result in the same set of tetrachords, but the number is still large and contains many degenerate examples. It is useful to classify the possibilities according to the number and kind of interval duplications required for each, since an excess of interval duplications results in impracticable examples. Table 13.2 lists nine such classes (A–I) for $u = 12$. Class A, with the fewest possible duplicated intervals, includes the example just discussed plus five others. In most cases the duplicated interval is ic3, but duplications of ic1 and ic5 are also possible. Class A *Tonnetze* are all based on one of the all-interval tetrachords, (0137) or (0146). Classes B, C, and D duplicate two intervals and omit one. They are subdivided on the grounds of what kinds of intervals are duplicated, whether they are what we might call primary (those belonging to all tetrachord types) or secondary (those belonging to just two types). In classes B and C, both duplicated intervals occur once

Table 13.2. List of three-dimensional *Tonnetze*.

Class	Properties	Tetrachords	Duplications	Omits	Opt. Space
A	One symmetric trichord	(0125) (0126) (0146)	ic1		$Ph_{1,2,6}$
		(0237) (0157) (0137)	ic5		$Ph_{2,5,6}$
		(0136) (0236) (0146)	ic3		$Ph_{1,2,4}$
		(0147) (0236) (0137)	ic3		$Ph_{2,3,4}$
		(0147) (0258) (0146)	ic3		$Ph_{2,3,4}$
		(0136) (0258) (0137)	ic3		$Ph_{2,4,5}$
B	Two symmetric trichords	(0124) (0125) (0135)	ic1, ic2	ic6	$Ph_{1,3,6}$
		(0135) (0237) (0247)	ic5, ic2	ic6	$Ph_{3,5,6}$
C	One duplicated trichord	(0126) (0127) (0157)	ic1, ic5, (016)	ic3	$Ph_{1,2,6}/Ph_{2,5,6}$
D	Duplicated tetrachord	(0134) (0236)×2	ic1, ic3, (013), (014)	ic5	$Ph_{1,4,6}$
		(0145) (0125)×2	ic1, ic4, (014), (015)	ic6	$Ph_{1,3,5}$
		(0156) (0126)×2	ic1, ic5, (015), (016)	ic3	$Ph_{1,2,6}$
		(0156) (0157)×2	ic1, ic5, (015), (016)	ic3	$Ph_{2,5,6}$
		(0235) (0135)×2	ic2, ic3, (013), (025)	ic6	$Ph_{1,3,5}$
		(0235) (0136)×2	ic2, ic3, (013), (025)	ic4	$Ph_{1,2,4}/Ph_{2,4,5}$
		(0347) (0147)×2	ic3, ic4, (014), (037)	ic2	$Ph_{2,3,4}$
		(0158) (0237)×2	ic4, ic5, (015), (037)	ic6	$Ph_{1,3,5}$
		(0358) (0258)×2	ic3, ic5, (025), (037)	ic1	$Ph_{4,5,6}$

(*Continued*)

Table 13.2. (*Continued*)

Class	Properties	Tetrachords	Duplications	Omits	Opt. Space
E	Duplicated tetrachord	(0145) (0148)$\times$2	ic4 ($\times$3), ic1, (014), (015)	ic2, ic6	Ph$_{1,3,5}$
	with augmented triad	(0347) (0148)$\times$2	ic4 ($\times$3), ic3, (014), (037)	ic2, ic6	Ph$_{2,3,4}$
		(0158) (0148)$\times$2	ic4 ($\times$3), ic5, (015), (037)	ic2, ic6	Ph$_{1,3,5}$
F	Duplicated tetrachord	(0134) (0124)$\times$2	ic1, ic3, ic2, (013), (014)	ic5, ic6	Ph$_{1,3,6}$
	with symmetric trichords	(0358) (0247)$\times$2	ic5, ic3, ic2, (025), (037)	ic1, ic6	Ph$_{3,5,6}$
G	Duplicated symmetric	(0123) (0124)$\times$2	ic1 ($\times$3), ic2, (012), (013)	ic5, ic6	Ph$_{1,3,6}$
	trichord	(0257) (0247)$\times$2	ic5 ($\times$3), ic2, (025), (027)	ic1, ic6	Ph$_{3,5,6}$
H	Quadrupled trichord	(0167) (0127)$\times$2	ic1, ic5, ic6, (016)$\times$4	ic3, ic4	Ph$_{1,2,5}$
I	Tripled tetrachord	(0123)$\times$3	ic1$\times$3, ic2$\times$3	ics4,5,6	Ph$_{1,3,4}$
		(0257)$\times$3	ic5$\times$3, ic2$\times$3	ics1,4,6	Ph$_{3,4,5}$

as a primary and once as a secondary interval (they differ only in how the intervals are distributed among trichords). In class D, one duplicated interval occurs twice as a primary interval while the other occurs twice as a secondary interval. This group, characterized by duplicated tetrachords, is especially large. Class E and G *Tonnetze* both have a duplicated secondary interval in addition to a tripled interval, while classes F and H have three different duplicated intervals. Classes E and G differ in that the duplicated interval is secondary in class E and primary in class G, while class H has two duplicated primary intervals and class F only one. The list includes every *Tonnetz* with no 0 intervals and with $\mathbf{Z}_{12}$ as its minimal universe.

Table 13.2 also gives one or two optimal spaces for each *Tonnetz*, using a modification of the procedure described above for the two-dimensional case. When there are three variants of an interval (classes E and G), two dimensions must be included that minimize the DFT of that interval (maximize its spread in that dimension).

A musical interpretation of any of these possible *Tonnetze* may begin by identifying a musical distinction that can justify and take advantage of the interval duplications. For example, an (0235)-(0135) *Tonnetz* (class D) has duplications of ic2 and ic3 and two forms of (013) and (025) trichords depending on which form of ic2 and ic3 they contain. The different possible arrangements of major seconds also lead to two possible forms of (0135) and make the inversions of (0235) distinguishable. This is all true also of the traditional just intonation scales: they have two forms of major second, a larger "Pythagorean" whole step defined by the frequency ratio 9/8, and a smaller "just" whole step of ratio 10/9. When combined with a semitone, these make two kinds of minor third, Pythagorean (32/27), and just (6/5). Using abbreviations P and J for the two forms of whole tone and *s* for a semitone, the tetrachords are, in consecutive intervals, P*s*J, J*s*P, *s*JP, *s*PJ, PJ*s*, and JP*s*. The traditional just major and minor scale patterns, PJ*s*PJP*s* and P*s*JP*s*PJ, contain all of these possible strings plus one other type that defines the bad fourth, P*s*P. Just scales may therefore be represented as paths in this *Tonnetz* that produce the sequence of overlapping tetrachords between the bad fourth and augmented fourth.

Duplications may also be used to fold the spaces. The *Tonnetze* of classes D–G, for instance, can be folded to equate their duplicated tetrachords. Such a folding of the (0258)-(0358) *Tonnetz* produces a seventh-chord *Tonnetz* described by Jack Douthett in an unpublished letter written in 1997 [9]. The folding is over the whole-tone planes that house (026) trichords, turning these into boundaries. One way to imagine this space is to draw two (026) *Tonnetze*, as in Fig. 13.1, parallel to one another in a fattened 2-torus, one using the even whole-tone scale and one using the odd. Then connect all pitch classes from one plane to the other related by a consonant interval (ic3 or ic5, four connections for each pitch class) so as to link each pitch class to two adjacent (026)s on the opposite boundary, making overlapping dominant and half-diminished sevenths. For example, D on the even boundary connects to {G, B, F} and {B, F, A} on the odd one. According to Table 13.2, the optimal space for the (0358)-(0258) *Tonnetz* is $Ph_{4,5,6}$. The Ph_6 dimension is included for disambiguating the duplicated ic3s and ic5s. This dimension is therefore the one whose cycles are removed by the folding, so that it becomes simply a means of separating the two whole-tone planes between the boundaries of the fattened 2-torus. Therefore, this optimal realization of Douthett's *Tonnetz* can be accurately pictured through the projection onto $Ph_{4,5}$ space in Fig. 13.9. Each point shows the position of one tetrahedron, with dominant and half-diminished sevenths alternately pointing up and down, depending on their whole-tone affinity, and minor seventh chords connecting a major second on one plane to an obliquely directed major third on the other. Lines connect chords that share a face. The minor sevenths are fully connected, while the (0258)s each have one face on a boundary. Douthett's schematic diagram (Fig. 13.10) illustrates this.

After folding, the only duplications remaining in this *Tonnetz* are of the tritones on the boundaries. Cutting across the space are (036) *Tonnetze* with boundaries, making two-dimensional *Tonnetz*-bands like the one in the middle panel of Fig. 13.6. Identifying the duplicated tritones in these (036) *Tonnetze* make them spherical,

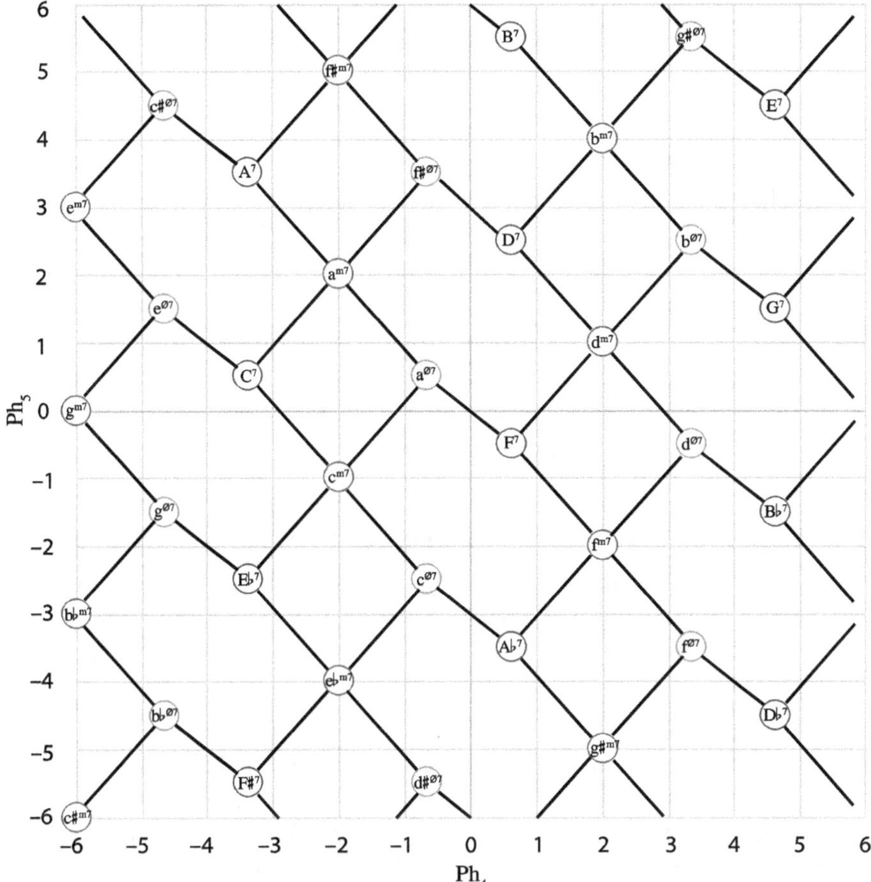

Fig. 13.9. The dual of Douthett's seventh-chord *Tonnetz* projected onto Ph$_{4,5}$-space.

as also shown in Fig. 13.6. This identification eliminates the last of the interval duplications. It also turns the boundaries into a new set of tetrachords: note in Fig. 13.1 that the (026) *Tonnetz* actually connects every note of the whole-tone collection to every other. The distinction between tritones is the only reason that tricords like {F♯, G♯, C} and {C, D, F♯} do not combine into a single tetrachord. Identifying the tritones folds each boundary into a set of three (0268) tetrahedra. The resulting space is the one described by Tymoczko [18] as the voice-leading *Tonnetz* for four-note chords, topologically

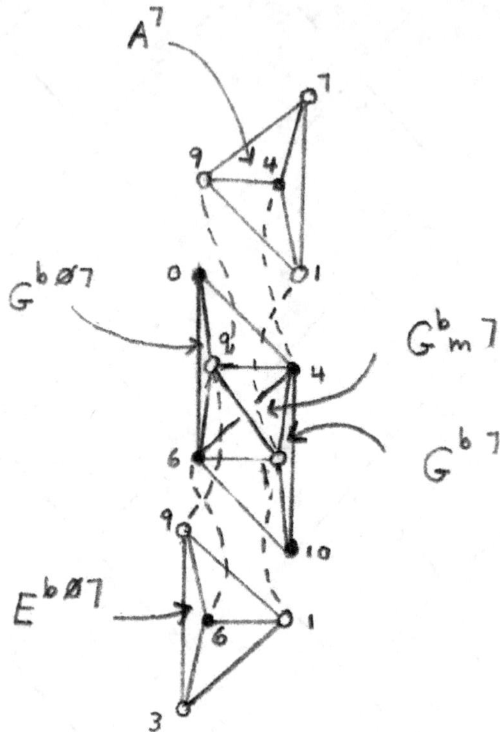

Fig. 13.10. Douthett's drawing [9] of the construction of a seventh-chord *Tonnetz*.

the direct product of a sphere and a circle. As with the augmented triad in Tymoczko's voice-leading *Tonnetz* for three-note chords (Fig. 13.2), the (0369) tetrachords have a special status in this space. Although they are fully connected in the network created by the triangulation, they are not properly elements of the triangulation as tetrachords. For Tymoczko they are essential intermediaries between dominant and half-diminished sevenths sharing three notes, so that the voice leading is entirely semitonal. But in the triangulation their status is equivalent to that of the two-dimensional planes that house trichordal *Tonnetze* made up of similar 2-faces of the triangulation, not of the tetrahedral (3-simplex) elements of the triangulation. The situation is analogous to that of Tymoczko's two-dimensional triadic *Tonnetz*, where the augmented triad, which he counts as an essential triangle (2-simplex), is, in the toroidal triangulation, an intervallic axis that happens to be limited to three elements. This distinction is essentially homological: the elements of the simplicial decomposition can be contracted to points in the space. Other maximal cliques, included in Tymoczko's networks but not in the simplicial decomposition, cannot.

13.7. The n-Dimensional Generalization

The construction defined for two and three dimensions above can be generalized to n dimensions:

(1) Choose n distinct, linearly independent line segments, $a, b, c, \ldots$, extending from pitch-class 0 to some other pitch class, which do not pass through any other pitch classes.
(2) Translate these u times (to each pitch class), making a skewed regular hypercubic lattice partitioning the space into u regions.
(3) Add the $\binom{n}{2}$ line segments obtained by taking vector sums of any two of $a, b, c, \ldots$ and extending these from pitch-class 0. These will triangulate each 2-face of the hypercube. Translate these to each pitch class to triangulate all of the planes of the hypercubic lattice.
(4) Continue this process for each m, $1 < m \le n$, taking vector sums of m of $a, b, c, \ldots$, adding these from each pitch class to complete a simplicial decomposition of each m-face of the hypercubic lattice.

Note that the pitch classes used in the process need not be unique — one of the line segments may even connect a pitch class to itself. Therefore degenerate examples such as *Tonnetze* on a one-pitch-class universe are possible. The number of edges defined in this process is

$$e = \sum_{i=1}^{n} \binom{n}{i} u = (2^{n-1} - 1)u, \tag{13.1}$$

where $\binom{n}{i}$ gives the number of i-faces of the hypercube incident upon a given vertex.

For example, in the three-dimensional case we can define a skewed cubic lattice with three intervals, a, b, c, triangulate the faces with intervals $a + b, b + c, a + c$, then complete the triangulation of the 3-faces, the skewed cubes themselves, with $a + b + c$. The tetrahedra correspond to all the permutations of a, b, and c. To get the augmented-second *Tonnetz* described above, let $a \to 3$ (a points from 0 to 3), $b \to 4$ (etc.), and $c \to 11$. Then $a + b \to 7$, $b + c \to 3$, $a + c \to 2$, and $a + b + c \to 6$. The tetrahedra may be constructed by stringing together a, b, and c in any order. In our example, *abc* and *cba* correspond to tetrachord type (0147), *acb* and *bca* to (0236), and *bac* and *cab* to (0137). We get the same *Tonnetz* by either negating all of a, b, and c, or exchanging any of a, b, or c with $(-a - b - c)$.

For $n = 4$, we get a *Tonnetz* of $24u$ 4-simplexes representing pentachords (12 types occurring in all transpositions and inversions). The number of intervallic axes is 15, so the number of duplications for $u = 12$ would be necessarily quite large.

Acknowledgments

Thanks to Mariana Montiel and Robert Peck for their efforts in organizing an excellent special session at the 2016 Joint Mathematics Meetings in Atlanta that motivated the development of this chapter, and also to Emmanuel Amiot for his mathematical insights and conversations about the DFT and Richard Cohn for pointing me towards additional examples of analysis with generalized *Tonnetze* and related concepts.

Bibliography

[1] E. Amiot, The torii of phases, in *Mathematics and Computation in Music: 4th Int. Conf., MCM 2013*, eds. J. Yust, J. Wild and J. A. Burgoyne (Springer, Heidelberg, 2013), pp. 1–18.

[2] E. Amiot, *Music Through Fourier Space: Discrete Fourier Transform in Music Theory* (Springer, Heidelberg, 2016).

[3] L. Bigo, M. Andreatta, J.-L. Giavitto, O. Michel and A. Spicher, Computation and visualization of musical structures in chord-based simplicial complexes, in *Mathematics and Computation in Music: 4th Int. Conf., MCM 2013*, eds. J. Yust, J. Wild and J. A. Burgoyne (Springer, Heidelberg, 2013), pp. 38–51.

[4] L. Bigo, D. Ghizi, A. Spicher and M. Andreatta, Representation of musical structures and processes in simplicial chord spaces, *Comput. Music J.* **39**(3) (2015) 9–24.

[5] S. Brown, Some instances of ic1/ic5 interaction in post-tonal music (and their Tonnetz representations), *Gamut* **6**(2) (2013) 17–50.

[6] M. J. Cantazaro, Generalized Tonnetze, *J. Math. Music* **5**(2) (2011) 117–139.

[7] R. Cohn, Neo-Riemannian operations, parsimonious triads and their 'Tonnetz' representations, *J. Music Theory* **41**(1) (1997) 1–66.

[8] R. Cohn, Tonal pitch space and the (Neo-)Riemannian *Tonnetz*, in *Oxford Handbook of Neo-Riemannian Theory*, eds. E. Gollin and A. Rehding (Oxford University Press, New York, 2011), pp. 322–348.

[9] J. Douthett, 3-Dimensional tiling, Unpublished letter to John Clough Working Group (1997).

[10] E. Gollin, Some aspects of three-dimensional Tonnetze, *J. Music Theory* **42**(2) (1998), pp. 195–206.

[11] E. Gollin, From matrix to map: *Tonbestimmung*, the *Tonnetz*, and Riemann's combinatorial conception of interval, in *Oxford Handbook of Neo-Riemannian Theory*, eds. E. Gollin and A. Rehding (Oxford University Press, New York, 2011), pp. 271–293.

[12] D. Lewin, Notes on the opening of the F♯ minor fugue from WTCI, *J. Music Theory* **42**(2) (1998) 235–239.

[13] G. Mazzola, *Topos of Music: Geometric Logic of Concepts, Theory, and Performance* (Birkhäuser, Basel, 2002).

[14] D. Muzzulini, Musical modulation by symmetries, *J. Music Theory* **39**(2) (1995) 311–327.

[15] J. R. Reed and M. N. Bain, A tetrahelix animates Bach: Revisualization of David Lewin's analysis of the opening of the F♯ minor fugue from WTCI, *Music Theory Online* **13**(4), 2007.

[16] M. Siciliano, Toggling cycles, hexatonic systems, and some analyses of early atonal music, *Music Theory Spectrum* **27**(2) (2005) 221–248.

[17] D. Tymoczko, *A Geometry of Music: Harmony and Counterpoint in the Extended Common Practice* (Oxford University Press, New York, 2011).

[18] D. Tymoczko, The generalized Tonnetz, *J. Music Theory* **56**(1) (2012) 1–52.

[19] P. C. Van den Toorn and J. McGinness, *Stravinsky and the Russian Period: Sound and Legacy of a Musical Idiom* (Cambridge University Press, Cambridge, 2012).

[20] J. Yust, Schubert's harmonic language and Fourier phase space, *J. Music Theory* **59**(1) (2015) 121–181.

[21] J. Yust, Applications of DFT to the theory of twentieth-century harmony, in *Mathematics and Computation in Music, 5th Int. Conf., MCM 2015*, eds. T. Collins, D. Meredith and A. Volk (Springer, Cham, 2015), pp. 207–218.

[22] J. Yust, Special collections: Renewing set theory, *J. Music Theory* **60**(2) (2016) 213–262.

[23] J. Yust, Harmony simplified, in *Organized Time: Rhythm, Tonality, and Form.* (Oxford University Press, 2018), Chapter 10.

Chapter 14

Deterministic Geometries: A Technique for the Systematic Generation of Musical Elements in Composition

Brent A. Milam

School of Music, Georgia State University,
Atlanta, GA 30303, USA
bmilam1@gsu.edu
www.gsu.edu

14.1. Introduction: Deterministic Methodologies

The methodologies presented in this chapter are purposefully formative rather than analytical. To some degree, deterministic methodologies for the generation of musical material have been employed for centuries as exemplified in isorhythmic motets and canonic imitation [3, 9]. In the last century, the concept of determinism was applied most conspicuously to pitch with the use of Schoenberg's 12-tone serialism and expanded to include multiple parameters in the total serialism of Babbitt, Stockhausen, and Boulez. The level of determination can vary greatly depending on the technique employed and the degrees of freedom allowed by the composer. At some level, all such methods must employ predefined sets of elements that rely on arbitrary decisions of the composer.

This chapter explores the use of mathematically deterministic games as novel generative processes for material in musical composition. It is not the intent of the author to remove all arbitrary pre-compositional decisions, but rather to minimize such decisions

279

and link the deterministic output more directly with patterns arising from natural phenomena such as astronomy, biology, geology, and physics.

At every level, nature produces a vast array of stunning and surprisingly interconnected behaviors. Such natural processes are best understood and even predicted through the language of mathematics [11]. The use of natural systems as material for musical structure, a process called "sonification", provides countless possibilities for unexpected yet coherent relationships. The technique presented herein is a preliminary attempt to coordinate the outcome of purely mathematical processes, such as group theory transformations, vector calculus, or geometric relationships, with various elements of musical experience to produce a deterministic composition with inherently logical and yet unique relationships. Ultimately, these efforts seek to mold an artistic statement that mimics the unexpected and astonishing beauty underlying complex natural systems while minimizing the arbitrary decisions of human intuition.

It should be noted that due to the preliminary and proof-of-concept nature of this study, the results reported herein were obtained primarily by hand without rigorous proof and without the use of computational software other than Microsoft Excel to calculate linear values and plot data. In order to compensate for this limitation and provide sufficient confidence in the present conclusions, numerous scenarios were run for each case using multiple initial conditions. In the end, well over 60 individual runs with different initial points were made for each boundary condition (triangle/square). For further study, it is easy to see how software could readily calculate a myriad of possible outcomes using different geometric boundaries conditions, multiple initial conditions, and defined element-spaces. Such applications are currently in development.

14.2. Geometric Patterns

14.2.1. *The Sierpinski triangle (Game I)*

The mathematical games explored include various modified versions of the Sierpinski Triangle [14]. In this game, an initial point is selected within the boundary of an equilateral triangle with subsequent points determined by a proportional translation of the distance between the previous point and an associated vertex of the geometric shape. These games produce either apparently random sequences or repeating geometric patterns depending on the selection of an initial seed but in either case, the outcome is unambiguously deterministic. The original version of this game will be called, "Game I". Rules of Game I:

Rule 1. *Choose any initial point, p_0, within an equilateral triangle (Fig. 14.1).*

Rule 2. *Double the distance from that point to the nearest corner (vertex, v_a) along a straight line beginning from that corner to arrive at the next point, p_1.*

Rule 3. *Repeat this process of doubling the distance until a point, p_i, falls outside the perimeter of the triangle (Fig. 14.1).*

After selecting the initial point, the resulting pattern is completely deterministic.

14.2.1.1.   *The Sierpinski Gasket*

Is there an initial point that would prolong the game indefinitely? These initial points lie along a series of lines forming a fractal pattern known as the Sierpinski Gasket shown in Fig. 14.2. This pattern is produced by the iterative process of connecting the midpoints of each

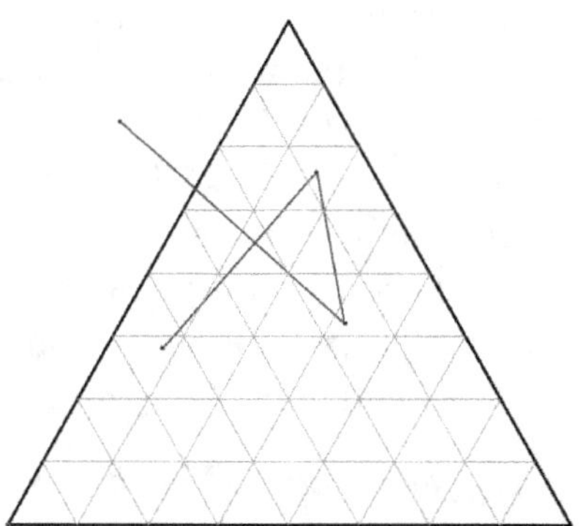

Fig. 14.1.   Deterministic Game I.

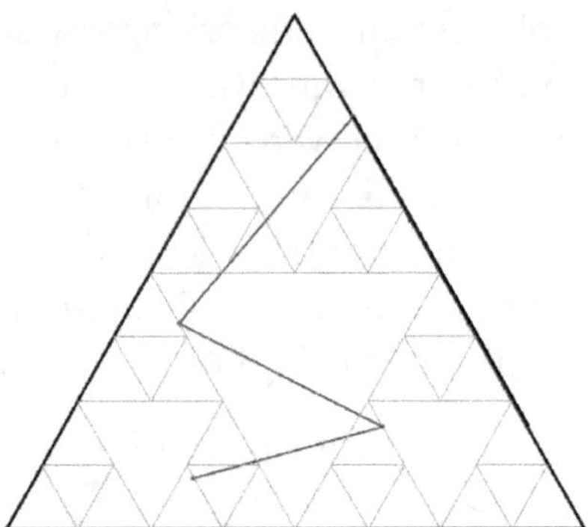

Fig. 14.2. The Sierpinski Gasket.

segment in the group of equilateral triangles pointing in a particular direction (in the case of Fig 14.2, those pointing "up"). This process will produce a smaller triangle in the center of each equilateral triangle with segments half the length of the parent and pointing in the opposite direction. Figure 14.2 demonstrates this process through three levels of triangles but the process could be repeated indefinitely. Any initial point selected along the boundary of these self-similar triangles will eventually land on the initial boundary condition itself and thereafter iterate indefinitely.

14.2.2. *Modified version of the game (Game II)*

Due to the relatively short sequence produced by most initial conditions, the applications to musical composition are limited. To provide compositional flexibility, it would be desirable to have a greater variety of sequence lengths. With this consideration, the rules of the original Sierpinski game were modified to ensure a continuous iteration apart from the fractal solutions provided by the Sierpinski Gasket. This was accomplished by incorporating a reflection across the boundary to return any out-of-bounds point to the game. This new process adds an additional rule to the original game.

Rule 4. *If a point, p_i, falls outside the boundary triangle, reflect the point back across the boundary condition equidistant along a line perpendicular to that boundary.*

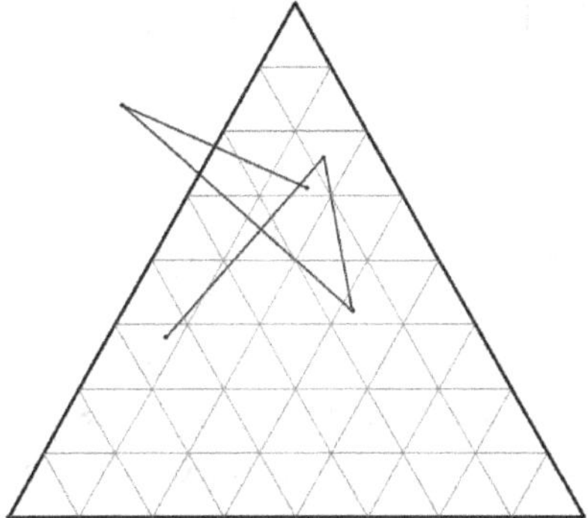

Fig. 14.3. Reflection across boundary condition.

Figure 14.3 demonstrates this reflection for the original out-of-bounds point given in Fig. 14.1.

By this process, any initial condition will now produce a continuous series of points incorporating both in-bound points and special out-of-bounds "events". This modified version of the original game will be called "Game II".

14.2.2.1. *Repeating geometric patterns*

The results of Game II were quite unexpected. It was discovered that most randomly selected starting points would eventually generate repeating geometric patterns referred to as "deterministic geometries" in this chapter. Three general categories of divergent outcome were observed:

A. the *Sierpinski Gasket* (**SGP**) resulting in repeated iteration along a boundary segment,
B. a *repeating geometric pattern* (**RGP**) of two or three-fold symmetry, and
C. *"deterministic chaos"* [14] (**DCP**) resulting in an apparently chaotic orbit.

It was anticipated that an initial starting point not along the Sierpinski Gasket (condition **A**) would produce a continuous but otherwise chaotic orbit, as described in **C**; however, it was observed that any initial starting point with rational coordinates that was not along the Sierpinski Gasket would actually produce condition **B**. Condition **A** was the predictable solution of the original game, but **B** was thoroughly unexpected and by far the most common outcome.

Another surprising result was that in both **A** and **B**, the final repeating geometric pattern was always preceded by a short series or "random" walk of one to five points as if the game was searching for the appropriate geometric pattern. It is conceivable that one might select an exact point of the repeating pattern as the initial starting point in which case no initial walk would occur; however, without prior knowledge of the eventual repeating geometric pattern (RGP), this seems a highly improbable selection. The occurrence of an initial "random" walk raises two important questions that have yet to be answered: (1) Why are the initial points not members of the eventual pattern? (2) If indeed any set of rational coordinates may be selected as the initial starting point, why does it take so few initial points, in most cases less than three, to "find" the RGP?

Figure 14.4 shows the final outcome of Game II for the starting point used in Figs. 14.1, 14.3, and 14.6. In this case there are two initial points, p_0 and p_1. Points p_2 through p_{11} are all members of the formal RGP with the reflection of the out-of-bounds point, p_{11}, across the boundary segment AC yielding p_2 once again. Figure 14.4 demonstrates an RGP with two-fold symmetry while Fig. 14.5, with

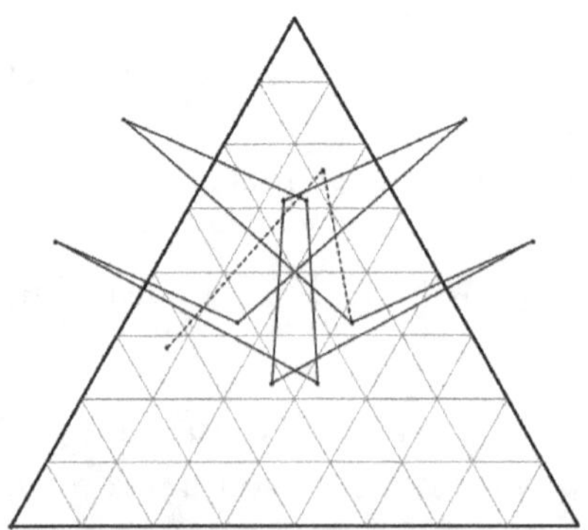

Fig. 14.4.   2 + 10 Point two-fold symmetry.

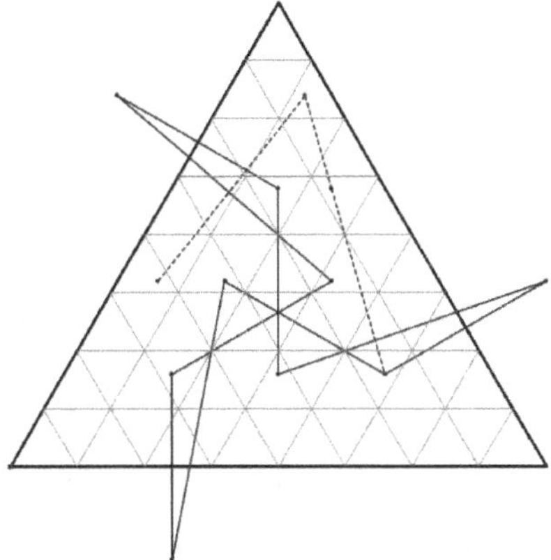

Fig. 14.5. $3 + 10$ point three-fold symmetry.

a slightly different initial starting point, demonstrates an RGP of three-fold symmetry.

14.2.2.2. *Asymptotic convergence (Game III)*

The SGP, RGP, and DCP outcomes are all products of the original divergent version of the game wherein doubling the distance from the nearest vertex is always pushing the points away from the vertices of the triangle. Another modification of the game incorporates a convergent solution by changing rule 4 as follows:

Rule 4. *If a point, p_i, falls outside the boundary triangle, return to the previous point, p_{i-1}, and halve the distance from that point to the furthest corner (vertex, v_f) along a straight line to arrive at the next point, p_i.*

With this convergent process, any initial condition will once again produce a continuous series of points but no longer any out-of-bounds "events". This version of the game will be called "Game III". The results of this convergent modification give a fourth category of outcome:

D. an *asymptotic convergence* (**ACP**) into one of two possible triangular arrangements.

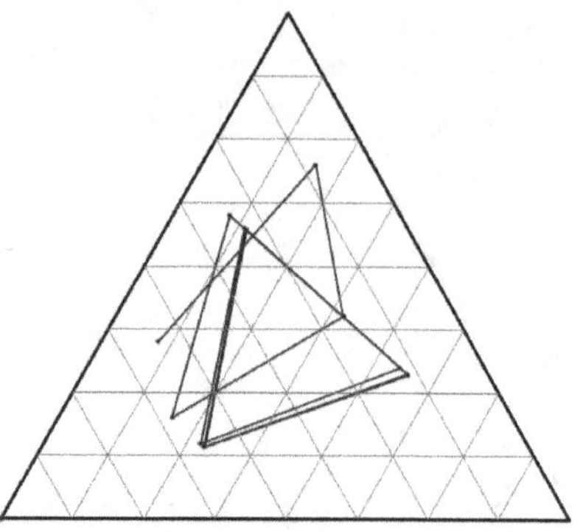

Fig. 14.6. ACP orientation 1.

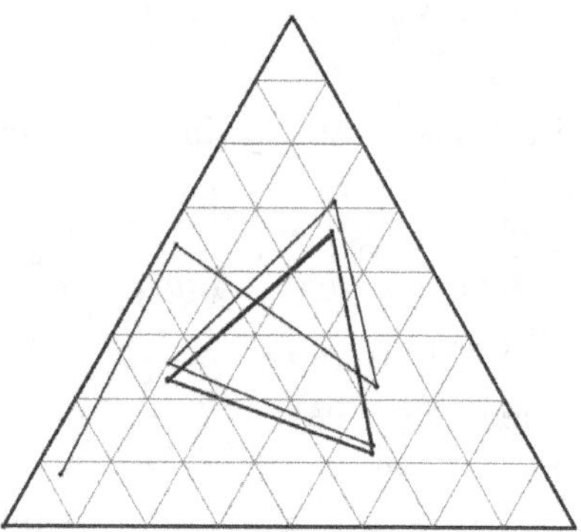

Fig. 14.7. ACP orientation 2.

Figure 14.6 employs the same three initial points found in Fig. 14.1; however, rather than using Rule 2 to double the distance and thereby produce an out-of-bounds event, the new Rule 4 is used to half the distance from the furthest vertex. Figure 14.6 shows that this new version of the game (Game III) will produce a continuous series of points that asymptotically converges to the three vertices of a smaller equilateral triangle slightly offset from the original boundary in a counterclockwise direction.

There are two possible equilateral arrangements for this convergent modification. Figure 14.7 uses a different starting point to demonstrate convergence into the other equilateral arrangement offset in the clockwise direction.

14.2.2.3. *Deterministic chaos*

As described in Sec. 14.2.2.1, not all initial points will produce a repetitive geometric pattern. The DCP is the only result that appears to iterate indefinitely without a repeating series of points. In order to create this DCP, the starting point must have irrational coordinates. These irrational coordinates give rise to a series of points that resemble a random walk.

Although the DCP outcome may appear completely chaotic, the succession of points is still absolutely deterministic. It depends only on the coordinate values of the initial starting point. Manfred Schroeder describes these outcomes as "deterministic chaos" [14]. Schroeder states that "the rules governing the 'game' are unambiguously deterministic, but the results are ultimately unpredictable".

14.2.3. *Rectangular boundary*

It should be noted that another obvious adaptation of the original game is to employ other geometric boundary conditions, such as a rectangular boundary. Many iterations with rectangular boundaries were investigated as part of this study; however, as none of the results affect the basic proposition of the study, the specific outcomes and considerations for alternative geometries will be left for future publication.

14.3. Elements of Sonority

In previous studies, the present author, in collaboration with Nickitas J. Demos, has attempted to categorize parameters by which a listener may describe and analyze any sonic event the listener considers "musical". Once such elements of musical experience are defined, the evolution of those elements can be specifically manipulated in isolation or in conjunction with other elements to shape the compositional process and serve musical objectives. The methods of analysis associated with traditional common practice and post-tonal theory are inadequate to describe numerous sonic events outside, and in many cases even inside, the tradition of western music. Many books on music appreciation, fundamentals, and musicology seek to categorize the elements of music in the useful terms of pitch, rhythm, texture, timbre, etc.; however, the basic level of such presentations is too imprecise and incomplete to capture the specifics of many sonic events [6, 8, 12, 16]. Furthermore, by what criteria might a listener systematically analyze or describe the timbre or orchestration of a musical event? This latter question was the primary concern raised by Demos in 1994 and continues to be a major focus of his efforts to develop a comprehensive analysis of sonority in the context of advanced orchestration [4]. Jan La Rue has also made an attempt to list the parameters by which one might explain musical experience; however, La Rue's focus is on the development of a comprehensive methodology for the analysis of musical styles and periods [7].

For the purposes of this chapter, the elements of sonority will be divided into four general classifications. These general classifications include time, pitch, timbre, and "other" (i.e., those characteristics not readily classified as time, pitch, or timbre). Each of these classifications is further divided into subcategories as necessary to isolate distinguishable characteristics, or "elements", of that general classification; however, it is important to recognize that none of the subcategories are truly isolated. Most of the subcategories easily overlap with and are influenced by multiple other subcategories, even across boundaries of the general classifications, such that absolute separation is at times impossible. Nevertheless, the characteristic elements of each subcategory are distinguishable to a degree that their contribution to the characterization of sonority is fundamental. It should also be noted that definitions for the following classifications and characteristics are necessarily vague for the scope of this chapter with specific details regarding perception of such characteristics as tempo, pitch, and timbre left to the numerous sources in music theory, musicology, and cognitive science that detail such characteristics [5–8, 10, 12, 16].

14.3.1. *Elements of time*

The classification of time concerns those elements of sonority directly dependent on the passage of time for their perception. As such, time is divided into the four subcategories of tempo, meter/pulse, rhythm, and form. Tempo describes the perceived speed of the basic pulse which will in many ways correspond to the speed with which sonic events occur. Meter is an implied periodicity in the sense of pulse which can be steady, free, symmetric, or asymmetric. Rhythm is essentially the density of attack events and the duration of each. Other considerations for rhythm include the concepts of simultaneity and syncopation (i.e., the way in which a given set of events frustrates the expected meter and pulse). Finally, form refers to the sense of structure created by such features as phrase, cadence, and sectional delimitation. Other considerations for form include the perception of length, balance, similarity, contrast, and return.

14.3.2. *Elements of pitch*

The classification of pitch concerns those elements related to the perception of frequency and how that frequency changes over time. In this case, the term "pitch" is misleading as it does not specifically refer to either tempered scale degrees or pitch classes. Nevertheless, the association of the musical term "pitch" with the technical term "frequency" in a musical context is helpful and the two terms will be used interchangeably for the purpose of this study. The four subcategories of pitch include the linear or horizontal dimension which is directly related to the passage of time, the vertical or harmonic dimension related to the perception of static collections, the compass of the sound, and the transformation of one pitch collection into another. The linear aspect of pitch corresponds to the perception of melody, contour, phrase structure, counterpoint, imitation, and the envelop of a sustained event. The vertical aspect of pitch includes the perception of pitch centricity and the association of various pitch events into a collection (tempered or microtonal), otherwise called "harmony". The vertical dimension also concerns the placement of that collection in the frequency domain, i.e., "voicing". The compass of a pitch collection is defined by the register and range/span of its frequencies. Transformation refers to the means by which one pitch collection changes over time through the process of progression, sequence, modulation, and voice-leading into a new collection [2, 13, 15].

14.3.3.　*Elements of timbre*

The classification of timbre defines those elements of sonority related to the quality of a given sound. The four subcategories of timbre include the method of sound production, orchestration, articulation, and special effects. The method of sound production (i.e., instrumentation) takes into account the materials, size, construction, resonant qualities, and initializing vibration/attack for a given sonic event including the use of electronics. Orchestration describes the manner by which multiple sonic events are combined over a given period of time. Articulation is the means of attack for the initializing vibration. Special effects include those events wherein the sound is produced by non-idiomatic means such as pizzicato, tremolo, sul ponticello, flutter tongue, multiphonics, key clicks, etc.

14.3.4.　*Other elements of sonority*

Those elements of sonority that clearly affect the perception of a musical event but are not principally classifiable under the perception of time, frequency, or timbre include texture, event density, dynamics, musical gesture, and spatialization. The textur of a sonic event concerns the interaction between instruments or sound sources. The layering of these interactions is often defined as monophonic, homophonic, chordal, polyphonic, or heterophonic. Another consideration of texture is the method of subordinate accompaniment to a more predominate sonic event. Event or ensemble density concerns the number of simultaneously sounding instruments or sonic events in a given period of time. Dynamics is the perception of amplitude and the degree of comparative volume between events. A musical gesture is a relatively short yet salient, musically meaningful, and structure-defining event that gives anchor points to the perception of time (i.e., form). Examples of musical gesture could include a repeated pitch, rhythm, chord, timbre, dynamic, texture, effect or even a singular defining event. Finally, spatialization concerns the

Table 14.1. Elements of sonority analysis.

Time	Pitch	Timbre	Other
Tempo	Linear/Horizontal	Sound Production	Texture
Meter/Pulse	Vertical/Harmonic	Orchestration	Density
Rhythm	Compass	Articulation	Dynamics
Form	Transformation	Special Effect	Musical Gesture
			Spatialization

location and direction of sound sources. Other considerations in the category of spatialization include the use of panning between sources and the physical space or venue within which the sonic events occur. Table 14.1 provides a summary of the elements of sonority as defined above.

14.4. Assigning Values to Musical Elements

To relate the results of the deterministic games from Sec. 14.2.2 to a musical composition, quantitative values derived from the deterministic geometries (SGP, RGP, DCP, and ACP) must be applied to the various elements of sonority from Sec. 14.3.3. The predefined parameters selected by the composer include the shape of the geometric boundary, the version of the game (Game II, Game III, or both), the initial starting point, the internal subdivision of the boundary geometry, and the relationship of the elements within the subdivided cells, the "network".

There are multiple ways to subdivide the space within a given geometric boundary in a symmetrical fashion. The obvious and most logical solution would be to create geometric subdivisions of the same shape as the original boundary. For example, in the case of a triangular boundary, the parent triangle may be subdivided into 64 smaller versions of the original as demonstrated by Fig. 14.8. This arrangement overlaps the pattern of the Sierpinski Gasket shown in Fig. 14.2.

These subdivisions, called "cells", create a network of alternating upward and downward facing triangles and are labeled according to row and position. Each row will be assigned a letter beginning with

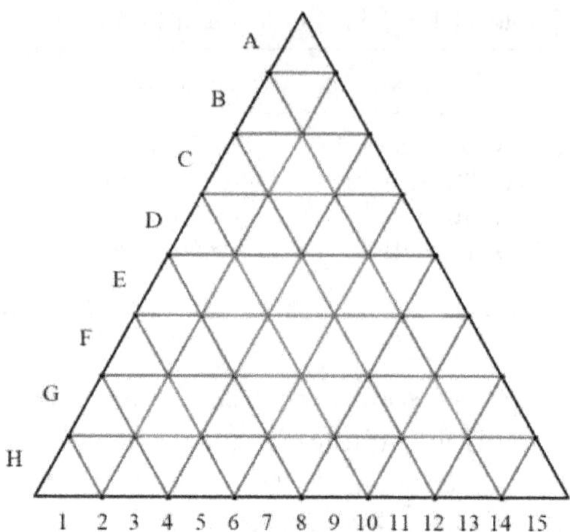

Fig. 14.8. Subdivision of boundary into cells.

A at the top and H at the bottom and each cell in a given row will be numbered accordingly, left to right, beginning at 1. Consequently, the apex of the triangle is cell A1 and the bottom right-hand corner is cell H15. Odd numbered cells correspond to triangles facing up and even numbers represent those facing down.

The quantitative values produced by the deterministic geometries (SGP, RGP, DCP, or ACP) include those associated with each line segment (x, y coordinates, slope, intercept, length, and angle of incidence), the geometric boundary (out-of-bounds values, vertex proximity, and vertex distance), and the geometric subdivision, i.e., network cells (cell value, trajectory path, and line intersection points). Some of these values for the RGP in Fig. 14.4 are listed in Table 14.2.

Point -2 and -1 of Table 14.2 correspond to the initial walk before the formal RGP. The actual RGP is represented by points 0–9. The trajectory following point 9 returns back to point 0 after which the pattern repeats. The values from this table may be assigned to the elements of sonority mentioned in Table 14.1: tempo (pulse), beats per measure (meter), attack, duration, simultaneity (rhythm), section length (form), pitch collection, pitch centricity, linear contour, harmony, voicing (pitch), register, range (compass), progression,

Table 14.2. Qualitative Values from RGP of Fig. 14.4.

Point	Coordinates			Slope	Angle	Intercept	Length	Nearest		Furthest	
	x	y	Cell					Vertex	Dist.	Vertex	Dist.
−2	275	350	F2	1.273		0	445.11	B	445.11	C	805.06
−1	550	700	C4	−6	47.61	4000	304.14	A	304.14	B	890.22
0	600	400	E7	−1	35.54	1000	565.69	C	565.69	B	721.11
1	200	800	OUT	−0.5	18.43	900	357.77	A	360.56	C	1131.4
2	520	640	C3	−18	119.74	10000	360.56	A	360.56	B	824.62
3	540	280	F7	−0.61	55.48	608.7	538.52	C	538.52	A	721.11
4	80	560	OUT	−0.5	4.78	600	357.77	B	565.69	C	1077.0
5	400	400	E3	1	108.43	−0.01	565.69	B	565.69	C	721.11
6	800	800	OUT	0.5	18.43	400	357.77	A	360.55	B	1131.4
7	480	640	C3	18	119.74	−8000	360.55	A	360.55	C	824.63
8	460	280	F5	0.609	55.48	0	538.50	B	538.5	A	721.11
9	920	560	OUT	0.5	4.78	100.04	357.77	C	565.71	B	1077.0
0	600	400	E7		108.43			C	565.71	B	721.11

modulation, voice-leading (transformation), instrumentation, articulation, special effect, gestures (timbre), layered interaction (texture), density, amplitude, envelop (dynamics), location, direction, and panning (spatialization). For example, associations could be made in the following ways, although numerous alternatives are possible:

Network cell	$\rightarrow$	Equal tempered pitch
Slope of line segment	$\rightarrow$	Dynamic level
Angle of incidence	$\rightarrow$	Direction
Length of line segment	$\rightarrow$	Note/event duration
Nearest boundary vertex	$\rightarrow$	Envelope, register, instrumentation, articulation, location
Line crossing	$\rightarrow$	Pitch centricity, gesture
Out-of-bounds value	$\rightarrow$	Rest, cadence, end of phrase

It is left to the discretion of the composer to define the degree to which such associations are employed from only one or two deterministic elements to total determinism. Once the initial conditions are established and the game begins, the various elements of sonority would thereby unfold in a deterministic fashion. Any remaining elements of sonority not assigned a deterministic value are left to the freedom of the composer's musical intuition.

Another possibility worth investigation is the use of quantitative values from the deterministic geometry in a feedback manner to inform the initial conditions of another run. The values from Table 14.2 could determine the geometric boundary, the elements associated with network cells, the version of the game (divergent or convergent), and/or the initial starting point for a future deterministic run. Multiple possibilities of this feedback process have been considered and marginally explored; however, due to the computational demands, results remain incomplete at this time.

14.4.1. *Pitch space: Triangular Tonnetz*

The simplest assignment for the purposes of this chapter is the designation of pitch. The location of line segments and terminal points in the RGP of Fig. 14.4 will be used as qualitative values.

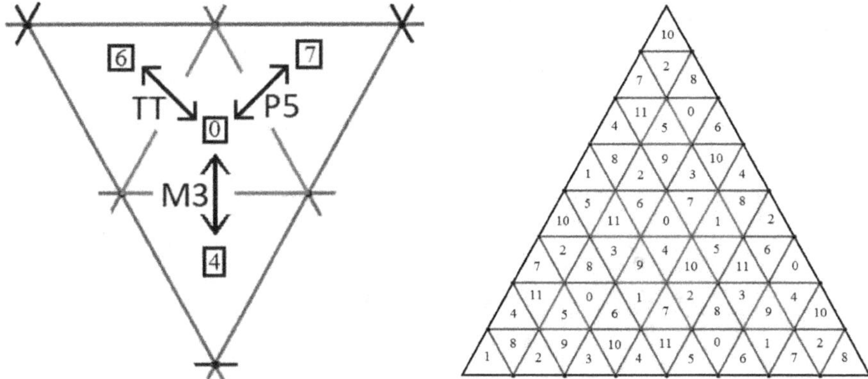

Fig. 14.9. Cell-centric *Tonnetz*.

The network of cells, as suggested above, will be assigned equal-tempered pitch values. Using the network of cells to define a pitch space presents two possibilities. The simplest would be to assign a pitch class to each individual cell; however, there must be a systematic relationship between those pitches, i.e., a *Tonnetz* or tonal network, for the cells to produce a coherent and repeatable pattern as shown in Fig. 14.9. In this case, a pitch class of 0 is specified for the center of the triangle; however, this decision is completely arbitrary.

Another possibility is to assign pitch classes to the vertices of each cell. In this case, each cell would represent a trichordal harmony. Using the same intervallic relationships suggested in Fig. 14.9, one might produce the vertex-centric *Tonnetz* shown in Fig. 14.10.

14.4.1.1. *Pitch sets from terminal points*

Two methods were used to assign quantitative values from the RGP to the equal-tempered pitch classes in the *Tonnetz*. The first was to associate the terminal points of each line segment, i.e., the series of x and y coordinates, with the cell in which it lands. The linear pitch collection is thereby determined by the series of cells into which the successive points of the RGP fall. The second would be to assign pitch according to the trajectory of the line segment from one point to the next. In this case, each cell through which the line segment

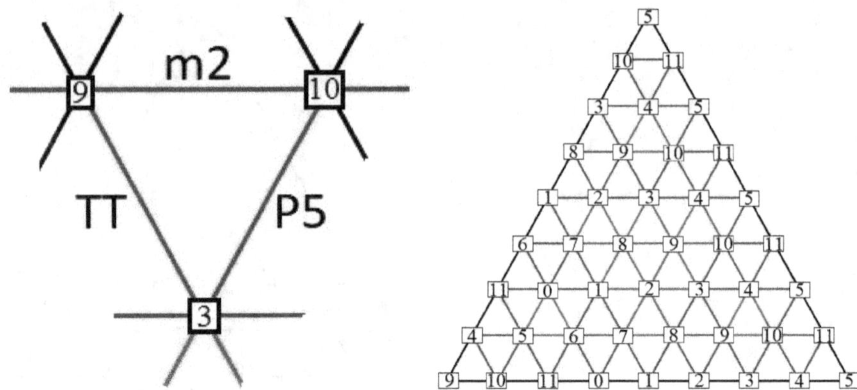

Fig. 14.10. Vertex-centric *Tonnetz*.

Table 14.3. Pitch sequences.

Deterministic geometry	Pitch sequence
RGP of Figure 14.4	(2 0) 1 × 5 T × E × 5 9 ×
RGP of Figure 14.5	(T 8 T) 3 E × 0 1 × 9 7 ×
ACP of Figure 14.7	(2 0 1 0 5) E 6 2

crosses would be the next pitch in the sequence. This variation would naturally create longer sequences. In either case, one question remains, what value should be assigned when the point "escapes" the geometric boundary? It was previously suggested that out-of-bounds events might be associated with a rest, cadence, or end of the phrase. For the sake of simplicity in our present example, we will insert a rest.

The pitch collections in Table 14.3 are the result of using the specified deterministic geometries with the cell-centric *Tonnetz* of Fig. 14.9. The pitch classes in parenthesis represent the initial points before the formal repeating pattern and the value "*x*" represents a rest based on an out-of-bounds condition.

If the vertex-centric *Tonnetz* of Fig. 14.10 is used instead, the result will be a series of trichords. Using the RGP of Fig. 14.4 and the cells associated with the terminus points of each line segment as indications of harmonic collections, the sequence of trichords would be:

(7)	(5)	T		T	9		8		T	8
(6)	(4)	9	×	9	3	×	7	×	9	2
(0)	(0)	4		4	2		3		4	1

Once again, the values in parenthesis represent the initial trajectories before the formal RGP.

14.4.1.2. *Pitch sets from trajectory path*

Using the second method whereby pitch is assigned according to the trajectory of a line segment from one point to the next and the cell-centric *Tonnetz* of Fig. 14.9, the first segment in Fig. 14.4, corresponding to the initial walk, would be as follows:

$$\underline{2} \quad T \quad 5 \quad E \quad 6 \quad 2 \quad 9 \quad 5 \quad \underline{0}$$

The underlined values represent the terminal points of the previous method.

The second segment of the initial trajectory crosses the vertex point of six cells. The values of these cells are represented by angle brackets and may be handled in multiple ways, such as with hexachordal harmony, two successive trichordal harmonies, or completely ignored.

$$\underline{0} \quad \langle 056, 39\,T \rangle \quad 3 \quad 7 \quad \underline{1}$$

Previously, it was suggested that the length of a line segment could correspond with the duration of a pitch or event. This is a logical association to make with the previous method of pitch determined by terminus points. In the trajectory case, the length of the segment within a given cell could be associated with the duration of that specific pitch.

If the vertex-centric *Tonnetz* of Fig. 14.10 is used in conjunction with the trajectory path, each successive trichord will change by only one pitch. This will yield the familiar parsimonious voice-leading associated with neo-Riemannian transformations [11]. The trajectory of the third line segment for the RGP of Fig. 14.4 would yield:

$$
\begin{array}{cccccc}
9 & 9 & & 9 & 9 & 9 \\
T & 3 & 3 & 3 & 8 & 8 \\
4 & 4 & & 2 & 2 & 3
\end{array}
$$

Just as in the cell-centric version, the third segment passes through the vertex of six connected cells; however, in the vertex-centric *Tonnetz*, this vertex is assigned a specific value. Consequently, the resulting pitch is unambiguous. This vertex crossing is represented by the single pitch class of 3 above.

14.4.2. *Resulting musical composition*

For the sake of completeness, a sample composition using the ACP from Fig. 14.6, the *Tonnetz* from Fig. 14.9, and the terminal point of pitch assignment is shown in Fig. 14.11.

Fig. 14.11. Sample composition "Lattice I". Copyright © 2016 Milam Caravan Music All Rights Reserved.

Acknowledgments

I am deeply indebted to Nickitas J. Demos, my teacher, mentor, and colleague, for his insight, direction, and refinement of my ideas concerning the elements of sonority. My table entitled "Elements of Sonority Analysis" (Table 14.1) is derived directly from Demos' own presentation entitled "Dimensions of Sonority" used in conjunction with his composition and orchestration classes. His ideas regarding the perception of timbre, sound density, spatialization, and most particularly the concept of musical gesture have proven invaluable to my own understanding.

Bibliography

[1] R. Cohn, Neo-Riemannian operations, parsimonious trichords, and their "Tonnetz" representations, *J. Music Theory* **41**(1) (1997) 1–66.

[2] R. Cohn, *Audacious Euphony: Chromaticism and the Triad's Second Nature* (Oxford University Press, New York, 2012).

[3] D. Cope, *New Directions in Music*, 7th edn. (Waveland Press, Inc., Long Grove, IL, 2001).

[4] N. J. Demos, The role of sonority in György Ligeti's *Melodien*, DMA monograph, The Cleveland Institute of Music, Case Western University, Ohio, USA, (1994).

[5] J. Giangrande, The tritone paradox: Effects of pitch class and position of the spectral envelope, *Music Percep.* **15**(3) (1998) 253–264.

[6] R. Kamien, *Music: An Appreciation*, 5th brief edn. (McGraw-Hill, New York, 2006).

[7] J. La Rue, *Guidelines for Style Analysis* (W. W. Norton, New York, 1970).

[8] J. Machlis, *Enjoyment of Music*, 6th edn. shorter (W. W. Norton, New York, 1991).

[9] J. B. Mailman, Agency, determinism, focal time frames, and processive minimalist music, in *Music and Narrative Since 1900*, eds. M. L. Klein and N. Reyland (Indiana University Press, Bloomington, Indiana, 2013), p. 128.

[10] K. Miyazaki, Absolute pitch identification: Effects of timbre and pitch region, *Music Percep.* **7**(1) (1989) 1–44.

[11] *NOVA "The Great Math Mystery: Invention or Discovery,"* DVD, produced and directed by D. McCabe and R. Reisz (WGBH Educational Foundation, 2015).

[12] L. G. Ratner, *Music: The Listener's Art* (McGraw-Hill, New York, 1957).

[13] S. Rings, *Tonality and Transformation* (Oxford University Press, New York, 2011).

[14] M. R. Schroeder, *Fractals, Chaos, Power Laws* (W. H. Freeman and Company, New York, 1991).

[15] D. Tymoczko, *A Geometry of Music: Harmony and Counterpoint in the Extended Common Practice* (Oxford University Press, New York, 2011).

[16] J. Yudkin, *Understanding Music*, 2nd edn. (Prentice-Hall, Inc., Upper Saddle River, NJ, 1999).

Section III

Chapter 15

Flamenco Music and Its Computational Study

Francisco Gómez

Department of Applied Mathematics,
Polytechnic University of Madrid,
carretera de Valencia, km. 7, s/n, 28031, Madrid, Spain
fmartin@etsisi.upm.es

15.1. Introduction

Music is a phenomenon that has always fascinated researchers in a wide range of disciplines. Music is complex, multidimensional and universal. Music can be seen as a phenomenon of people, society, and culture, as studied in anthropology and sociology [16]; as a physical phenomenon, and as such studied by physicists and engineers [1]; as a phenomenon of the mind, which is then an object of study of musical cognition [11, 21, 22]; as a purely musical phenomenon, and then questions such as melody, rhythm, and harmony organization are relevant; as an affective phenomenon, thus entering in the realm of psychology [4, 20]. However long this enumeration may seem, music has also sparked the interest of researchers in seemingly farther fields such as mathematics and computer science. The interest of mathematicians in music is by no means new; it can be traced back to ancient Greece. That interest has been renewed over the centuries as mathematics has progressed itself. New understanding in mathematics has sooner or later led to new interpretations of musical structures. Music is full of patterns and structures and that inexorably have attracted mathematicians' interest. This interest, however, should not be understood as an irrepressible desire to find patterns and structures in music irrespective of the object itself. That would provide an amiss, unfair impression of mathematics and its methods. In the mathematical research of music there has been a genuine interest to understand the nature of music. The reader can find a reasoned discussion about the scope and purpose of mathematics and computer science in music in [25]. Since computers allow such formidable ways to process musical data, a great wealth of mathematical research in music has served as theoretical foundations in musical technology. Music technology can be defined as the interdisciplinary science of computational description, analysis and treatment of music and music data. An important subfield of music technology is music information retrieval (MIR), the scientific discipline of retrieving data from music.

Over the last few years the use of computational tools in music research has grown rapidly, in music technology and especially in MIR. For example, Meinard [15] identifies four broad areas where audio-signal MIR techniques have intensively been applied to music, namely, audio retrieval, music synchronization, structure analysis, and performance analysis (although he himself acknowledges "this list only scratches the surface"). However, those tools are not intended to substitute for traditional methods of research in music. Nowadays computational tools are thought of to assist, complement, and increase the power of analysis that uses traditional methodologies, both in quantitative and qualitative terms.

The number of applications of music technology one can find in the literature is simply spectacular; just look at the papers presented at ISMIR [14], the main conference in the field, to obtain an idea of such variety of results. Surprisingly enough, most of the research and applications were done for Western music, either popular music or classical music from the common practice. In 2010 Cornelis *et al.* [3] found that the problem persists, and in spite of *all* the substantial amount of research undertaken over the past few years, the amount of research done to understand and process ethnic music (non-Western, traditional, and folk music) is still scant and scattered. The challenges posed by the research in ethnic music are significant because of the particular musical features of a given tradition, which in many cases are markedly different from the Western one.

Research in music technology has mainly focused on written Western music, and only recently researchers have begun paying attention to oral traditions. In this respect, flamenco music is a case in point. Flamenco music is a non-Western, oral tradition with very particular musical features whose analysis certainly requires an interdisciplinary approach. A simple question, such as musical transcription, is by no means solved in flamenco music (see discussion below), just to name a central problem in musical analysis. Furthermore, other fundamental problems in flamenco music are still open, such as melodic and rhythmic similarity, style classification, singer identification, among others.

The main purpose of this chapter is to make known the efforts of COFLA [23] (Computational Analysis of Flamenco Music), a group of researchers from several disciplines who study flamenco music, as well as its results; the group was founded by the authors of this chapter. Its goal is to analyze flamenco music from different disciplines, incorporating music technology in that analysis. To accomplish that goal, COFLA is composed of an interdisciplinary team including experts from areas such as Musicology, Ethnomusicology, History, Literature, Education, Sociology, but also Mathematics, Engineering, and Computer Science. The structure of this chapter is as follows. The next section gives a brief overview of the main musical features of flamenco music. Afterwards three problems on which the COFLA group is currently working are presented. Next, the problem of melodic similarity in flamenco a cappella songs is addressed from an interdisciplinary perspective (which is original research).

15.2. Flamenco Music

Flamenco music is an eminently individual yet highly structured form of music. Indeed, on the one hand, in flamenco music there is a high degree of improvisation and spontaneity; on the other hand, there is an extremely stable organization of the musical material without which improvisation would not work. Flamenco music is the result of reciprocal influences of several cultures through time, which originated a unique combination of singing, dancing, and guitar playing. Among these influences are Jewish and Arabic cultures, but also the culture of the Andalusian Gypsies, but also from the culture of the Andalusian Gypsies, who decisively contributed to its form as we know flamenco today. We refer the reader to the books of Blas Vega and Ros Ruiz [2], Navarro and Ropero [19], and Gamboa [6] for a comprehensive study of styles, musical forms and history of flamenco.

According to Gamboa [6], flamenco music was mainly developed from the singing tradition. Therefore, the singer's role is dominant and fundamental in flamenco music. In the flamenco jargon, singing is called *cante*, and songs are termed *cantes*; in this chapter we use this terminology. Next, we describe the main general features of flamenco *cante* (after [17]).

- Instability of pitch. In general, notes are not clearly attacked. Pitch glides or *portamenti* are very common.
- Sudden changes in volume (loudness). Those sudden changes are very often used as an expressive resource.
- Short melodic pitch range. It is normally limited to an octave and characterized by the insistence on a note and those contiguous to it.
- Intelligibility of voices. Lyrics are important in flamenco, and intelligibility is then desirable. For that reason, contralto, tenor, and baritone are the preferred voice *tessituras*.
- Timbre. Timbre characteristics of flamenco singers depend on the particular singers. As relevant timbre aspects, we can mention breathiness in the voice and absence of high frequency (singer) formants.

Since they will be used later for illustration purposes, we describe briefly a cappella *cantes*, which constitute an important group of styles in flamenco music. They are songs without instrumentation, or in some cases with some percussion. From a musical point of view, a cappella cantes retain the following properties.

- Conjunct degrees. Melodic movement mostly occurs by conjunct degrees.
- Scales. Certain scales such as the Phrygian and Ionian mode are predominant. In the case of the Phrygian mode, chromatic rising of the third and seventh degrees is frequent.
- Ornamentation. There is also a high degree of complex ornamentation, melismas being one of the most significant devices of expressivity.
- Microtonality. Use of intervals smaller than the equal-tempered semitones of Western classical music.
- Enharmonic scales. Microtonal intervallic differences exist between enharmonic notes.

These features are not exclusive to a cappella *cantes* and can be found to various degrees in other flamenco styles.

15.3.   The COFLA Group

The COFLA group has as its main goal the interdisciplinary study of flamenco music. In this section and as illustration, we present three problems the COFLA group is currently working on.

Musical transcription of flamenco music. As mentioned earlier, in many other musical traditions the problem of transcription does not exist or is tractable. In flamenco music, unfortunately, it is not the case. As an oral tradition, performers never had the need to transcribe. Furthermore, flamenco has started to be studied from academia recently and available scores are scant, limited entirely to guitar. In fact, a consensus about which is the best method to transcribe flamenco music has not been reached among flamenco scholars. Whereas Donnier [5] advocates for the use of Gregorian neumes to annotate flamenco singing, Hurtado and Hurtado [12] and Hoces [10] argue in favor of Western classical notation. Emilia Gómez (COFLA) and collaborators have developed algorithms to automatically transcribe flamenco a cappella singing from audio input (see [7]). These transcriptions can be further tailored to obtain a melodic contour or detailed expressive features related to vibrato or ornamentation. A harder problem is that of the transcription of accompanied singing in polyphonic recordings, requiring a source separation method [8].

Melodic similarity in flamenco music. It is with regard to melodic similarity that flamenco proves to be an extraordinarily hard type of music to analyze. In other musical traditions, musical similarity is assessed through the sequence of notes and the order they appear in. Two *cantes* belonging to the same style may sound very different to an unaccustomed ear. Underlying each *cante* there is a melodic skeleton. Donnier [5] has termed that melodic skeleton as the "*cante*'s melodic gene". The singer may intersperse all kind of melismas, ornamentation and other expressive resources between the notes of the melodic skeleton. A flamenco aficionado would recognize two different fillings of the skeleton as being the *same song* as long as the skeleton is not changed. In order to help the reader understand this point, in Figs. 15.1 and 15.2 we show a transcription of two version of the same *cante* to Western musical notation; the scores have not been transposed and reflect the actual pitches used by the singers. A flamenco aficionado recognizes both versions as the same *cante* because certain notes appear in a certain order. What happens between two of those notes does not matter regarding style

Fig. 15.1. Mairena's rendition of *En el barrio de Triana*.

Debla: "En el barrio de Triana"

Chano Lobato

Trans.: J.M.R.

Fig. 15.2. Lobato's rendition of *En el barrio de Triana*.

classification, but does matter for assessing a performance or the piece itself. The main notes that the aficionado must hear have been highlighted in both figures. See [17] for further information. By style here we mean pieces that share certain musical features; in flamenco every piece belongs to a specific style (sometimes a piece can be composed of sections in several styles).

In these transcriptions many melismas from the actual recording were removed for ease of reading. It is clear now that the problem in musical similarity in flamenco is quite complex and intricate. A line of research COFLA is pursuing consists of defining a melismas lexicon so that melismas could be discarded, and thus main notes could be identified.

Detection of distinctive melodic patterns in flamenco music. Since flamenco is orally transmitted, melody plays a central role. However, the way flamenco musicians memorize melodies is not note by note. For each style they learn a few melodic patterns, out of which later they will add their personal touch. Therefore, the identification of those patterns is of paramount importance. There are two approaches to the problem of finding distinctive patterns. The first one is an inductive method. Flamenco corpora are automatically analyzed in search of patterns that appear often or in certain positions in the piece. For further details on how this problem has been addressed by the COFLA group, see [9]. The other method, which is more interesting from a interdisciplinary point of view, is deductive. Flamenco experts define patterns that they consider meaningful or distinctive of a given style. Then their hypotheses are tested for searching those patterns in large corpora of flamenco music. Both approaches lead to many fascinating open questions; here there is a short list: (1) What is the melodic pattern common to authoritative recordings of flamenco masters? (2) What is the characteristic ornamentation of a given style? (3) Which ornaments are mandatory in a given style?

15.4.   An Example of Interdisciplinary Research on Flamenco Music

As an example of the interdisciplinary research carried out by the COFLA group, we will study the problem of melodic similarity for style classification. One of the problems the COFLA group chose to tackle was that of a cappella *cante* classification. Three substyles, *deblas*, *martinete-1*, and *martinete-2*, from a style called *tons* were selected to carry out this study. *Tons* are *cantes* sung in free rhythm. Scale and melody type are modal. Frequent modes are major, minor, or Phrygian, though alternation of modes is also common. The lyrics of these songs range widely. This *cantes* possess a high degree of ornamentation; the *cantes* annotated in Figs. 15.1 and 15.2 above are actually *deblas*.

Since scores were not available, the first step we took was to extract an automatic transcription from the audio files. The corpus was composed of 72 *tons*, which included *deblas, martinete-1*, and *martinete-2*. We used the system proposed in [7], for the transcription task. Melodic transcription is usually structured into three different stages: low-level feature extraction, frame-based descriptor extraction (e.g., energy and fundamental frequency), note segmentation (based on location of note onsets), and note labeling. For more technical details, see [7] and the references therein.

The output of this first step is a MIDI-like file containing the sequence of notes and durations, representing the melodic contour. The corpus employed for our research can be found in [24].

In geometric terms, that sequence can be interpreted as a plane polygonal chain. The x-coordinate represents the duration of notes while the y-coordinate represents the pitch. From this point on, many mathematical techniques can be applied to the problem of determining the distance between two *cantes*. The problem of determining how similar two *cantes* has been transformed into the problem of determining the distance between the polygonal chains given by the melodic contours. There is a plethora of algorithms to compute the distance between two such chains (edit distance, n-grams, vector correlation). One of the most popular is the edit distance. Given two polygonal chains, we consider three basic operations to transform one chain into another: insertion of a note, deletion of a note, and substitution of a note. The edit distance between two polygonal chains is then defined as the minimum number of operations required to transform one chain into another. As a matter of fact, those operations have weights assigned, which depends on the specific problem.

However, most of those similarity measures lack perceptual validity, that is, they have not tested on subjects. In a 2004 paper, Müllensiefen and Frieler [18] addressed this problem. Their first step was to establish some ground truth for melodic similarity under certain conditions; secondly, they analyzed 34 similarity measures (or dissimilarity distances) found in the existing literature to determine the most adequate measures in terms of perceptual validity. The best similarity measure was a linear combination between the raw edit distance and the n-grams, specifically $3.355 \cdot ED + 2.852 \cdot NG$, where ED is the edit distance and NG the n-grams distance. We followed their results to compute our similarity measures between *cantes*. The distance matrix obtained was used to feed an algorithm that computes phylogenetic graphs. Given a distance matrix from a set of objects, a phylogenetic graph is a graph whose nodes are the objects in the set and such that the distance between two nodes in the graph corresponds to the distance in the matrix; see [13] for more information. If possible, the phylogenetic graph algorithm tries to output a tree (a connected graph with no cycles). If that is not possible, then the algorithm introduces new nodes, which may create cycles, and outputs a more general graph. However, phylogenetic graphs are always planar graphs (Fig. 15.3). Also, along the graph the algorithm outputs and index, *LSFit*, which is expressed as a percentage. It indicates how accurate the correspondence between the distances in the graph and the distances in the set of objects is. The higher the index is, the more accurate the correspondence between matrix and graph distances is. For our set of *cantes* that index is 99.19%; the resulting graph can be seen in Fig. 15.4.

Because of the problem of ornamentation mentioned earlier, the phylogenetic graph produced poor results (there were many misclassified *cantes*). The *cantes* were classified a priori by the flamenco experts in the COFLA group. Then, once the distance matrix was computed, a *cante* was classified according to its k nearest neighbors (several values of k were tested, but they all gave poor performance). For example, if $k = 1$, a *cante* is classified as the style of its nearest neighbor. The main problem lies in the fact that two *cantes* may have the same main notes and very different ornamentations between them. However, an edit distance would give a high distance between them. More knowledge was required to obtain better results.

Typically, musical descriptors fall in three categories: low-level descriptors, mainly related to properties of the audio signal such as frequency, spectrum, or intensity; mid-level descriptors, associated to pitch, melody, chords, timbre, beat, meter, or rhythmic patterns; and finally high-level descriptors, typically linked to meaning and expressiveness such as mood, motor and affective responses. To circumvent the problem of ornamentation in the *cantes*, we defined a distance based on mid-level descriptors. The mid-level descriptors used for all *tonás* in our study were the following: (1) Initial note of the piece; (2) Symmetry of the highest degree in the second hemistich (a hemistich is half of a line of poetry); (3) The frequency of the highest degree in the second hemistich; (4) Clivis at the end of the second hemistich (a melodic movement consisting of two descending notes); (5) Final note on the second hemistich; (6) Highest degree in the *cante*; (7) Duration of the *cante*. As a matter of fact, just using features very peculiar to a given style would distort the

Fig. 15.3.   The phylogenetic graph for the melodic contour distance.

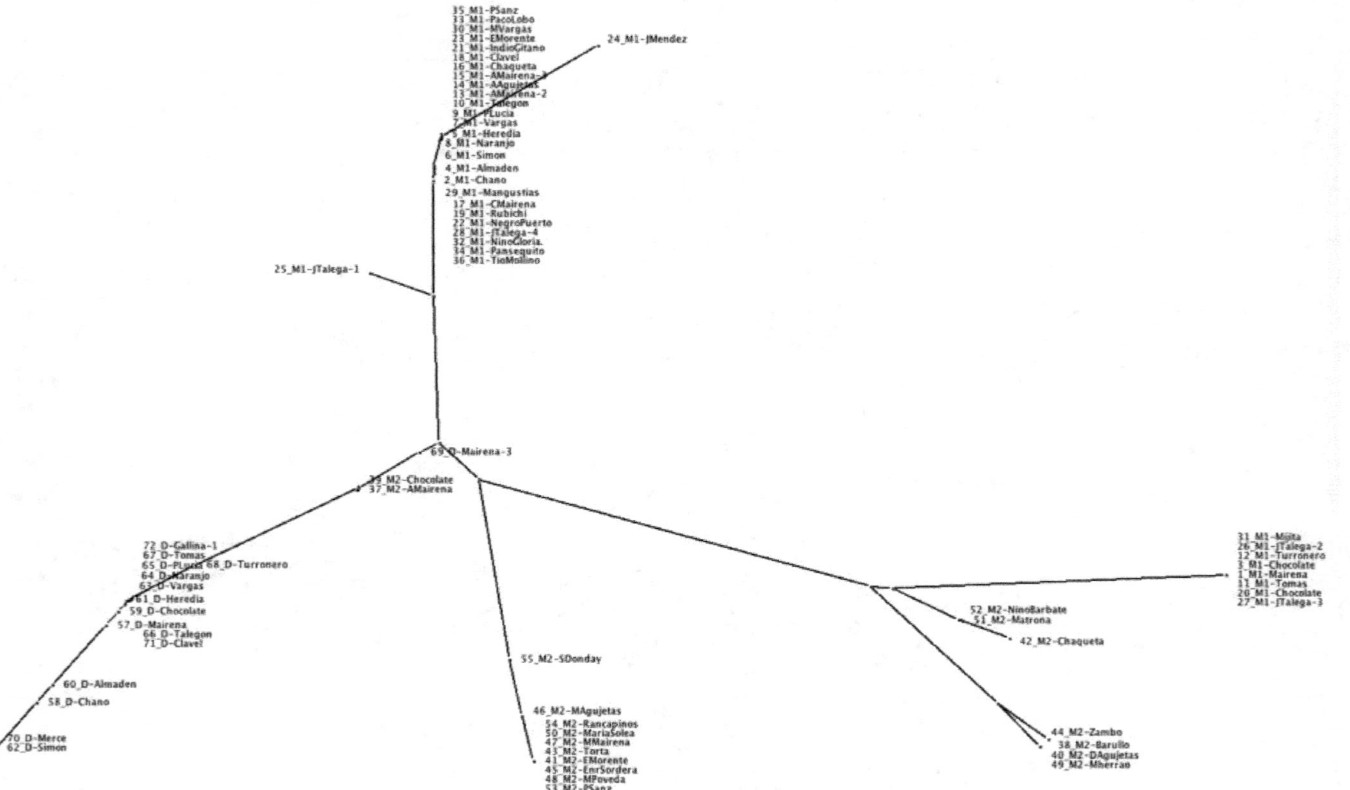

Fig. 15.4. The phylogenetic graph for the combination of distances.

analysis, as their discriminating power would be very high. Our intention was to select a set of a few features capable of discriminating between different *cantes*. Note that the variables chosen here are of a more musical nature. The computation of these variables requires that ornamentation is taken into account, and initially was done manually by flamenco experts. We are working on how to compute them automatically (some very hard computational problems arise here).

We defined a new distance that was a linear combination of the melodic contour and this mid-level distance. The results improved dramatically. The melodic contour accounted for local changes and the mid-level distance measured global changes. The clustering showed by the combined distance made sense according to flamenco experts and the number of misclassified *cantes* decreased to normal levels. Also, the distance allowed the study of intra-style characteristics; that study was not possible with the melodic contour distance.

15.5.   Conclusions

In this chapter we have presented an example of interdisciplinary research through the work of the COFLA group. In the group there are researchers of various disciplines, including mathematics, engineering, and computer science, who strive for understanding a musical tradition as complex and rich as flamenco music. At the outset we warned against doing mathematics for the sake of it (*the irrepressible desire to find patterns and structures*) and we believe we have escaped the danger. Indeed, in the example given, the problem of melodic similarity in a cappella *cantes*, we have seen that interdisciplinary methodology at work. The corpus was selected by flamenco experts; engineers, computer scientists, and mathematicians designed the algorithms to extract the melodic sequences from the audio files; mathematicians studied the problem of the similarity distance. As results were not satisfactory enough when the melodic contour distance alone was used, more research was in order. Flamenco experts designed the mid-level distance; again the scientists in the team were in charge of the computations and the assessment of results.

Many open problems remain to be considered. Our distance, although giving good results, can be certainly improved. In particular, we are working on the generalization of mid-level distance to other flamenco styles (the variables were specifically design for that corpus). An interesting question is what is the minimum set of variables that can be used to satisfactorily measure melodic similarity in flamenco music. The problem of automatically identifying ornaments in flamenco styles is still open and seems computationally hard. More research has to be done in the problem of the automatic detection of distinctive patterns. Although there are some studies, rhythmic similarity in flamenco music is a field waiting to be explored, particularly in hand-clapping rhythms.

Bibliography

[1] R. Berg and D. G. Stork, *The Physics of Sound* (Addison-Wesley, 2004).

[2] J. Blas Vega and M. Ríos Ruiz, *Diccionario enciclopdico ilustrado del flamenco* (Cinterco, Madrid, 1988).

[3] O. Cornelis, M. Lesaffre, D. Moelants and M. Leman, Access to ethnic music: Advances and perspectives in content-based music information retrieval, *Signal Process.* **90**(4) (2010) 1008–1031.

[4] D. Deutsch, *The Psychology of Music* (Academic Press, 1998).

[5] P. Donnier, Flamenco: elementos para la transcripcin del cante y la guitarra, in *Proc. III Congr. of the Spanish Ethnomusicology Society*, Spain (1997).

[6] J. M. Gamboa. *Una historia del flamenco* (Espasa-Calpe, Madrid, 2005).

[7] E. Gómez and J. Bonada, Towards computer-assisted flamenco transcription: An experimental comparison of automatic transcription algorithms as applied to a cappella singing, *Comput. Music J.* **37** (2013) 73–90.

[8] E. Gómez, F. Cañadas, J. Salamon, J. Bonada, P. Vera and P. Cabañas. Predominant fundamental frequency estimation vs. singing voice separation for the automatic transcription of accompanied flamenco singing, in *13th Int. Society for Music Information Retrieval Conf. (ISMIR 2012)*, Porto (08/10/2012).

[9] F. Gómez, A. Pikrakis, J. M. J. Mora, Daz-Báez, E. Gómez and F. Escobar, Automatic detection of ornamentation in flamenco, in *Proc. 4th Int. Workshop on Machine Learning and Music. NIPS Conf.*, Granada (2011).

[10] R. Hoces, La transcripcin musical para guitarra flamenca: anlisis e implementacin metodolgica. Ph.D. thesis, Universidad de Sevilla (2011).

[11] H. Honing, *Musical Cognition: A Science of Listening* (Transaction Publishers, 2013).

[12] D. Hurtado and A. Hurtado, El arte de la escritura musical flamenca, in *Bienal de Arte Flamenco*, Sevilla (1998).

[13] D. H. Huson and D. Bryant, Application of phylogenetic networks in evolutionary studies, *Mol. Biol. Evol.* **23**(2) (2006) 254–267.

[14] International Society for Music Information Retrieval (ISMIR), http://www.ismir.net/ (accessed in January, 2014).

[15] M. Meinard, New developments in music information retrieval, in *Audio Engineering Society 42nd Int. Conf.*, Ilmenau, Germany (July 2011).

[16] B. Merker and S. Brown. *The Origins of Music* (MIT Press, Cambridge, 2000).

[17] J. Mora, F. Gōmez, E. Gōmez, F. Escobar Borrego and J. M. Daz Bez, Characterization and melodic similarity of a cappella flamenco cantes, in *ISMIR* (*Int. Sympo. Music Information Retrieval*), Utrecht, The Netherland (August 2010).

[18] D. Müllensiefen and K. Frieler, Cognitive adequacy in the measurement of melodic similarity: Algorithmic vs. human judgments, *Computing in Musicology* **13** (2004) 147–176.

[19] J. L. Navarro and M. Ropero, (eds.), *Historia del flamenco* (Ed. Tartessos, Sevilla, 1995).

[20] R. E. Radocy and D. J. Boyle, *Psychological Foundations of Musical Behaviors* (Charles C. Thomas, Springfield, IL, 2003).

[21] J. Sloboda, *Exploring the Musical Mind: Cognition, Emotion, Ability, function* (Oxford University Press, 2005).

[22] D. Temperley, *The Cognition of Basic Musical Structures* (MIT Press, Cambridge, MA, 2001).

[23] The COFLA group, http://mtg.upf.edu/research/projects/cofla.

[24] The COFLA group, Corpus tonás, http://mtg.upf.edu/download/datasets/tonas (accessed in January, 2014).

[25] A. Volk and A. Honingh, Mathematical and computational approaches to music: challenges in an interdisciplinary enterprise, *J. Math. Music* **6**(2) (2012) 73–81.

Chapter 16

Examining Fixed and Relative Similarity Metrics Through Jazz Melodies

David J. Baker and Daniel Shanahan

Louisiana State University,
Baton Rouge, LA 70803, USA

16.1. Introduction

Similarity, defined here as the degree to which multiple items might share overlapping properties, inherently requires a comparison to be made between the two objects [5]. How the comparison of two or more musical ideas can be carried out is, at times, a perplexing question. A musical idea might be considered similar by specific parameters such as its rhythmic, melodic, harmonic features, or broader, more difficult to define aspects, such as contour and overall shape. Because of this, we might argue that the question of musical similarity occupies the space where music theory and music psychology intersect. Loosely defined, musical similarity is the extent to which two musical objects are thought to share similar characteristics. In this definition there are two parameters to consider. Firstly, which two musical objects are to be compared to one another? Secondly, which criteria should be used in order to make that comparison?

A special issue of *Musicae Scientiae* published in 2007 focused specifically on questions of musical similarity, and provides a many resources on the topic. For example, Typke, Weiring, and Veltcamp suggested that using the earth mover's distance can be used as a way to classify music that is in agreement with expert human judgments [22], Müllensiefen and Frieler modeled expert notions of melodic similarity and concluded that experts are, in fact, able to consistently make similarity judgments across different musical stimuli. They additionally matched algorithms that mirrored the expert judgments and argued that these methods could be used for real world problems such as folk song categorization and musical analysis [15]. Additional work has also shown that listeners are unconsciously sensitive to various statistical properties in melodies such as pitch distribution and the number of notes, and that these can be used to predict perceived similarity, although listeners may be unconsciously aware of it [9].

16.1.1.   *Fixed and relative similarity*

At this point, it might be worth distinguishing between two modes of similarity measurements, one we might call a fixed, and a second we could refer to as relative. The first is an fixed, stable, and somewhat objective view of similarity, in which a mathematical measurement provides some theoretical understanding of a relationship between X and Y, and is simply asking "how similar are these two objects given a fixed parameter?" A second conceptualization of similarity, which we will refer to as "relative similarity", asks "when accounting for a humans response to this questions what external information helps guide a choice of similarity between objects X, Y, and maybe even Z, in a given context?" Although objective (fixed) measurements have been used a great deal for mathematical and computational expediency, psychologists argue that perceived similarity is, by its nature, perceived in relation to a network of the objects that inform each contextual decision. For example, Cambouropoulos [4, p. 10] writes that "[s]imilarity is by definition relational and, therefore, cannot be objective".

There are, however, benefits to both approaches. On the one hand, two objects are able to be seen as similar regardless of whether or not a third object is there. For example, if one is drinking out of the same type of paper coffee cup as a colleague, they are able to understand that those paper coffee cups are quite similar, regardless of whether a third friend is present with an ornate ceramic mug. This is because an individual brings their own memories and beliefs to any situation where they are made to make a choice. On the other hand, it could be argued that the only reason the two paper cups are seen as similar is due to the internalized understanding of what dissimilar coffee cups would look like. These ideas are very much at the core of prototype theory, which claims judgments of similarity are informed by the distance one member of a category is to its central, defining member [8, 19].

It would follow that depending on which set of musical objects one chooses, as well as which characteristics one chooses to focus on in a relative context, the answer to the question of "how similar is X to Y?" is subject to change. Contrastingly, if modeling this question mathematically as an objective measurement of similarity, the answer will be the same each time the computation is run. While tackling this question from a strictly mathematical vantage point has its advantages (such as classifying folksongs, or searching for repeated structures in a large corpus of music [12] or building computational models to formalize arguments), questions of psychological similarity will involve some sort of human judgment, variance, and error, which are missing in this sort of computation.

16.1.2. *Considering human judgment*

A large amount of studies exploring musical similarity have employed similarity measurements in their research that juxtapose two melodies and ask participants to rate the degree to which they are similar (for example, see [9, 18]). While this approach has its advantages, one drawback is that these measures can only capture the sets of perceptual qualities of the two items in question. One point of interest that has been raised is that the context of a similarity judgment can influence ratings of similarity even by the same participant [21]. For example, if given the context two separate natural harmonic minor melodies and a major key melody, an individual might select the two minor melodies as most similar in one context. Though if those two same minor melodies are present in a subsequent trial with the addition of a different, yet novel harmonic minor melody, the two harmonic minor melodies might be then rated as more similar — perhaps even to a larger extent. We chose to investigate how the context of melodies might in fact impact judgments of similarity in both humans and computers in two ways:

(1) An experiment asks participants to gauge how similar two objects are when given *three* choices of melody.
(2) A computational model is given the same three choices and by using an objective measurement of similarity, also makes decisions about which two objects are the most similar.

We then compare results between the listeners behavior and the computational model, examining the degree to which a distance metric can be used to predict human choices when given a paradigm based on *relative* similarity. The process of modeling with computational tools how humans perform when carrying out behavioral tasks is increasing, but there is still a great deal of word to be done. Müllensiefen and Pendzich [17] examined the accuracy of similarity metrics in predicting the jury responses in musical plagiarism cases and found that while certain measures such as edit distance performed somewhat well, they were surpassed by Tversky's [21] model of similarity which takes into account the number of shared categorical features between two sets to be compared. Given the success of edit distance in this and other contexts, we choose to use this as a stepping off point for investigating the effects of context on musical similarity.

16.2. Distance as Transformation and Similarity

Looking at musical distances can be operationalized a handful of ways depending on how the question being asked. In this chapter we have chosen to view musical distance as a transformation, conceiving the distance between two objects as a result of the number of operations that are needed to move from one object to another. This method has been recently successful in modeling musical similarity [18] and here we take a related approach.

One of the most commonly-used transformation metrics is the Damerau–Levenshtein distance measure (also commonly referred to as edit distance). In some ways, the Damerau–Levenshtein distance could be thought of as a proxy for measuring human behavior similarity judgments in that it makes a calculation of the number changes needed to transform one object into another. Although the Damerau–Levenshtein distance measure is a rather simple transformative model, it has been applied successfully in music information retrieval tasks [5, 13] and has predicted melodic similarity well in previous research by Müllensiefen *et al.* who suggested that the distance that best matches human judgments of melodic similarity can be modeled with combination of the edit distance and an *n*-gram distance. We therefore chose to implement the edit distance as an entry point to implement and reasoned that once a paradigm was designed that could map human judgments on to those of the computer, it would then be possible in future research to explore other options of comparison.

Unlike prior research that often relies on participants making continuous ratings between two separate musical stimuli against one another [18], we opted for using a forced choice task that would provide an insight into how humans behave when confronted with having to pick between stimuli. Using the three way alternative forced choice (3AFC) paradigm listed above not only provided the opportunity to do analyses like those found in the aforementioned papers, but also created data that counts the number of times a participant selected a certain stimuli when presented alternative options. This not only allowed for our analyses to look at trial by trial inter-rater agreement, but also provided a paradigm where an algorithm could hypothetically partake in the experiment as if it were a participant. While a computation almost by definition will not display the same sort of variability as a human responses, this paradigm affords analyses that are easily interpretable when comparing human judgments of similarity to those of the computer.

16.3.   Experiment

We designed an experiment with the goal of exploring how different computational and algorithmic measures can be used to mirror human behavior using a contextual paradigm. This goal led to employing a design that was not dependent on any sort of explicit understanding of Western musical notation. While these aspects of the music were necessary in order to carry out some of the above studies, it could be argued that some of the results depend heavily on an individual's musical background as well as their training with the context of Western music notation which could cause the listeners to consciously or unconsciously focus on aspects of the music more linked to the notation than the actual aural events.

Work in the field of music psychology has suggested that many factors involved in music perception do not even require any sort of formalized training [3] and that factors outside of musical training can be used to model performance on perceptual tasks [2]. In light of these considerations, we set out to investigate if listeners, regardless of any sort of domain specific expertise, were able to perform consistent and reliable similarity judgments in a context that is not dependent on anything but auditory information.

In order to investigate these aspects of musical similarity, we employed a paradigm that that did not rely on any sort of musical formal training expertise [1]. This approach has been used successfully with naive listeners to investigate questions of genre perception where the authors concluded that based on similarity judgments alone, listeners were able to pick out the hypothesized salient features [10].

16.3.1.   *Methodology*

Participants ($N = 44$, 13 women, $\bar{X} = 201.16$, SD $= 1.45$, R $= 18$–24) for this experiment were selected from the subject pool at Louisiana State University's School of Music. All participants were given partial course credit for taking part in the experiment. Stimuli from this experiment were randomly selected from a subset of the database of encoded bebop transcriptions. It was decided *a priori* that seven phrases, or licks, of 24 notes would be chosen as this number of notes would create three bar phrases. Each excerpt was converted into eighth notes, transposed to the key of C major, set to quarter note = 120, and presented as MIDI recordings with a uniform timbre in order to control for any musical features outside of melody. Stimuli from the experiment are shown in Figs. 16.1–16.7. Software

Fig. 16.1. From Charlie Parker's "Warming Up on a Riff".

Fig. 16.2. From Charlie Parker's "Passport".

Fig. 16.3. From Charlie Parker's "Chasing the Bird".

Fig. 16.4. From Charlie Parker's "Cardboard".

Fig. 16.5. From Charlie Parker's "Bird Gets the Worm".

Fig. 16.6. From Charlie Parker's "Anthropology".

Fig. 16.7. From Charlie Parker's "Another Hairdo".

for the experiment was provided by the authors of Farrugia *et al.* [10].

16.3.1.1. *Procedure*

Participants first completed forms relating to their general demographic information, as well as their musical background as measured by the Goldsmiths Musical Sophistication Index [16]. After completing these forms, the experimenter directed their attention to an interface in Max/MSP where participants were told they needed to make a subjective choice about which two musical excerpts they found to be most similar. Participants were not prompted to focus on any sort of musical parameters and if they inquired about what to pay attention to, the experimenter would tell to focus on whatever aspects they found to be most salient. Participants were allowed to listen to any stimuli in any order, as many times as they would like to. The amount of times a participant listened to a track was recorded, unknown to them.

Additionally before running the experiment, we ran an *a priori* effect size calculation using the *pwr* package in [6] to identify approximately how many subjects we needed in order to generate a large effect size (0.5) [7]. Calculations indicated that with $df = 2$, $\alpha = 0.05$, $\beta = 0.8$ we would need approximately 39 participants.

16.3.2. *Algorithm implementation*

Mirroring the experimental design, we used the Damerau–Levenshtein distance measure from the Humdrum Toolkit and the Humdrum Extras package [11, 20] to simulate what the Damerau–Levenshtein distance would "select" in each permutation of the stimuli in our experiment. To do this, we computed Damerau–Levenshtein distance measures between each three stimuli for every unique trial and then used the smallest distance between the three melodies to select the two most similar stimuli. Computing a three way distance measure between each triad of stimuli and selecting the two which were most similar enabled use to somewhat anthropomorphize the algorithm. By doing this, we were able essentially have an algorithmic participant that would be able to always select the same melody in each context.

16.3.3. *Analysis*

Our initial analyses sought out to investigate two hypotheses:

(1) H_1: Participants will identify at each three alternative forced choice (3AFC) in accordance with edit distance.
(2) H_2: Participant choices will most often match an algorithm when given identical choices, based on relative similarity.

Before comparing the human responses to any sort of algorithm, data from participants was examined separately. Since participants not only rated each permutation of seven different stimuli, but did so three times, first data was checked for reliability in order to investigate the aforementioned question about if non-expert listeners are able to provide an acceptable degree of reliability between the rating blocks. Reliability between blocks was checked at the block level by running a Spearman rank order correlation between the trials selected most often between participants. A high degree of correlation was found between blocks 1 and 2 ($r = 0.88$), 2 and 3($r = 0.89$), and 1 and 3($r = 0.90$).

Because of the high degree of consistency between participants, deemed it was appropriate to pool participant responses and examine the degree to which human responses were in agreement with edit distance.

After computing all edit distance measures, we then tallied the amount of times that each participant was either in agreement or disagreement with what Damerau–Levenshtein distance selected and performed a chi-square test on each of the unique trials to determine if humans agreed with the Damerau–Levenshtein distance measures a significant amount of times to investigate H_1. Type I error inflation rates were controlled by using a Bonferroni correction. Results of H_1 suggest that, within a set of trials, there was a fair degree of agreement within some of them, but after correcting for multiple tests any sort significant effects disappeared due to the extremely conservative alpha ($p < 0.001$).

Table 16.1. Block comparison.

H_2 analysis
Block 1: $\chi^2(2, 35) = 0.25, p > 0.05$
Block 2: $\chi^2(2, 35) = 0.03, p > 0.05$
Block 3: $\chi^2(2, 35) = 0.71, p > 0.05$

After this step, we then looked at unique trial sets on the whole and computed a second, experiment level chi-square test on unique sets of trials to determine if Damerau–Levenshtein distance on the whole aligned with human choices a significant amount of times. While Damerau–Levenshtein distance did correctly predict human choices in some sets of trials, they did not predict human decisions on a global level as demonstrated in Table 16.1.

Although edit distance was not able to mimic human behavior in our experimental design, we used this as a stepping off point to inspect melody sets where humans did and did not agree with edit distance. The truncated Table 16.1 shows a listing of the highest and lowest agreements between our participant data and that of the edit distance.[a]

Upon inspecting the data, it appeared that similarities in contour variation (see stimuli 3 and 7) as well as the outlier status of stimuli 4 lead to a fair degree of consistency in ratings between edit distance and the participant judgments of similarity. This outlier status was also apparent when looking at the amount of times that a stimuli at the global level of the experiment was chosen as the oddball.

16.3.4. *Discussion*

In this study we sought out to investigate the extent to which Damerau–Levenshtein is able to mirror how humans would behave when presented with forced choice tasks. Before evaluating how well both algorithms and humans faired when compared to one another, it is first worth noting that within this dataset most

---

[a]The term "oddball" refers to the stimuli not selected in the group of three, thus selecting the other two as most similar.

participants exhibited a high degree of reliability in their ratings. This is in contrast to earlier research that screened participants for the reliability [14], but our findings may be due to the fact that our participant pool consisted of trained musicians. Alone this finding is important in that it suggests that, when prompted, participants are depending on a consistent and reliable process in their decision making. It is additionally worth noting that this degree of consistency was achieved via listening, and was not dependent on any sort of symbolic notation which carries many preconceived notions of Western music education.

Given this degree of consistency, it follows that it should be possible to come up with some sort of mathematical or statistical model that could accurately mirror this behavioral process. We began to explore how measures of edit distance might compare to how human selections by looking for agreements between edit distance and our rating data. While the results from the H_2 analysis table did not yield any significant results, it may have been that the way in which we decided to compare edit distance to our human measures was too stringent of a test for edit distance to accomplish given this context. A significant p-value for this test, if found, would have suggested that edit distance follows our participant data to an extreme degree. Secondly, the statistics reported in the H_2 analysis table are based on a second-order level of statistical tests meaning that the chi square value reported here was derived from a family of other statistical tests in which the critical values at the participant level were transferred one higher in order to generalize to the entire experiment. This type of analysis is extremely conservative and protects for type I errors almost to a limiting degree. Upon examining the data *post-hoc*, the amount of individual trials where participants did agree with what edit distance would have chosen is impressive and may indicate that our initial parameters that we chose were too conservative and would not again be adopted if this type of study was run again (Table 16.2).

Inspecting the distribution of trials where a high degree of agreement was achieved, we additionally suggest that there may be

Table 16.2. Subject agreement.

	Set	Oddball stimuli	Trials agreed
1	437	4	96
2	714	4	86
3	675	5	85
4	134	4	77
5	674	4	76
6	367	6	75
7	136	6	73
.	.	.	.
29	642	2	24
30	265	2	24
31	573	7	20
32	352	2	19
33	563	6	18
34	531	1	14
35	245	5	11

different types of appraisals happening for different melody contexts. Given the non-uniform distribution of what sets of trials were agreed upon as displayed in Fig. 16.8, some salient features may have been used in order to make the each judgments. Of particular interest in this setting would be repeated opening notes from Melody 4, which may have caused significant attention from the participants as well as the global contours of pairs of melodies that were paired together frequently (Fig. 16.9).

The question of what salient features an individual attends to also brings up the question of if in the 3AFC task that we used, were participants making judgments of similarity, as they were prompted? Or were participants, much like the way that we describe our analysis, treating this as some sort of oddball task. Though the experimental instructions explicitly asked for participants to make a rating of similarity, it might have been that they were treating the task as picking out the melody that is not like the other. Returning to our previous example of the one ceramic and two paper coffee cups, the issue becomes more apparent. Is it that the two paper cups are seen as separate from the ceramic and share more features in common, or

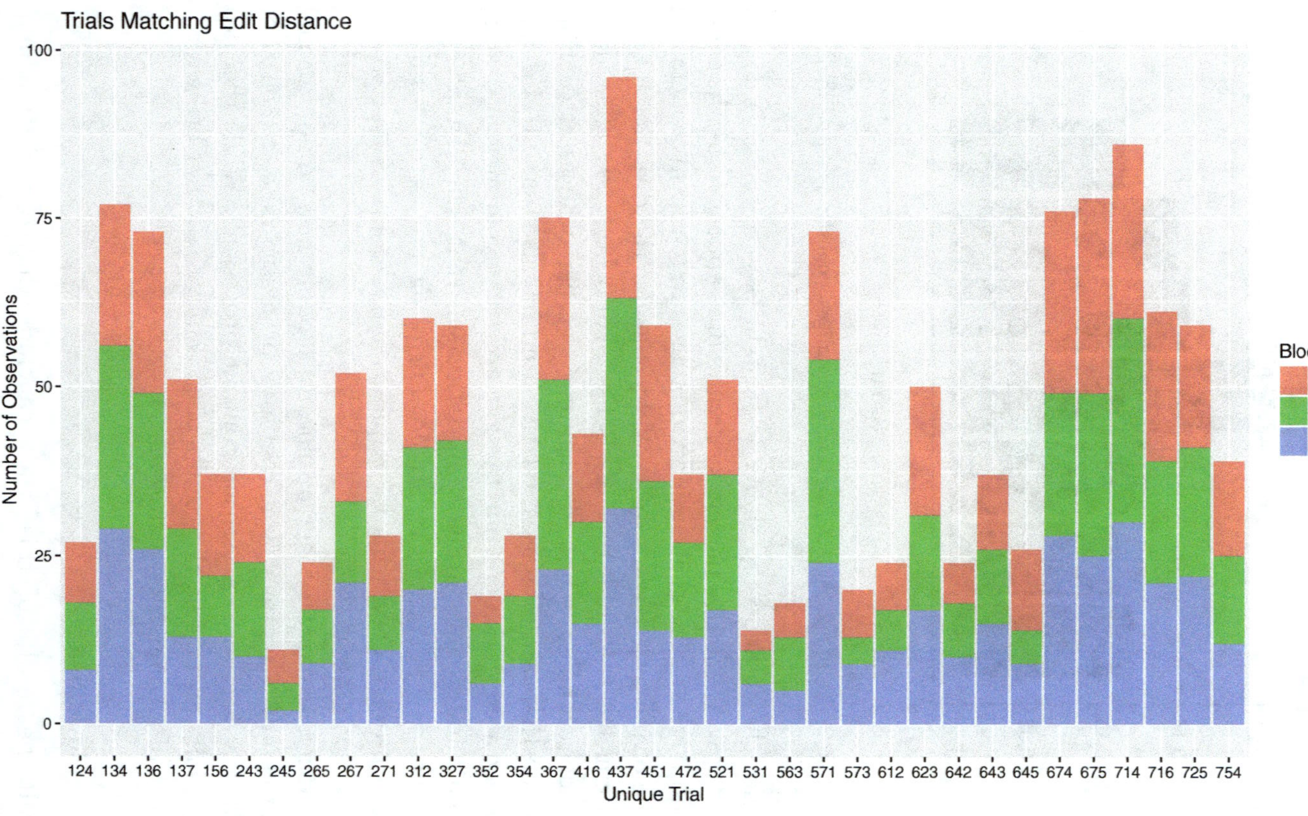

Fig. 16.8.   The number of observations each participant was in alignment with the choices made using edit distance.

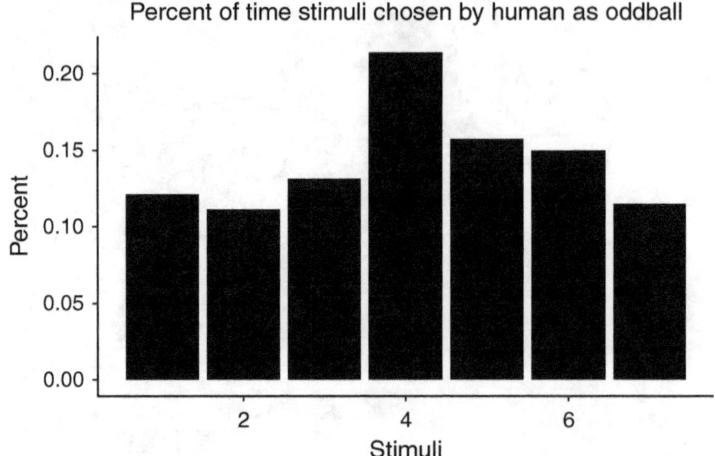

Fig. 16.9. The percentage of time each stimuli was chosen as an oddball.

is it that the ceramic cup shares less qualities with each of the two paper cups? While this pattern of results may appear to be the same end result, the means that an individual uses to reach this conclusion may be of interest. Future work might consider the directionality of paradigms like this. This type of discussion only bolsters claims made by Cambouropoulos *et al.* [4] mentioned earlier, who asserted that these questions are by nature relative, and in this case, directional.

If one were to accept our the logic of this experimental design, we could additionally argue that the reason that edit distance did not fair as well is because we are not mirroring any sort of cognitive decision process, only using a computational measure as a proxy for how someone might think.

Further, exploring other avenues of looking how computational measures might better reflect cognitive processes we can then turn to our *post-hoc* investigation on contour processing. While the idea was initially promising, the results of the analysis did not bring forth any sort of meaningful conclusions.

16.4. Conclusions

Overall the results of this study conclude that participants in a 3AFC reliably perform ratings of similarity across various conditions.

While this is not a new finding, it supports the idea that individuals use a consistent mental process when making similarity decisions about short melodies. In terms of modeling the degree to which humans make choices in accordance with the Damerau–Levenshtein edit distance, our data analysis did not yield any significant results. This could be in large due to the conservative preset alpha level decided *a priori*. Future work exploring this might consider how other algorithms might fare in this task as well as alternative ways of analyzing the dataset. The dataset used is publicly available at request of the authors.

Bibliography

[1] H. Allan, D. Müllensiefen and G. A. Wiggins, Methodological considerations in studies of musical similarity, in *ISMIR* (2007), pp. 473–478.

[2] D. J. Baker and D. Müllensiefen, Perception of leitmotives in richard wagner's der ring des nibelungen, *Front. Psychol.* **8** (2017).

[3] E. Bigand and B. Poulin-Charronnat, Are we experienced listeners? a review of the musical capacities that do not depend on formal musical training, *Cognition* **100**(1) (2006) 100–130.

[4] E. Cambouropoulos, How similar is similar? *Musicae Scientiae* **13**(1_suppl) (2009) 7–24.

[5] E. Cambouropoulos, T. Crawford and C. S. Iliopoulos, Pattern processing in melodic sequences: Challenges, caveats and prospects, *Comput. Humanities* **35**(1) (2001) 9–21.

[6] S. Champely, *Pwr: Basic Functions for Power Analysis. R Package Version 1.1. 1* (The R Foundation, Vienna, 2009).

[7] J. Cohen, Statistical power analysis, *Current Directions Psychol. Sci.* **1**(3) (1992) 98–101.

[8] I. Deliege, Grouping conditions in listening to music: An approach to lerdahl & jackendoff9s grouping preference rules, *Music Percept.* **4**(4) (1987) 325–359.

[9] T. Eerola, T. Jäärvinen, J. Louhivuori and P. Toiviainen, Statistical features and perceived similarity of folk melodies, *Music Percept.* **18**(3) (2001) 275–296.

[10] N. Farrugia, H. Allan, D. Müllensiefen and A. Avron, Does it sound like progressive rock? a perceptual approach to a complex genre, (2016), pp. 197–212.

[11] D. B. Huron, *The Humdrum Toolkit: Reference Manual* (Center for Computer Assisted Research in the Humanities, 1994).

[12] D. Meredith, Analysing music with point-set compression algorithms, in *Computational Music Analysis* (Springer, 2016), pp. 335–366.

[13] M. Mongeau and D. Sankoff, Comparison of musical sequences, *Comput. Humanities* **24**(3) (1990) 161–175.

[14] D. Müllensiefen and K. Frieler, Cognitive adequacy in the measurement of melodic similarity: Algorithmic vs. human judgments, *Comput. Musicology* **13** (2003) 147–176.

[15] D. Müllensiefen and K. Frieler, Modelling experts notions of melodic similarity, *Musicae Scientiae* **11**(1_suppl) (2007) 183–210.

[16] D. Müllensiefen, B. Gingras, J. Musil and L. Stewart, The musicality of non-musicians: An index for assessing musical sophistication in the general population, *PLoS One* **9**(2) (2014) e89642.

[17] D. Müllensiefen and M. Pendzich, Court decisions on music plagiarism and the predictive value of similarity algorithms, *Musicae Scientiae* **13**(1_suppl) (2009) 257–295.

[18] M. Pearce and D. Müllensiefen, Compression-based modelling of musical similarity perception, *J. New Music Res.* **46**(2) (2017) 135–155.

[19] Rosch, E. Principles of categorization, *Concepts: Core Readings* **189** (1999).

[20] C. Sapp, Humdrum extras, (2008), http://extras.humdrum.org/, http://extras.humdrum.org/ (accessed: 2017-11-1).

[21] A. Tversky, Features of similarity, *Psychol. Rev.* **84**(4) (1977) 327.

[22] R. Typke, F. Wiering and R. C. Veltkamp, Transportation distances and human perception of melodic similarity, *Musicae Scientiae* **11** (2007) 153–181.

Chapter 17

In Search of Arcs of Prototypicality

Daniel Shanahan

*Ohio State University,
Columbus, OH 43210, USA*

17.1. Introduction

In his 1947 book, *On Human Finery*, Quentin Bell argued that fashions change because of imitation: we imitate people that belong to social groups of which we would like to be a part [3]. At the same time, we attempt to identify ourselves as part of a social group to exclude those that we think should not be a part of that group. This implies that fashion is in a state of perpetual change. As more people acquire a certain fashion, the original members will seek to be more individualized and unique. Fashion, Bell would argue, is a perpetual motion machine, in which groups are constantly trying to assimilate, which incentivizes others to dissimilate, *ad infinitum.*

Some theories of musical change have also incorporated this concept. Meyer [15] argues that musical style, defined as "a replication of patterning, whether in human behavior or in the artifacts produced by human behavior, that results from a series of choices made within some set of constraints", can change gradually over time — the result of internal factors — or immediately and rapidly — a consequence of an event or an individual's actions. This mirrors much of the work done in sociolinguistics on the cause of linguistic and dialectical change [13]. Within this notion of change, music theorists often discuss how there is a process of assimilation and dissimilation that occurs across the lifespan of musical styles. For example, Gjerdingen [9, 10] discusses how musical ideas undergo an "arc of prototypicality", which seem to follow the trajectory of a bell curve, and are correlated with the usage of the musical idea. As a musical gesture is used more frequently, it is used more consistently, and as it begins to be used less, it undergoes mutates and splinters into new ideas. Much like Bell's notion of fashion, the musical ideas that are quite commonplace are replaced with those that are increasingly differentiated from the norm, which, in turn, becomes the norm to which others assimilate.

Corpus studies, however, have tended to focus on arcs of *usage*, rather than arcs of *typicality* (for example, see [5]). Although Gjerdingen [9] hypothesizes that the two are inextricably linked, the relative dearth of studies looking at such shifts might lead one to question why this might be the case. It is quite likely that there are practical challenges to a computational examination of typicality in general: although we can discuss melodies as similar or dissimilar (see Chapter 16), discussing their relative similarity in relation to the broader whole requires quite a bit of data. Additionally, what is analyzed as a marker of prototypicality (often termed a schema) can mutate in ways that are not always discernable through traditional search patterns. Searching for gestures based on metric regularity is possible, but is likely to miss quite a few options. One way around this might be to avoid the schema question entirely, instead focusing on "low-level" aspects such as entropy, variance, and correlation to a broader whole. Here I discuss some of these issues while examining a recently-curated corpus, the RISM-World Dataset. Specifically, I will look at eighteenth-century Italian melodies, as a great deal has been written on the schema — and prototype — driven nature of much of music in this repertoire. I explore the question of prototypicality by looking primarily at "high-level" and "low-level" features, and will discuss the barriers that still need to be crossed in order to more fully engage with a question of changing waves of prototypicality in music.

17.2. Using the RISM Corpus

Before discussing this, however, it might be valuable to discuss the types of data one might use to ask such a question. Music theory has recently seen an influx in corpora over the past decade or so. For quite a while, a significant proportion of corpus studies examined the Barlow and Morgenstern Dataset [2] when looking at WesternArt music, and the Essen Folksong Database [16] for investigating broader questions pertaining to folk music. Both datasets were encoded more than 20 years ago, and were converted to Humdrum Toolkit [12] by David Huron and the collaborators at the Center for Computer Assisted Research in the Humanities at Stanford (CCARH). Recent studies have included full transcriptions of Native American Folksongs [17], the harmonic progressions of jazz standards [5], Dutch folksongs [18], hip-hop transcriptions [6], and a curation of web-based online MIDI transcriptions and performances [19], among others.

This study employs the RISM-World Corpus, which has been curated from the broader RISM collection of data from libraries across the globe (Shanahan, Sapp, and Bell, *in progress*). The entire dataset contains more than one million incipits. Composer information, including nationality and birth and death dates were added by both human and algorithmic processes, and the algorithmic addition of metadata was corrected by hand. Specifically, an algorithm scraped online sources such as the Virtual International Authority File (VIAF, see [4, 14]), and searched for matching information. If information matched completely, it was labeled as accurate, while if there were two mismatched nationalities, it was checked by a human being. When removing duplicates, and when only examining the composers for which we could acquire metadata, the corpus contains roughly 600,000 melodic incipits. This study will only be using a selection from the eighteenth centuries written by Italian composers. This subset contains roughly 101,000 incipits, although each test was run on a sample of 10–15,000 melodies due to computational limitations.

17.2.1. *The issue of date affixation*

There is an ongoing debate in the field of corpus studies as to how to affix a date to a composition when analyzing pieces over time. The obvious method would be to simply use date of first publication or performance, but the abundance of posthumous pieces, as well as pieces performed or published significantly after composition casts some doubt on this method. Albrecht and Huron look at the year in which the composer turned 25, as they argue that most of the learning of tonal grammars and stylistic nuances will have taken place by then [1]. Daniele and Patel use the composer's midpoint year [7] and subsequent work has shown this to be a fairly reliable indicator [8].

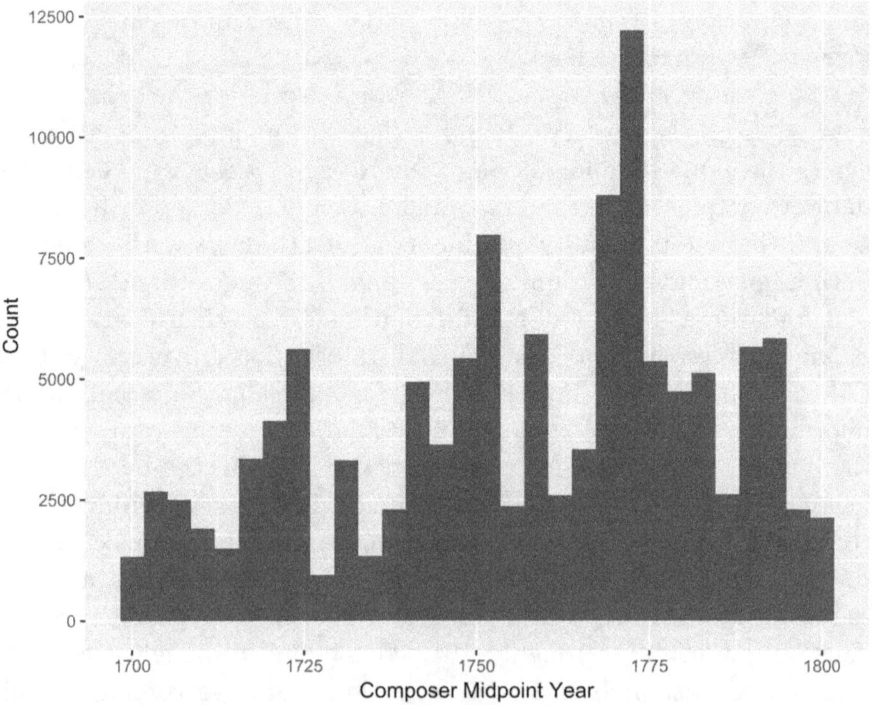

Fig. 17.1. Histogram of themes by composer midpoint year, $N = 176,335$. A number of these themes lacked key information, and were not able to be included in the final analysis, in which $N = 100,000$.

This study therefore employs composer midpoint (that is, the mean between the birth and death years of the composer) to affix dates for each composer. This can be seen as problematic: it ignores stylistic change within composers, instead affixing a single point to each individual. Given a large enough sample, however, we might expect to find broader change over time between individuals. As we are examining such large scale changes, we will focus on the between-composer differences, rather than the within-composer differences here.

As can be seen in Fig. 17.1, there is a healthy distribution of composers throughout the eighteenth century, with peaks in the latter half. Daniele and Patel argue that there were a number of nationalistic changes that took place in the eighteenth century as German nationalism began to rise, and composers turned away from more "Italianate" styles. By only examining Italian style, I hope to simply look at changes in compositional grammar over time, and at what rate these changes were adopted.

17.3. Examining the Changing Tonal Grammar Through "Low-Level Features"

One approach to examining "wave of style" might be to first examine the tonal grammar being used in a short period of time, such as a decade, and examine how that grammar changes. Once it has been established how this change has occurred (for example, perhaps $\hat{4}$ and $\hat{7}$ are used far less in the early decades of a century, rather than the later decades) the "key-profile" of each melody can be correlated with the key-profile of each decade. We would expect to find the correlation coefficient to increase as the decade approaches, and decrease as the decade passes. Albrecht and Huron used this approach to trace the evolution of the major and minor mode from 1400–1750, continually examining the key-profiles over time [1]. Similarly, examined how evolving key-profiles function differently when given examples of ambiguous key [20].

Using this approach can illustrate the change in compositional styles throughout the eighteenth-century, and would ideally demonstrate relationships of each melody to a more general tonal "norm". We might consider this a "bottom-up" approach, or one that focuses on "low-level" features such as overall pitch-class distribution, rather than "high-level" features such as melodic gestures or harmonic progressions. Given the pitch distribution of a certain period in time, will we would expect to see a change in the correlation of a given pitch distribution to the given pitch distribution.

Each melody was analyzed using the Humdrum Toolkit [12], and the key-profile of each decade was calculated. It would be likely that the second half of the century would exhibit the most stylistic change. The end of the Galant era, along with the increased output of iconoclasts such as Mozart and Haydn might suggest a shift in the usage of tonal grammars. Therefore, we might learn a great deal from examining the key-profile distribution relative to the distribution of the 1750s. Figure 17.2 shows a random sample of

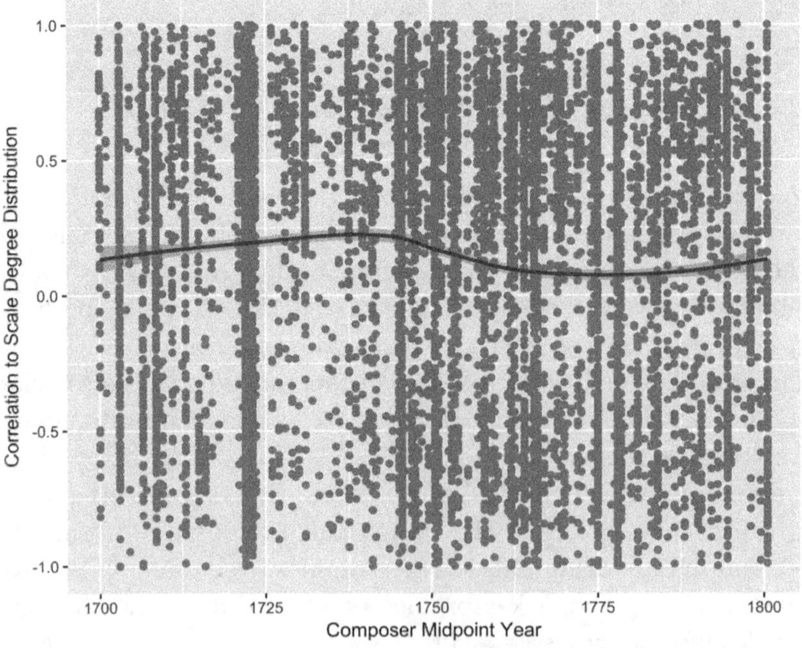

Fig. 17.2. A localized regression of the correlation coefficients of a random sample of 10,000 melodies (out of a total of 101,709 themes) to the 1750s prototype. Note how there is an incline leading int 1750, followed by a period of dissimilation throughout the decade.

10,000 themes by composers whose midpoint years fall within the eighteenth century, along with a localized regression line (LOESS) indicating the changing correlation of a theme to the scale degree distribution of the 1750s.[a] As would be expected, we do see an increased relationship as we move into the 1750s, and then a sharp decline throughout that decade. It would seem that, melodies began to employ different distributions throughout the course of the decade, moving away from the prototypical grammar, dissimilating as would be expected.

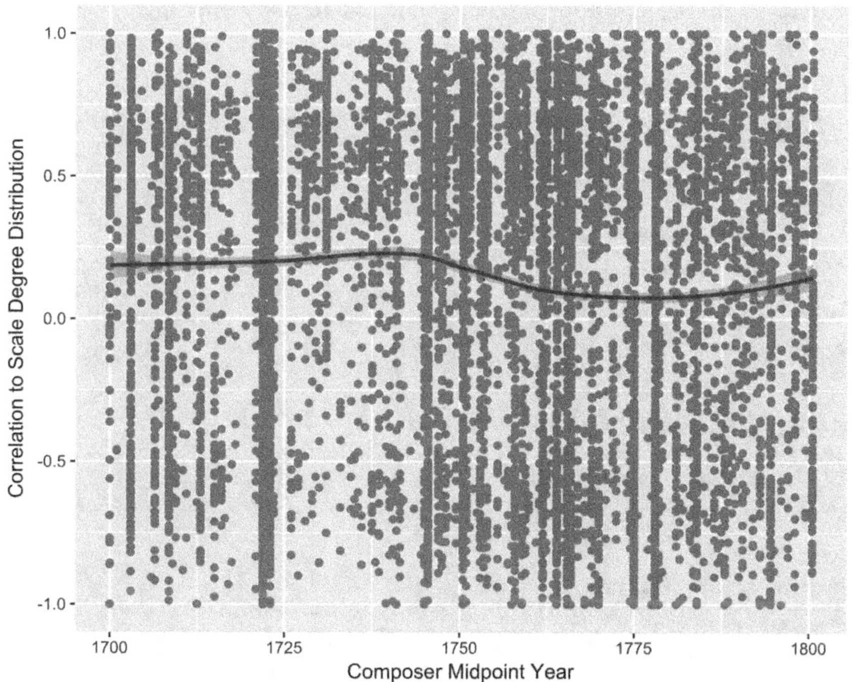

Fig. 17.3. A localized regression of the correlation coefficients of a *second* random sample of 10,000 melodies to the 1750s prototype.

---

[a]It should be noted here that, due to the large number of pieces in question, I'm avoiding discussing statistical significance. Significance allows us to understand if we have enough power to examine the question at hand, demonstrating whether our mathematical microscope provides enough resolution. With this many themes, most relationships are significant.

Examining a broader correlation, such as each melody to the scale degree distribution of the *entire century* shows that there is a fairly stable relationship. That is, if our window of prototypicality covers something as broad as an entire century, it would make sense that it would be difficult to see much change. This can be seen below in Fig. 17.3 on a subset of roughly 15,000 pieces.

We might then surmise that the tonal grammar changes in relatively small windows, but that is likely not the case. As Fig. 17.4 shows, instead of the tonal usage changing over time, we find an almost frustratingly stable distribution of pitches. Although there is a little bit of change in usage with $\hat{3}$ and $\hat{4}$, specifically around the 1760s and 1770s, it seems to be quite minor. How could this be? Surely the tonal language of the eighteenth-century, a period encompassing the height of the Baroque era, the entirety of the

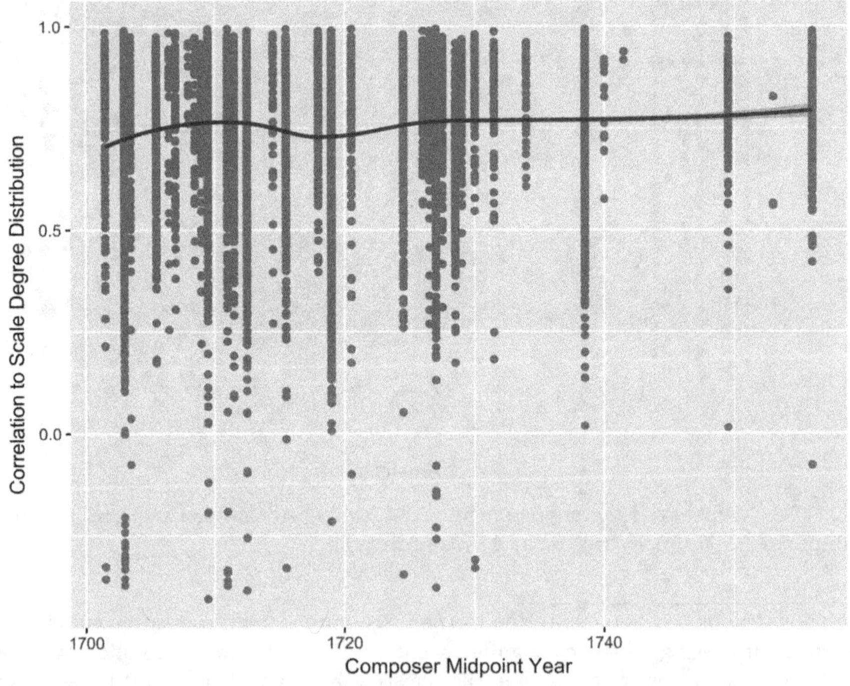

Fig. 17.4. The average correlation of a theme to the broader scale degree distribution of eighteenth-century Italian melodies ($n = 15,484$).

Galant era, and the majority of classicism, was anything but static. The answer might lie within the notion of style vs. structure. In languages, words come in and out of use, and slang terms are picked up and discarded quite readily. The syntax and grammar of a well-formed sentence, however, does not change: sentences still need verbs, and these verbs should be appropriately placed. While this is definitely a primary concern of this immediate research question, my goal in discussing this here is to also stress the disconnect between the computational convenience of low-level features and the often ineffable nature of musical prototypicality and similarity. For more discussion on this (see Chapter 16). Nevertheless, we can see *traces*

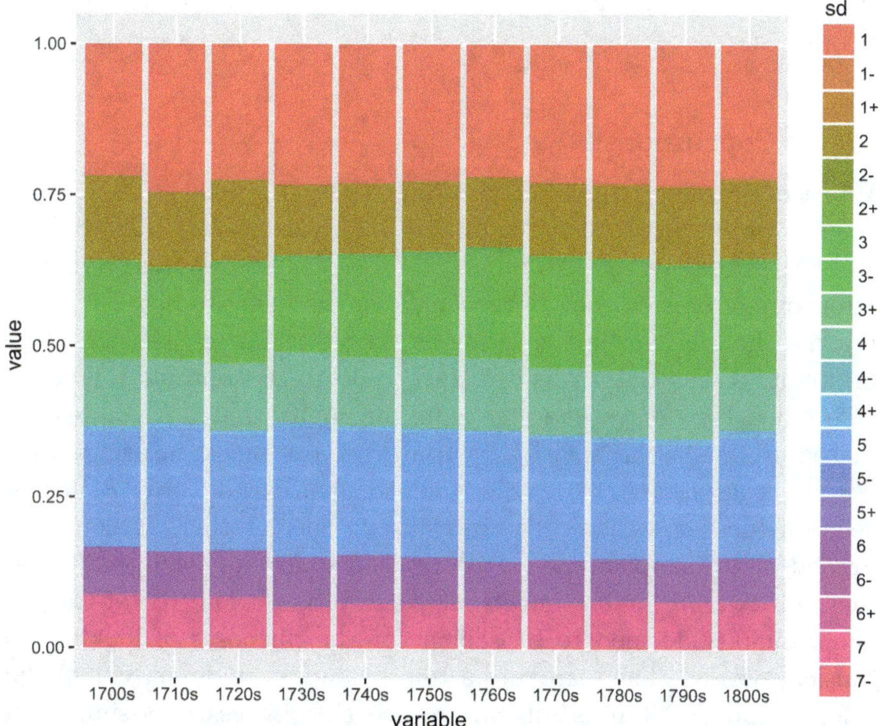

Fig. 17.5. The change in scale degree usage by decade in eighteenth-century Italian melodies in the RISM corpus. Note the astonishingly consistent distribution of scale degrees throughout the century.

of these stylistic changes when paired with a specific period, given enough data.

Examining shifts in prototypicality from the perspective of "low-level" features such as correlation of key-profiles would therefore seem to be leading us down a garden path. Studies that have focused on the changing nature of schemata have demonstrated how they can change quite rapidly. For example, Broze and Shanahan argue that harmonic schemata in Jazz standards change quite rapidly, correlated with the death of Charlie Parker in 1955, not the advent of modal jazz in 1959, as predicted [5]. Yet, low-level features — at least those pertaining to scale degree usage — seem to change at an almost glacial pace. We might extrapolate that this hierarchy of features can be mapped onto the dichotomy of "internal/external" and "gradual/rapid" stylistic change: grammar is relatively stable, but styles are fluid (Fig 17.5).

17.4.   Conclusions

When performing large-scale corpus studies, it is easy to sometimes rely on factors because of "computational convenience". Gjerdingen [11] writes that there are a number of assumptions we tend to make, often out of convenience, when engaging in such studies. First, we assume that the music is existing as a single auditory stream. Even a monophonic piece, such as a Bach cello suite, contains a number of implied harmonies and contrapuntal points that are not readily picked up by searching with a computer. Additionally, he argues that corpus studies tend to privilege contiguous events, often ignoring relationships at variable distances from one another. It could be argued that, in examining prototypicality through computational methods, we are falling into these same pitfalls. The RISM-Corpus short incipits deemed to be salient for the purposes of looking up musical works. They are often short, and provide only melodic information. That it is difficult to see the process of assimilation and dissimilation in tonal usage likely speaks to the likelihood that such changes occur primarily as "high-level" features, changing over a constant profile of tones.

This is not to say that corpus methods should not be used to investigate aspects of prototypicality, but one should focus on the type of corpus used. Ideally, entire pieces would be encoded, including as much harmonic information as possible. Ideally, much of this information would be annotated by humans, allowing for the labeling of schema that might fall outside of the score of traditional searching methods (for example, non-metrically regular schema). Unfortunately, there is not an abundance of these corpora at the moment. The Yale Classical Archives Corpus [19] provides one possibility, but its high proportion of performed MIDI transcriptions might mean that the rhythmic information is somewhat unreliable. Nevertheless, as musical corpora continue to proliferate, it is likely that we will be able to ask more musically-informed questions of them, and perhaps begin to understand more about the fashions of musical style.

Bibliography

[1] J. D. Albrecht and D. Huron, A statistical approach to tracing the historical development of major and minor pitch distributions, 1400–1750, *Music Percept.* (2014), http://mp.ucpress.edu/content/31/3/223.abstract.

[2] H. Barlow and S. Morgenstern, *A Dictionary of Musical Themes* (Crown Pub, 1948).

[3] Q. Bell, *On Human Finery* (Hogarth Press, London, 1949/1976).

[4] R. Bennett, C. Hengel-Dittrich, E. T. ONeill and B. B. Tillett, Viaf (virtual international authority file). Linking die deutsche bibliothek and library of congress name authority files, in *World Library and Information Congress: 72nd IFLA General Conference and Council* (2006). https://pdfs.semanticscholar.org/8a0a/3e8c91500c58fb06bffe94a757aa50da8ee4.pdf.

[5] Y. Broze and D. Shanahan, Diachronic changes in jazz harmony, *Music Percept.* **31**(1) (2013) 32–45, http://mp.ucpress.edu/content/31/1/32.abstract.

[6] N. Condit-Schultz, MCFlow: A digital corpus of rap transcriptions, *Empirical Musicol. Rev.* **11**(2) (2017) 124–147, http://emusicology.org/article/view/4961.

[7] J. R. Daniele and A. D. Patel, An empirical study of historical patterns in musical rhythm: Analysis of german & italian classical music using the nPVI equation, *Music Percept.* (2013) 10–18, http://www.jstor.org/stable/10.1525/mp.2013.31.1.10.

[8] J. R. Daniele and A. D. Patel, Stability and change in rhythmic patterning across a composers lifetime, *Music Percept.* **33**(2) (2015) 255–265, http://mp.ucpress.edu/content/33/2/255.abstract.

[9] R. O. Gjerdingen, The formation and deformation of Classic/Romantic phrase schemata: A theoretical model and historical study, *Music Theory Spectrum* **8** (1986) 25–43.

[10] R. O. Gjerdingen, *A Classic Turn of Phrase: Music and the Psychology of Convention* (University of Pennsylvania Press, 1988).

[11] R. O. Gjerdingen, "Historically Informed" corpus studies, *Music Percept.* **31**(3) (2014) 192–204, doi:10.1525/mp.2014.31.3.192, http://www.jstor.org/stable/10.1525/mp.2014.31.3.192.

[12] D. B. Huron, *The Humdrum Toolkit: Reference Manual* (Center for Computer Assisted Research in the Humanities, 1994).

[13] W. Labov, *Principles of Linguistic Change: Internal Factors*, Vol. 2 (Social, 1994).

[14] M. F. Loesch, VIAF (the virtual international authority file) — http://viaf. org, *Technical Services Quart.* **28**(2) (2011) 255–256, http://www.tandfonline.com/doi/full/10.1080/07317131.2011.546304.

[15] L. B. Meyer, *Style and Music: Theory, History, and Ideology* (University of Chicago Press, 1989), https://market.android.com/details?id=book-X44_zSrh0nYC.

[16] H. Schaffrath and D. Huron, *The Essen Folksong Collection in the Humdrum Kern Format* (Center for Computer Assisted Research in the Humanities, Menlo Park, CA, 1995).

[17] D. Shanahan and E. Shanahan, The densmore collection of native american songs: A new corpus for studies of effects of geography and social function in music, in *Proc. 13th Int. Conf. Music Perception and Cognition*, (2014), pp. 206–209.

[18] P. Van Kranenburg and B. Janssen, What to do with a digitized collection of western folk song melodies? in *Workshop on Folk Music Analysis (FMA2014)*, (2014), p. 117.

[19] C. White and I. Quinn, The Yale-Classical archives corpus, *Empirical Musicology Rev.* **11**(1) (2016) 50–58, http://search.proquest.com/openview/8df452287826234ad62e1fba5702d9e3/1?pq-origsite=gscholar&cbl=2030105.

[20] C. W. White, Changing styles, changing corpora, changing tonal models, *Music Percept.* **31**(3) (2014) 244–253, http://mp.ucpress.edu/content/31/3/244.abstract.

Index

R

S

T